CRUDE OIL EMULSIONS – COMPOSITION STABILITY AND CHARACTERIZATION

Edited by **Manar El-Sayed Abdel-Raouf**

Crude Oil Emulsions – Composition Stability and Characterization
Edited by Manar El-Sayed Abdel-Raouf

Published by InTech
Janeza Trdine 9, 51000 Rijeka, Croatia

Publishing Process Manager Vana Persen

Technical Editor Teodora Smiljanic

Cover Designer InTech Design Team

Additional hard copies can be obtained from orders@intechopen.com

Crude Oil Emulsions – Composition Stability and Characterization,
Edited by Manar El-Sayed Abdel-Raouf
p. cm.
ISBN 978-953-51-0220-5

For EU product safety concerns:
IN TECH d.o.o., Prolaz Marije Krucifikse Kozulić 3, 51000 Rijeka, Croatia,
info@intechopen.com or visit our website at intechopen.com.

Meet the editor

Prof Dr Manar El-Sayed Abdel-Raouf is a Professor of polymer chemistry for the Petroleum Application Department at the Egyptian Petroleum Research Institute. She studied at the faculty of science and got her MS. In 1995. The title of her Msc thesis was "Ethylene Oxide – Propylene Oxide Block Copolymers as Demulsifiers for Water-in-Oil Emulsions". She obtained her PhD in organic chemistry in 2000 with her thesis on "Alleviation of the Environmental Pollution by Recycling of Poly (Ethylene terephthalate) into useful Products". Prof. Dr Manar El-Sayed has published over 30 papers in international journals,attended 7 international conferences and, participated in two scientific projects for waste recycling. She is a member of an American-Egyptian joint project in Polyethylene terephthalate recycling, and a member in Tanks Services Center at EPRI.

Contents

Preface

We live in what has been called the Petroleum Age. This hydrocarbon-rich mixture of crude oil and gases runs our factories, our cars, heats homes and has changed the face of life on the earth since it's discovery on 1901.

Crude oil is an extremely versatile, naturally occurring, flammable liquid consisting of a complex combination of various molecular hydrocarbons weights and other organic liquid compounds that are found in geologic formations beneath the Earth's surface. Petroleum is recovered principally through oil drilling.

Refining it creates everything from asphalt and gasoline to lighter fluids and natural gas, along with a variety of vital elements such as sulphur and nitrogen. Petroleum products are also vital ingredients ("feedstocks") in the manufacture of medicines, chemicals and plastics.

The aim of this book is to discuss the many topics related to crude oil. It includes five sections. The first section deals with asphaltenes-these are natural surfactants that stabilize petroleum emulsions. The chapters of the second section discuss some advanced methods used in characterization of crude oil and it's components. The third section includes and publishes new discoveries and improvements in crude oil biology. The fourth section discusses the natural components presented in crude oil and those factors affecting crude oil stability. The final chapter involves some innovative ideas and concepts related to the crude oil.

Finally, we would like to thank all the authors who had participated in this book for their valuable contribution. While there is no doubt that this book may have omitted some significant findings in the petroleum field, we hope the information included in will be useful for both petroleum engineers and chemists, in addition to the academic researches working in this field.

Manar El-Sayed Abdel-Raouf
Professor of Petroleum Chemistry
Petroleum Application Department
Egyptian Petroleum Research Institute
Egypt

Part 1

Asphaltenes in Crude Oil

1

Asphaltenes – Problems and Solutions in E&P of Brazilian Crude Oils

Erika Chrisman, Viviane Lima and Príscila Menechini
Federal University of Rio de Janeiro/DOPOLAB
Brazil

1. Introduction

The history of oil in Brazil began in 1858, when the Marquis of Olinda signed Decree No. 2266 granting Barros Pimentel the right to extract mineral asphalt for the manufacture of kerosene, on land situated on the banks of the Rio Marau in the province of Bahia. But who really came to be known as the discoverer of oil in Brazil was Monteiro Lobato that on January 21, 1939, already under the jurisdiction of the newly created Department of National Production, began drilling the well DNPM-163 in Bahia.

Only in 1953 began the research of Brazilian oil by the government of Vargas, which established the state oil monopoly, with the creation of Petrobras.

Petrobras - Petroleo Brasileiro S/A was founded on October 3, 1953 and headquartered in Rio de Janeiro, now operates in 28 countries in the energy sector, primarily in the areas of exploration, production, refining, marketing and transportation of oil and derivatives in Brazil and abroad. Its current motto is "An integrated energy company that works with social and environmental responsibility" (http://pt.wikipedia.org/wiki/Petrobras).

Since its creation, Petrobras has discovered oil in several states, and in every decade, new oil fields are discovered. Oil production in Brazil grew from 750 m^3/day at the time of the creation of Petrobras to more than 182,000 m^3/day in the late '90s thanks to continuous technological advances in drilling and production on the continental shelf.

In 2006, Brazil managed to achieve sustainable self-sufficiency in oil production with the operations of the FPSO (Floating Production Storage Offloading) P-50 in giant Albacora East field, in northern of Bacia de Campos in the state of Rio de Janeiro (www.autosuficiencia.com.br).

In 2007, Brazil announced the discovery of oil in the so-called pre-salt, which later turned out to be a large oil field, extending over 800km off the Brazilian coast, from the state of Espírito Santo to Santa Catarina, below thick layer of salt (rock salt) and covering sedimentary basins of Espirito Santo, Campos and Santos. The first pre-salt oil extracted in 2008 (www.petrobras.com.br).

The year 2009 was marked by the beginning of production in pre-salt layer in the Santos Basin, and in August 2010, Brazil had a record of oil production, with 2, 078 million barrels per day, up 6% over the same period in 2009.

According to the National Petroleum Agency (ANP), the volume exceeds one thousand barrels the previous record from April 2010. Produced in the Tupi, Jubarte and Cachalote fields, the pre-salt oil contributed 43, 087 thousand barrels per day in volume in August (www.monitormercantil.com.br).

Fig. 1. Pre-salt layer. Source: www.anp.gov.br

The company ranks second among the largest publicly traded oil in the world. In market value, it is the second largest in the Americas and the fourth largest in the world in 2010. In September 2010, it became the second largest energy company in the world, always in terms of market value, according to Bloomberg data and the Agency Brazil.

It became internationally famous for having made in October 2010 the largest capitalization of all publicly traded history of mankind: US$ 72.8 billion (R$ 127.4 billion), almost double the record so far, the post office in Japan (Nippon Telegraph and Telephone), with US$ 36.8 billion capitalized in 1987. In August 2011 the company broke two more records for net income: US$ 10.94 billion in the second quarter, and also the record of R$ 21.9 billion in the first half of the year. (Http://pt.wikipedia.org/wiki/Petrobras).

2. Proved reserves

In 2010, the proved oil reserves in the world reached about 1.38 trillion barrels, after a 0.5% increase over 2009.

In South and Central America, the rise was driven by Colombia, Brazil and Peru, which saw its proved reserves grow 39.7%, 10.7% and 10.6% in that order.

With this increase, partly due to the findings in the pre-salt, the Brazilian proved reserves reached 14.2 billion barrels of oil, and placed the country in 15th position in world ranking of reserves.

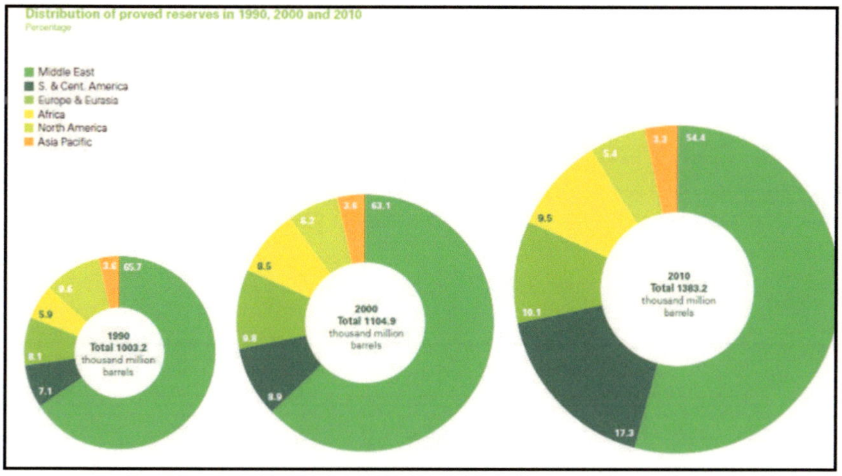

Fig. 2. Statistical Review of World Energy 2011. Source: www.bp.com

3. Consumption

In 2010, world oil consumption was 3.2% over 2009, totaling 87.4 million barrels/day. The oil was more consumed in the region of Asia-Pacific, with a total of 27.2 million barrels/day or 31.2% of the total. Consumption growth was 5.3% over 2009, especially to China which, after the United States was the country with the second largest consumer in the world, 9.1 million barrels/day, 10, 4% more than last year.

The Central and South America also recorded high in its consumption, as a result of increases in almost all countries, except Chile, which had a low of 6.2%. Thus, the increase in consumption in the region was 4.8%, reaching 6.1 million barrels/day or 7% of the world. Brazil was the country with the largest increase in consumption in the region - 8.6% - and reached 2.6 million barrels / day or 3% of world total. Thus, the country jumped to seventh in the ranking of the largest consumers of oil in the world.

Compared to 2009, the volume of oil produced worldwide in 2010 increased 2.3%, from 80.3 to 82.1 million barrels/day.

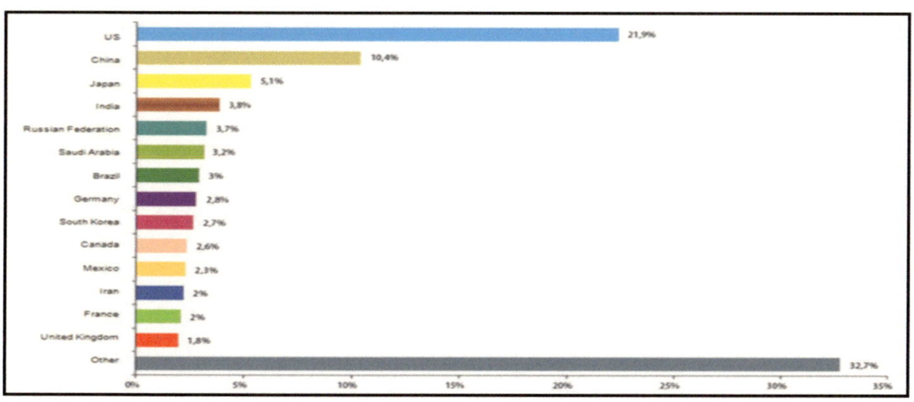

Fig. 3. World consumption of oil. Source: www.anp.gov.br

4. Production

Compared to 2009, the volume of oil produced worldwide in 2010 increased 2.3%, from 80.3 to 82.1 million barrels/day.

Oil production in Central and South America rose by 3.5%, driven mainly by increases in Colombia, Peru and Brazil, respectively, 16.8%, 8.2% and 5.3%. With the increase in the volume of oil produced, Brazil reached the 12th position among the world's largest producers of oil in 2010.

Oil: Production *												Change 2010 over 2009	2010 share of total
Thousand barrels daily	2000	2001	2002	2003	2004	2005	2006	2007	2008	2009	**2010**		
US	7733	7669	7626	7400	7228	6895	6841	6847	6734	7271	**7513**	3.2%	8.7%
Canada	2721	2677	2858	3004	3085	3041	3208	3297	3251	3224	**3336**	4.3%	4.2%
Mexico	3450	3560	3585	3789	3824	3760	3683	3471	3167	2979	**2958**	-0.8%	3.7%
Total North America	**13904**	**13906**	**14069**	**14193**	**14137**	**13696**	**13732**	**13616**	**13152**	**13474**	**13808**	**2.5%**	**16.6%**
Brazil	1268	1337	1499	1555	1542	1716	1809	1833	1899	2029	**2137**	5.3%	2.7%
Venezuela	3239	3142	2895	2554	2907	2937	2808	2613	2558	2438	**2471**	1.4%	3.2%
Total S. & Cent. America	**6813**	**6722**	**6619**	**6314**	**6680**	**6898**	**6865**	**6635**	**6676**	**6753**	**6989**	**3.5%**	**8.9%**
Kazakhstan	744	836	1018	1111	1297	1356	1426	1484	1554	1688	**1757**	4.4%	2.1%
Norway	3346	3418	3333	3264	3189	2969	2779	2551	2459	2358	**2137**	-9.4%	2.5%
Russian Federation	6536	7056	7698	8544	9287	9552	9769	9978	9888	10035	**10270**	2.2%	12.9%
United Kingdom	2667	2476	2463	2257	2028	1809	1636	1638	1526	1452	**1339**	-7.7%	1.6%
Total Europe & Eurasia	**14950**	**15450**	**16289**	**16973**	**17580**	**17542**	**17599**	**17815**	**17590**	**17745**	**17661**	**-0.4%**	**21.8%**
Iran	3855	3892	3709	4183	4248	4234	4286	4322	4327	4199	**4245**	0.9%	5.2%
Iraq	2614	2523	2116	1344	2030	1833	1999	2143	2428	2442	**2460**	0.6%	3.1%
Kuwait	2206	2148	1995	2329	2475	2618	2690	2636	2782	2489	**2508**	0.6%	3.1%
Saudi Arabia	9491	9209	8928	10164	10638	11114	10853	10449	10846	9893	**10007**	0.7%	12.0%
United Arab Emirates	2620	2551	2390	2695	2847	2983	3149	3053	3088	2750	**2849**	3.5%	3.3%
Total Middle East	**23547**	**23120**	**21858**	**23442**	**24981**	**25488**	**25675**	**25309**	**26338**	**24629**	**25188**	**1.7%**	**30.3%**
Angola	746	742	905	870	1103	1405	1421	1684	1875	1784	**1851**	3.8%	2.3%
Nigeria	2155	2274	2103	2238	2431	2499	2420	2305	2113	2061	**2402**	16.2%	2.9%
Total Africa	**7804**	**7897**	**8028**	**8411**	**9336**	**9902**	**9918**	**10218**	**10204**	**9698**	**10098**	**4.2%**	**12.2%**
China	3252	3306	3346	3401	3481	3637	3705	3737	3809	3800	**4071**	7.1%	5.2%
India	726	727	753	756	773	738	762	769	768	754	**826**	9.8%	1.0%
Indonesia	1456	1387	1289	1176	1130	1090	996	972	1003	990	**986**	-0.3%	1.2%
Total Asia Pacific	**7874**	**7811**	**7837**	**7742**	**7854**	**7959**	**7940**	**7951**	**8054**	**7978**	**8350**	**4.9%**	**10.2%**
Total World	74893	74906	74700	77075	80568	81485	81729	81544	82015	80278	82095	2.2%	100.0%
of which: OECD	21531	21314	21440	21174	20775	19870	19463	19114	18414	18471	**18490**	0.2%	22.1%
Non-OECD	53361	53592	53260	55900	59793	61616	62266	62430	63600	61807	**63605**	2.7%	77.9%
OPEC	31145	30640	29261	31020	33776	34951	35098	34757	35722	33365	**34324**	2.5%	41.5%
Non-OPEC £	35734	35606	35907	35556	35385	34695	34315	33991	33466	33699	**34287**	1.9%	41.7%
European Union #	3493	3285	3339	3128	2902	2659	2422	2388	2222	2088	**1951**	-6.5%	2.4%
Former Soviet Union	8014	8660	9533	10499	11407	11839	12316	12795	12827	13214	**13484**	2.0%	16.8%

* Includes crude oil, shale oil, oil sands and NGLs (the liquid content of natural gas where this is recovered separately).
Excludes liquid fuels from other sources such as biomass and coal derivatives.
^ Less than 0.05.
w Less than 0.05%.
£ Excludes Former Soviet Union.
\# Excludes Estonia, Latvia and Lithuania prior to 1985 and Slovenia prior to 1991.
Notes: Annual changes and shares of total are calculated using million tonnes per annum figures.

Table 1. World production of oil. Source: www.bp.com

5. The origin of oil

The oldest formations of the world's oil has about 500 million years and it is a result of the slow process of nature, which produced deposits of sediment in large depressions in the bottom of seas and lakes, accumulating for thousands of years, successive layers of sedimentary rock containing microorganisms, animals and plants.

The action of heat and weight of these layers on the deeper sedimentary deposits has been transforming the organic matter through thermochemical reactions in kerogen, the initial stage of oil, then by the action of higher temperatures and pressures, they were broken, making it on deposits of oil and gas.

5.1 Constituents of oil

Oil can be defined as to its chemical composition as a naturally occurring complex mixture consisting predominantly of hydrocarbons (up to more than 90% of its composition) and derived organic sulfur, nitrogen, oxygen and organometallic.

The oils from different oil reservoirs have different characteristics. Within this complex mixture, there is a fraction with high molecular weight components called asphaltenes that causes serious precipitation problems, since its production by refining (Carvalho, 2003).

The hydrocarbons present in oil can be classified into four main classes: saturated (alkanes and cycloparaffins), aromatics (hydrocarbons, mono, di and polyaromatic), resins (fractions consist of polar molecules containing heteroatoms such as N, O or S) and asphaltenes (they are molecules similar to the resins, but with a higher molecular weight and polyaromatic core). This classification is known as SARA (Wang et al, 2002; Speight, 2001; Tissot and Welt, 1978).

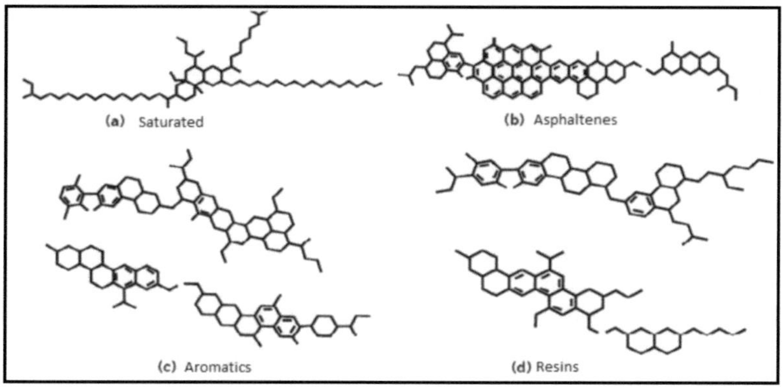

Fig. 4. Structures representing saturated, asphaltenes, aromatics and resins. Source: Bernucci et al., 2006.

Table 04 shows the elemental analysis of five Brazilian oils performed in DOPOLAB - Laboratory of Development and Optimization of Process Organic, showing considerable amounts of asphaltenes present.

Oil	Saturated (%)	Aromatics (%)	Resins (%)	Asphaltenes (%)
A	22.4	22.8	36.4	18.4
B	20.7	29.8	30.4	19.1
C	16.0	25.1	25.0	33.9
D	14.2	40.5	24.7	20.5
E	19.4	36.7	12.6	31.2

Table 2. Chemical composition of Brazilian oil. Source: DOPOLAB, 2011.

The relative amounts of individual compounds within each group of hydrocarbons is characteristic of each type of oil.

Metals can be found at levels ranging from 1 to 1200 ppm, the main being iron, zinc, copper, lead, molybdenum, cobalt, arsenic, manganese, chromium, sodium, nickel and vanadium, the latter two with the highest incidence.

5.2 Classification of oil

The classification of oils, according to their constituents, it has an interest from geochemists to the refiners. The first aim is to characterize the oil is to relate it to the rock and measure its degree of degradation and the refiners seek to know the amount of the various fractions that can be obtained, as well as its composition and physical properties.

Such information is important because: paraffinic oils are excellent for the production of aviation kerosene (jet fuel), diesel, lubricants and paraffins; naphthenic oils produce significant fractions of gasoline, naphtha, aviation fuel and lubricants; while the aromatic oils are best suited for the production of gasoline, solvents and asphalt (Thomas, 2001).

5.3 Petroleum refining

The processing of oil, called refining, begins by distillation, a unit operation, consisting of the vaporization and subsequent condensation fractional of its constituents by the action of temperature and pressure due to the difference in their boiling points. Thus, with the variation of the conditions of a heating oil, it is possible vaporization of compounds light, medium and heavy that can be separated when they condense. In parallel, there is the formation of a heavy residue that consists mainly of high molecular weight hydrocarbons, which under the conditions of temperature and pressure at which the distillation is performed, does not vaporize (Mariano, 2005).

Oil refineries are a complex system of multiple operations that depend on the properties of the oil that will be refined as well as the desired products. For these reasons, the refineries may be very different. Depending on the type of oil being processed and the profile of the refinery, i.e., the existing treatment units, you get larger or smaller portions of each type of fraction.

Oil: Refinery capacities												Change	2010
												2010 over	share
Thousand barrels daily *	2000	2001	2002	2003	2004	2005	2006	2007	2008	2009	**2010**	2009	of total
US	16595	16785	16757	16894	17125	17339	17443	17594	17672	17688	**17594**	-0.5%	19.2%
Canada	1861	1917	1923	1959	1915	1896	1914	1907	1951	1976	**1914**	-3.1%	2.1%
Mexico	1481	1481	1463	1463	1463	1463	1463	1463	1463	1463	**1463**	-	1.6%
Total North America	**19937**	**20183**	**20143**	**20316**	**20503**	**20698**	**20821**	**20964**	**21086**	**21127**	**20971**	**-0.7%**	**22.8%**
Brazil	1849	1849	1854	1915	1915	1916	1916	1935	2045	2095	**2095**	-	2.3%
Venezuela	1269	1269	1269	1269	1284	1291	1294	1303	1303	1303	**1303**	-	1.4%
Total S. & Cent. America	**6271**	**6246**	**6296**	**6353**	**6377**	**6405**	**6413**	**6502**	**6658**	**6688**	**6707**	**0.3%**	**7.3%**
Germany	2262	2274	2286	2304	2320	2322	2390	2390	2366	2362	**2091**	-11.5%	2.3%
Italy	2485	2485	2485	2485	2497	2515	2526	2497	2396	2396	**2396**	-	2.6%
Russian Federation	5655	5628	5590	5454	5457	5522	5599	5596	5549	5527	**5555**	0.5%	6.1%
United Kingdom	1778	1769	1785	1813	1848	1819	1836	1819	1827	1757	**1757**	-	1.9%
Total Europe & Eurasia	**25399**	**25276**	**25159**	**25005**	**25066**	**24999**	**25042**	**24966**	**24840**	**24761**	**24516**	**-1.0%**	**26.7%**
Iran	1597	1597	1597	1607	1642	1642	1727	1772	1805	1860	**1860**	-	2.0%
Saudi Arabia	1806	1806	1810	1890	2075	2100	2100	2100	2100	2100	**2100**	-	2.3%
Total Middle East	**6491**	**6746**	**6915**	**7039**	**7256**	**7284**	**7409**	**7522**	**7598**	**7818**	**7911**	**1.2%**	**8.6%**
Total Africa	**2897**	**3164**	**3228**	**3177**	**3116**	**3224**	**3049**	**3037**	**3171**	**3022**	**3292**	**8.9%**	**3.6%**
China	5407	5643	5933	6295	6603	7165	7865	8399	8722	9479	**10121**	6.8%	11.0%
India	2219	2261	2303	2293	2558	2558	2872	2983	2992	3574	**3703**	3.6%	4.0%
Japan	5010	4705	4721	4683	4567	4529	4542	4598	4650	4621	**4463**	-3.4%	4.9%
South Korea	2598	2598	2598	2598	2598	2598	2633	2671	2712	2712	**2712**	-	3.0%
Total Asia Pacific	**21478**	**21853**	**22444**	**22579**	**23037**	**23537**	**24693**	**25561**	**26094**	**27653**	**28394**	**2.7%**	**30.9%**
Total World	82473	83469	84183	84468	85355	86147	87427	88552	89446	91068	91791	0.8%	100.0%
of which: OECD	44761	44697	44900	45024	45169	45202	45422	45634	45784	45742	**45124**	-1.3%	49.2%
Non-OECD	37712	38771	39283	39444	40187	40945	42005	42918	43662	45326	**46667**	3.0%	50.8%
European Union #	15456	15540	15691	15729	15803	15811	15857	15784	15658	15553	**15240**	-2.0%	16.6%
Former Soviet Union	8574	8404	8133	7937	7940	7945	7961	7958	7961	7965	**8033**	0.9%	8.8%

Source: Includes data from Parpinelli Tecnon.

* Atmospheric distillation capacity on a calendar-day basis.
# Excludes Lithuania prior to 1985 and Slovenia prior to 1991.
Note: Annual changes and shares of total are calculated using thousand barrels daily figures.

Table 3. Oil refining capacity worldwide. Source: www.bp.com

Lighter oils produce greater volume of gasoline, LPG and naphtha, and so exhibit a higher commercial value, while heavy fuel oils produce higher volumes of oil yield and asphalt. The average derivatives, such as diesel and kerosene, are also particularly important for our country where the highway is intense and the demand for diesel is great.

In 2010, the effective refining capacity installed worldwide was 91.8 million barrels/day to a world oil production of 82.1 million barrels/day.

The United States retained its first place in ranking global refining capacity (19.2% of total), followed by China (11%), Russia (6.1%), Japan (4.9%) and India (4 %). Together, these five countries accounted for 39.6% of global refining capacity.

Brazil has climbed to ninth place in global refining capacity ranking, with 2.1 million barrels/day or 2.3% of world capacity.

6. Asphaltenes

Over the past year, there was a significant increase in studies of asphaltenes due to increased production of heavy oil and due to the diminishing reserves of oil lighter (Yarranton et al, 2000a, 2002; Calemma et al, 1995).

The asphaltenes are a mixture of high molecular weight aromatic components oil shale. Coal and oil can vary from 1% by weight in light oils, up to 17% in heavy oils. According to Leon et al (2001), the asphaltenes are considered the fraction of oil that has the highest number of aromatic rings and higher molecular weight.

According to several researchers (Ortiz et al 2010; Nordgard et al, 2009, Yasar et al, 2007; Ancheyta and Trejo, 2007; Deo et al, 2004; Mullins et al, 2003; Kilpatrick et al, 2003a and 2003b; Speight, 2001; Leon et al, 2001 and 2000; Bauget et al, 2001; Gafanova and Yarranton, 2001; Yarranton et al, 2000a and b; Rogel, 2000, Andersen and Speight, 1999; Murgich et al, 1999; Speight et al, 1994; Speight and Long, 1995), the most accepted definition for asphaltenes is related to their solubility and says that asphaltenes are insoluble in aliphatic hydrocarbons such as n-heptane or n-pentane and soluble in aromatic hydrocarbons such as toluene. According Shkalikov et al (2010) asphaltenes represent insoluble precipitates obtained from solutions of oil in alkanes of lower molecular weight such as pentane, hexane and heptane.

According Oseghale and Ebhodaghe, 2011; Mustafa et al, 2011; Ortiz et al, 2010; Nordgard et al, 2009; Gauthier et al, 2008; Yasar, 2007; Ancheyta et al, 2004; Kilpatrick et al, 2003; Mullins, 2003; Siddiqui, 2003; Murgich, 2002; Priyanto et al, 2001; Sheu, 2002; Speight, 2001; Leon et al, 2000; Rogel, 2000; Yarranton et al, 2000b; Murgich et al, 1999; Calemma et al, 1995, the asphaltenes consist of a heterogeneous complex mixture of molecules highly polydisperse in size and with a chemical composition poly condensed aromatic rings, aliphatic chains, naphthenic rings, and containing heteroatoms such as nitrogen, oxygen, sulfur, presenting itself in the form of carboxylic acids, amides, amines and alcohols, and metals such as iron, nickel and vanadium. Different types of metals (Ni, V, Fe, Al, Na, Ca, Mg) are present in crude oils and tend to accumulate in the asphaltenic fraction in trace amounts. Vanadium and nickel are the most commonly found and in most cases present as porphyrin complexes, being responsible for the poisoning of catalysts in the improvement of oil (Mustafa et al, 2011; Nordgard et al, 2009; Ancheyta and Trejo, 2007).

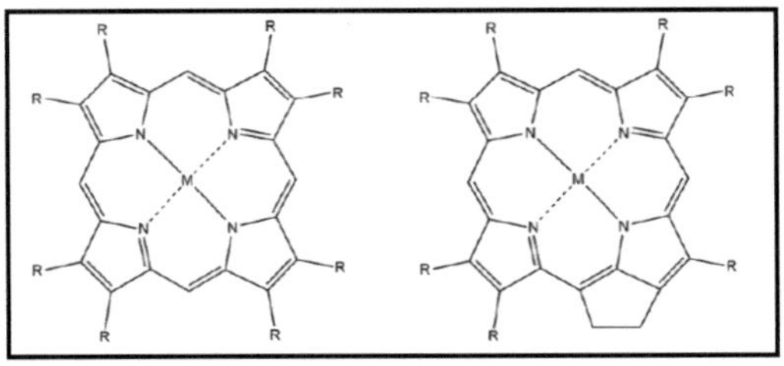

Fig. 5. Structures of metal porphyrins. Source: Mustafa et al, 2011.

The asphaltenes are arguably the most complex fraction of oil. In variations of pressure, temperature or composition of oil, asphaltenes tend to associate and precipitate causing several costly operational problems from transport to refining (Trejo et al, 2007; Lira-Galeana and Duda, 2006).

Among all the oil fractions, the molecular structure of asphaltenes is the least understood (Trejo et al, 2007). Several researchers (Carvalho, 2003; Speight, 1999 a; Speight, 1999 b; Andersen and Speight, 1999; Speight et al, 1984) have concentrated their efforts on improving the information about this mixture, deepening their knowledge of the chemical structures involved, characterizing the functions and establishing their behavior against solvents. They also seek to explain the way their molecules are stabilized and dispersed in the oil.

Overall, the researchers concluded that the level and nature of asphaltenes in a sample is due to a series of parameters such as the origin of oil, the flocculating agent, the time used for precipitation, temperature, procedure used and ratio oil / flocculant agent. All these combined parameters not only influence the amount of asphaltene precipitated, but also in its composition, which can be obtained from an asphaltene solid dark brown to a black (Silva, 2003).

The elemental composition of asphaltenes varies in a ratio of C/H of $1.15 \pm 0.05\%$, however, values outside this range are sometimes found, according to Speight (1999 a).

Fig. 6. Illustration of Asphaltenes. Source: Lima, 2008.

Notable variations may occur, particularly in the proportions of heteroatoms such as oxygen and sulfur, but they are always exposed in a very characteristic manner. For example, nitrogen occurs in the asphaltenes in various heterocyclic chains, the oxygen can be identified as carboxylic, phenolic and ketone (Speight et al, 1994); while sulfur is in the form of thiols, thiophenes, benzothiophenes, dibenzothiophenes and naphtebenzothiophenes as well as in systems such as sulphide, alkyl-aryl, aryl-aryl and alky- laryl (Speight et al, 1984).

Examples of structures present in the asphaltenic fraction based on data from infrared, ^{1}H NMR and ^{13}C with Venezuelan oils (A, B, C, D) are shown in Figure 07.

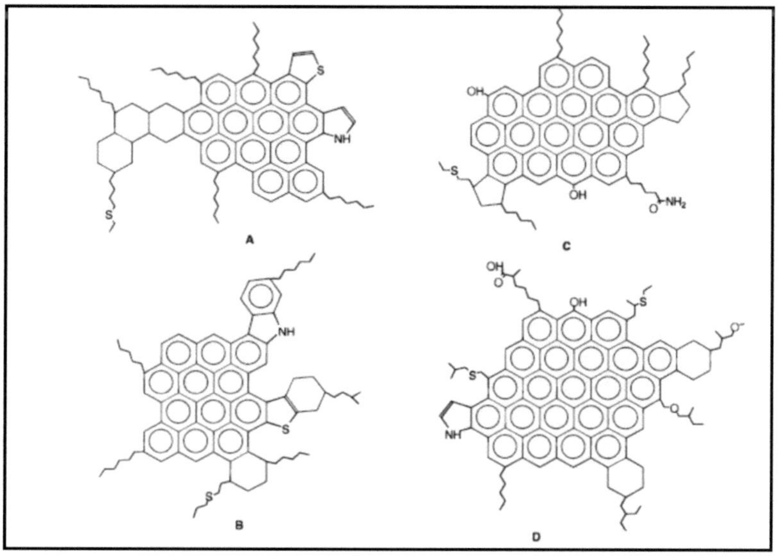

Fig. 7. Hypothetical structures of asphaltenes. Source: Leon et al, 2000.

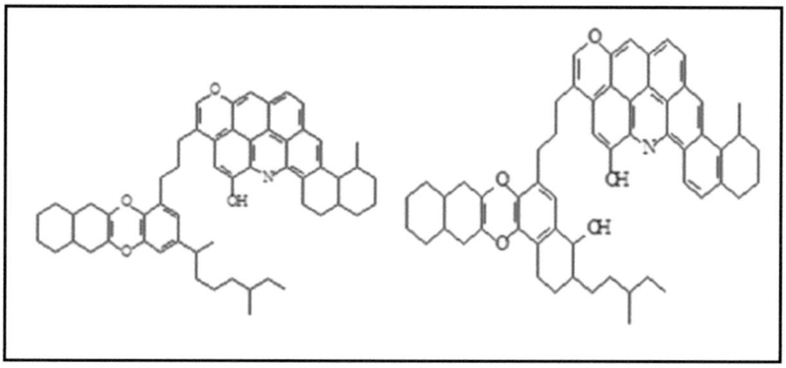

Fig. 8. Structures representing asphaltenes. Source: Lima, 2008.

The basic structure of asphaltenes consists of a number of rings polycondensates, replaced by aliphatic or naphthenic groups, and there may be between 6 and 20 rings. These structures gather in piles at the level of the aromatic rings forming, then, particles, each particle made up 4 to 6 structures. These particles can come together to form a structure called the aggregate. The size of an aggregate is clearly dependent on the structures involved in the same (Caldas, 1997).

To Merdrignac and Espinat (2007), asphaltenes contain aromatic molecules variables and with different amounts of heteroatoms, metals and functional groups. Such structures can not be represented by a single model of the molecule. Several models are proposed in the literature to describe them, among the main continental and archipelago. Figure 09 presents some of these models.

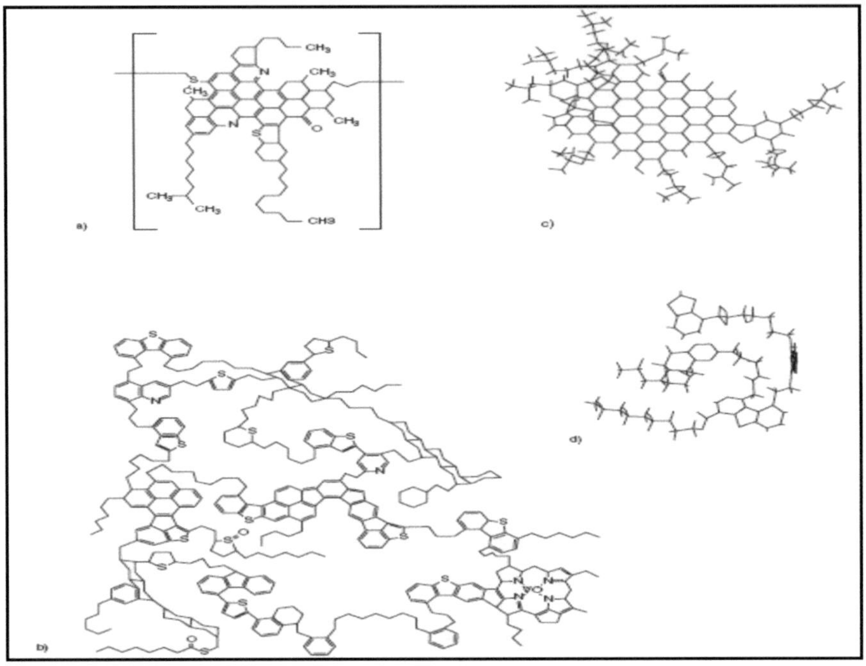

Fig. 9. Examples of structures of asphaltenes: a) crude Venzuelan oil b Athabasca c) continental type structure d) archipelago type structure. Source: Merdrignac and Espinat (2007).

According to Merdrignac and Espinat (2007), the continental structure is defined by asphaltenes with a large central region of aromatic rings while the archipelago structure describes asphaltene molecules with various aromatic regions linked by smaller alkanes.

6.1 Deposition of asphaltenes

Variations in temperature, pressure and chemical composition can cause precipitation of asphaltenes from crude oil. Rainfall and subsequent deposition of asphaltenes can cause problems in all stages of production, for example, transportation and processing, causing the loss of efficiency equipment in steps of production of crude oil. In the reservoir rock, seal can cause partial or complete its pores, resulting in the loss of oil recovery.

The formation of the asphaltenes deposit is one of the most studied phenomena in the production and processing of crude oil. Researches are looking increasingly for the improvement about the chemical structure of molecules present in the asphaltene fraction and, consequently understand its behavior in oil.

According to Leon et al (2000), the main features observed in asphaltene of petroleum that present deposition problems are: high atomic ratio carbon/hydrogen, high aromaticity and high condensation of aromatic rings.

Asphaltenes are found in heavy oil and, therefore, distillation residues, affect refining operations. The asphaltenes act as coke precursors and lead to deactivation of catalysts. They are the main contributors to the formation of deposits in refinery equipment, and because of this, refinery units must be taken out of service for removal of deposits, thus increasing the costs (Gonçalves et al, 2007).

The cost associated with the asphaltene deposition during production and refining operations is in the order of billions of dollars a year. For this reason, the prevention or minimization of precipitation of asphaltenes is an important goal for many oil companies (Rogel et al, 2010).

6.2 Extraction of asphaltenes

Boussingault (1837) and Marcusson (1931) did a remarkable job on asphaltenes of oil that was used to establish a procedure for the separation of these, developed by Nellensteyn (1933), based on the solubility of asphaltenes in carbon tetrachloride. This procedure converged to the method known today for the separation of asphaltenes using n-heptane or n-pentane as a flocculating agent.

Nellensteyn (1933) proposed not only a method of separation, but also suggested a conceptual outline of the structure of asphaltenes in oil. He proposed that asphaltenes are formed by high molecular weight hydrocarbons that form a colloidal system that can be adsorbed on a surface. This revolutionary idea at the time, is so incredibly precise that although questionable, persists today (Sheu, 2002).

Due to the proposed Nellensteyn (1933), there was an enormous effort in studying the fundamental properties of asphaltene molecules, such as molecular weight, structure and characteristics related to these properties, as well as the influence of extraction method on the type of asphaltene fraction obtained (Silva, 2003).

The Institute of Petroleum of London (Standard Methods for Analysis and Testing of Petroleum and Related Products - vol.1, IP-143) developed a methodology which is a standardized test which consists in the precipitation of part of the oil with n-heptane and then dissolving the precipitate with toluene. The precipitate is soluble in toluene and is then called asphaltenes. This methodology, as well as its American version (ASTM 6560-00), are commonly used by the oil industry for the quantification of asphaltenes.

There are several methods of extraction of asphaltenes, and although these are well accepted, there are questions, due to be a fraction of asphaltene a solubility class. There is debate about the different extraction methods and changes that these procedures can generate on the properties of this asphaltenic fraction.

According Shkalikov et al (2010) the yield of asphaltenes depends on certain factors such as temperature, pressure, ratio sample/solvent, performance of preparation steps such as filtration, repeated washing of the precipitated asphaltenes with solvents and drying. All these variations certainly complicate the comparison of results by generating different asphaltenes. Currently, the most active researchers in this area are already talking about the search for a unique and standards methodology to ensure uniformity of concepts.

To better understand the aggregates generated by asphaltenes during the separation processes, it is also important to characterize the resins, as these are also part of the composition and can also act as surfactants to stabilize emulsions.

6.3 Resinas

The dispersion of asphaltenes is mainly attributed to the resins (polar aromatic). The resin molecules play a role of surfactants in stabilizing colloidal particles of asphaltenes in oil. There are concepts about precipitation of asphaltenes and the most widely accepted says that the dissolution of resins is followed by precipitation of asphaltenes (Shkalikov et al, 2010). On this basis, the stability of oil can be represented by three phase systems: asphaltenes, aromatics (including resins) and saturated, which are delicately balanced (Speight, 1992).

The presence of resins in oil prevents the precipitation of asphaltenes by keeping the same particles in colloidal suspension. When a solvent is added to oil, resins are dissolved in the liquid, leaving active areas of asphaltene particles, which allow the aggregation of the same and, consequently, precipitation (Andersen and Speight, 1999).

According to Oseghale and Ebhodaghe (2011) the stability of asphaltenes in oil depends on the ratio resin/asphaltene in the oil. In contrast to the asphaltenes, resins are soluble in n-alkanes as n-heptane and n-pentane (Shkalikov et al, 2010). According to Speight (1992) is the criterion of solubility which allows setting them and consider that they have a similar structure of asphaltenes, however, with a molecular weight less than these (Andersen and Speight, 1999) (Figure 10).

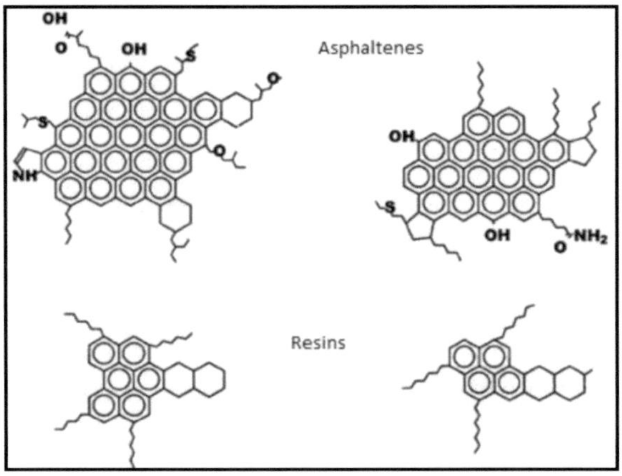

Fig. 10. Structures of hypothetical resins. Source: Rogel (2000).

Resins compared to asphaltenes have a lower content of aromatics, but are rich in heteroatoms, mainly oxygenates (Faria, 2003).

It should be noted that during the process of purification and precipitation of asphaltenes due to the existence of clusters, always exists the possibility of a certain amount of resin precipitate with the asphaltenes.

6.4 Characterization of asphaltenes

The need for knowledge of molecular structure of asphaltenes is the key to developing treatments to prevent their precipitation and may help in understanding its function as a stabilizer of emulsions.

According Merdrignac and Espinat (2007), the detailed structural characterization of heavy fractions is generally difficult to achieve, mainly due to the large complexity of fraction and limitations of analytical techniques. Structural information has been obtained, but can not represent all the chemical and structural variety that these complex mixtures of asphaltenes may contain. Another problem that hinders the characterization of aggregation is the phenomenon presented by the asphaltenes. That is, the asphaltenes have a tendency to form aggregates of high molecular weight, whose distribution depends on the solvent employed, pressure and temperature, making it difficult to know their true distribution in the original oil (Carvalho, 2003).

In recent decades, NMR has been used as a tool for the characterization of mixtures, especially the ^{13}C NMR, providing relevant information about the structure of complex systems such as asphaltenes. Simultaneous use of ^{1}H and ^{13}C NMR allows the determination of a series of structural parameters such as fraction of aromatic carbon, the average number of carbons in an alkyl attached to aromatic systems and the percentage replacement of this system (Skoog et al, 2002).

The NMR technique, in particular, provides reliable molecular parameters about characteristics of aromatic rings and aliphatic chains of asphaltene structures (Speight, 1999).

6.5 The impact of asphaltenes in petroleum refining

In the activities of the oil industry, the deposition of organic compounds is frequent. Among the deposits that cause operational problems, we can identify two predominant groups: paraffins and asphaltenes. Therefore, it is necessary to determine the conditions under which these deposits occur and the way in which they can be avoided in order to generate the least possible damage to the process (Smith, 2003).

Deposits may occur in reservoir rock and source rock for oil. This impairs the production of the well by causing the blocking of pores of the rock and by changing a very important property of the reservoir rock, its wettability, which is the tendency of a fluid to spread or adhere to a solid surface in the presence of non-miscible fluids, and can be modified by adsorption of polar compounds and/or deposition of organic material and thus affect the migration of oil. This is an extremely serious problem, since it can lead to the loss of the well (Faria, 2003; Menechini, 2006).

The phenomenon of deposition can also occur on the production lines. It is known that the use of any method, chemical (injection of solvents, for example) or mechanical (using scrapers), to remove this type of deposit is an expensive operation and requires a lot of security because any accident can lead to line loss (Carnahan, 1989).

The deposition of asphaltenes can also happen in separators during the final stage depressurization of oil (Almehaideb and Zekri, 2001), as well as in almost all stages of production, processing and transportation of oil and is an extremely serious problem that affects significantly the costs of oil industry.

In refining, these constituents may lead to catalyst deactivation and the formation of waste during the thermal and thermo-chemical processing of heavy residues of oil (Speight, 2001).

Due to the economic impact of this problem, the existing literature about asphaltenes is vast, complex and inconclusive (Chinligarian and Yen, 2000).

The biggest challenge associated with this kind of deposit is: what is chemically known as asphaltenes. For this reason, the asphaltenes have been studied mainly with regard to the identification of chemical structures present in this complex mixture. The molecular knowledge of heavier fractions of oil is not conclusive because of complexity of the molecules involved and the families of molecules that are part of these fractions. Thus, it is recommended that all development work to assist in identifying and characterizing properties of complex molecules present in this fraction, as well as studying the stability of the same over the physical and chemical processes by which oil is to generate products commercial interest that undoubtedly sustain the world economy.

7. Case study

As oil undergoes a series of processes involving heating and atmospheric distillation in order to raise its energy potential Chrisman and Lima, 2009 studied the influence of cutting temperature in the asphaltenic fractions.

In this work, the goal was to identify the differences in the average molecular parameters of asphaltenes obtained during the simulation of atmospheric distillation in the laboratory on five different temperatures. Significant changes were observed in the structures of each of the fractions obtained from two Brazilian oils called A and B, especially in the higher cut temperatures.

The extraction and quantification of content of asphaltenes were performed using the ASTM 6560-00 and characterization of asphaltenes was performed using the analytical techniques: elemental analysis, infrared and nuclear magnetic resonance of ^{1}H and ^{13}C. Significant changes were observed in almost all molecular parameters during the distillation at different temperatures, using as standard asphaltene of crude oil. The results obtained confirm the occurrence of oxidation with the increase of cut temperature, probably because of aromatization of naphthenic rings and closing of lateral chains. Examples of events are presented in proposals for representative structures of these fractions.

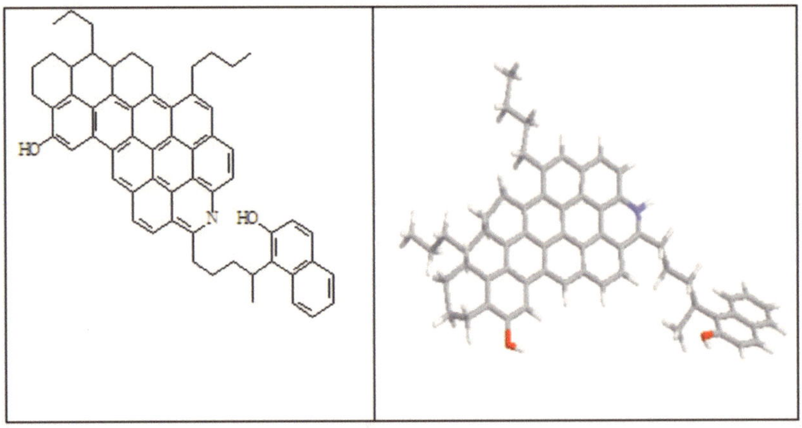

Fig. 11. Structure of asphaltene (crude oil A). Source: Chrisman and Lima, 2009.

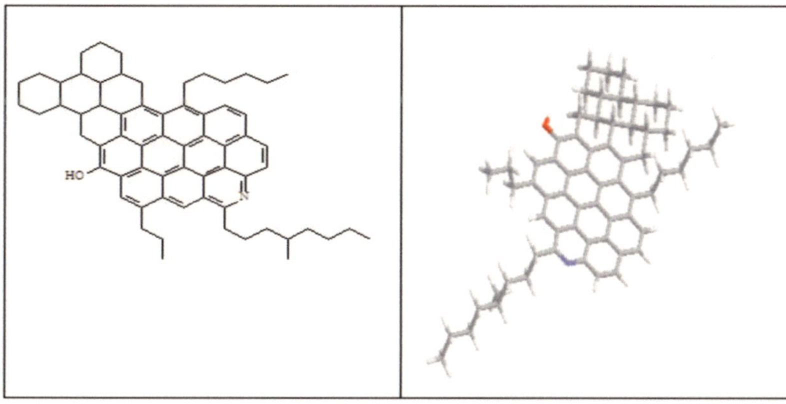

Fig. 12. Structure of asphaltene (residues of cutting to 420°C oil A). Source: Chrisman and Lima, 2009.

It can be observed in the structures shown in Figures 11 and 12 that with increasing temperature, naphthenic rings were formed from lateral chains and even being in some cases, the oxidized aromatic rings. At the temperature of 420°C it was observed aromatization of naphthenic rings.

8. Emulsions

Emulsions are defined as a heterogeneous liquid system consisting of immiscible liquids with one another where one liquid is dispersed in another in the form of drops. The emulsions are distinguished by the amount of liquid dispersed in another one.

An emulsion consists basically of two phases: a continuous phase (external), where the droplets are dispersed, and a dispersed phase (internal or discontinuous), which are themselves dispersed droplets.

Three conditions are necessary for the formation of an emulsion:

- immiscibility between liquid of emulsion
- Shaking to disperse one liquid in another
- Presence of emulsifying agents (surfactants)

The characteristics of an emulsion are constantly changing since the beginning of formation until their complete resolution and vary with temperature, pressure, degree of agitation and time of formation. From a purely thermodynamic point of view, an emulsion is considered an unstable system due to a natural tendency of the system liquid/liquid separation and to reduce their interfacial area and thus their interfacial energy. However, most of the emulsion is stable for a period of time, have kinetic stability that is due to smaller drop sizes and the presence of an interfacial film around the drop.

8.1 Stability of an emulsion

The stability of emulsion can be determined by the type and amount of surface active agents or surfactants that can occur naturally in crude oil, for example, the asphaltenes. These surfactants tend to concentrate in the water/oil interface where form interfacial films stabilizing the emulsion by reducing the interfacial tension (IFT) and promotion of emulsification and dispersion of droplets (Lee, 1999). When energy is added to the system, the particles are broken down into smaller parts and with higher energy, become smaller and, consequently, greater its stability, it is more difficult to treat (Silva, 2008).

Some fine solid particles present in crude oil are able to stabilize emulsions by diffusion into the oil/water interface to form rigid structures that can sterically prevent the coalescence of droplets. To act as stabilizers, the particles must be much smaller than the size of emulsion droplets. They must present themselves at the interface and be sprayed with two phases (aqueous and oily) to stabilize the emulsion. Examples of wet solids in oil are wax and asphaltenes and examples of wet solids in water are inorganic compounds such as, for example, $CaCO_3$ and $CaSO_4$, clay and sand.

The temperature can modify the physical properties of oil, water, interfacial films and the solubility of surfactants in oily and aqueous phases affecting the stability of an emulsion. When increasing the temperature, it is observed a decrease in viscosity of emulsion caused primarily by a decrease in oil viscosity. The temperature increases the thermal energy of drops and, consequently, increases the frequency of drop collisions. This also reduces the interfacial viscosity and results in a rate of faster drainage of the film, thus increasing the coalescence of droplets. The temperature increase leads to a gradual destabilization of interfacial films.

The drop size distribution affects the viscosity of emulsion, which is larger when the droplets are smaller. Usually the emulsion with smaller droplet size is more stable and the time for separation of water must be larger. The viscosity of emulsion will also be higher when the droplet size distribution is narrow (ie, the droplet size is fairly constant).

The pH of aqueous phase has a strong influence on the stability and the type of emulsion formed, it affects the rigidity of the interfacial film. The low pH (acid) generally produces emulsions W/ O (corresponding to wettability in oil, solid films) and high pH (basic) produces emulsion O/W (corresponding to the water wettability, mobile movies).

The emulsions are stabilized by films that are formed around the drops of water in water/oil interface. These films are the result of adsorption of polar molecules of high molecular weight that are interfacial active, i.e. show behavior similar to surfactants. These films increase emulsion stability by reducing the IFT and increased interfacial viscosity. Highly viscous interfacial films act as a mechanical barrier to coalescence. The characteristics of interfacial films vary depending on the type, composition and concentration of polar molecules present in crude oil, temperature and pH of the water. (Kokal, 2005 and Ortiz et al, 2010).

8.2 Viscosity of emulsions

The viscosity of an emulsion is directly proportional to the viscosity of continuous phase and is defined as the relationship between stress and shear rate. Highly viscous oils usually form more stable emulsions These oils cause emulsions difficult to treat, because they decrease the movement of droplets, retarding the coalescence.

The volume fraction of the dispersed phase is the most important factor that affects the viscosity of emulsions. With increasing volume fraction of dispersed phase, the internal circulation is reduced, and the viscosity of emulsion increases.

The effect of particle size distribution on the viscosity of emulsions is very important for high values of concentration of the dispersed phase. For lower concentrations, however, the effect is much smaller. When the average size of water droplets dispersed is lower, higher is the residence time of emulsion.

The shear rate influences the viscosity of emulsions only when it has characteristics of a non-Newtonian fluid. For low values of concentration of dispersed phase, the emulsion exhibits characteristics of Newtonian fluid and, consequently, the shear rate does not affect the viscosity of emulsion. For high values of concentration of dispersed phase, the emulsions exhibit features non-Newtonian (pseudoplastic fluids) and the apparent viscosity decreases significantly with an increasing shear rate (Kokal, 2005 and Silva, 2008).

8.3 The impact of asphaltenes in oil emulsions

The formation of an emulsion W/O can be a serious obstacle for the production of oil, and some oils are particularly prone to form more emulsions than others (Muller et al, 2009). According to Ortiz et al (2010), treatment of these emulsions W/O is still a challenge in the oil industry due to the high stability versus coalescence.

Phenomenological investigations of physical and chemical properties related to the strength of emulsion are described in the literature in terms of mechanisms, properties and classes of potential compounds that stabilize emulsions.

Characteristics commonly linked to w/o emulsions are the API gravity, total acidity index (TAI) and asphaltene content (Muller et al, 2009). In the petroleum industry, most of emulsions produced is of type A/O. Figure 13 illustrates an emulsion W/O, where water droplets are dispersed in oil. It is observed from this figure that stable emulsions are characterized by properties that prevent the coalescence of small drops of water, while in unstable emulsions the water droplets coalesce rapidly.

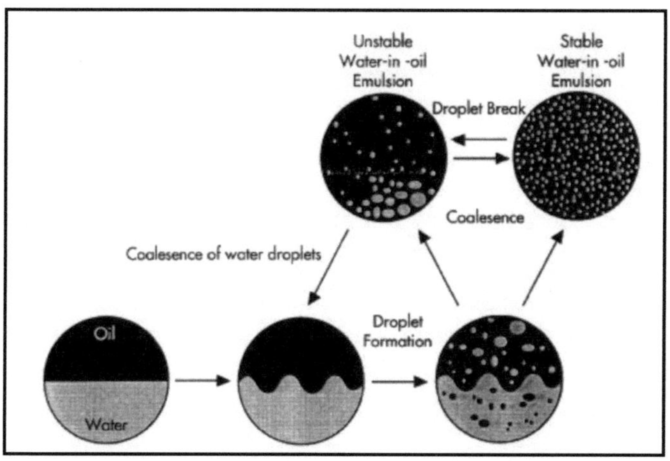

Fig. 13. Formation of emulsions A / O. Source: Lee, 1999.

A stable emulsion W/O consists of an aqueous phase, an oil phase and an emulsifying agent. Certain compounds and particles found in crude oil can act as emulsifying agents (surfactants) and thus promote and stabilize these emulsions. Surfactants have hydrophilic and hydrophobic regions so as to fall within the oil-water interface and stabilize emulsions. If the concentration of particles and surfactants are sufficiently high, then the coalescence of water droplets is prevented, leading to stable emulsions. Figure 14 shows the stabilization of a drop of water in an oily continuous phase by the presence of surfactant.

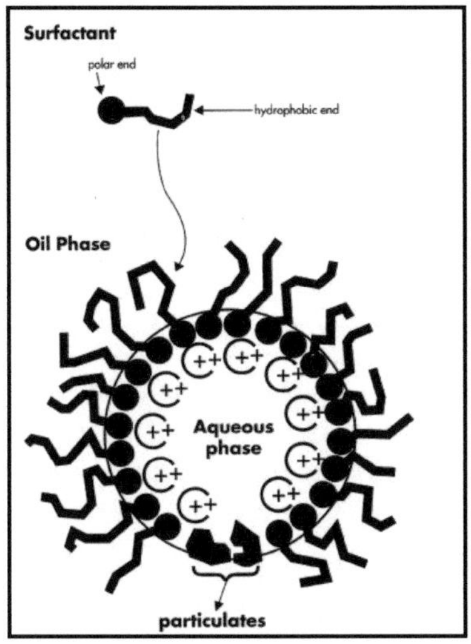

Fig. 14. Emulsion A/O stabilized by surfactants. Source: Lee, 1999.

Certain fractions of polar and high molecular weight exist in crude oil, natural surfactants considered, contribute to the formation of emulsions A / O. These fractions include waxes, asphaltenes and resins and can be dissolved or particulate form (Wei et al, 2011; Ortiz et al, 2010; Kokal, 2005; Lee, 1999). These compounds are seen as the main constituents of interfacial films, where they accumulate and thus stabilize the droplets and, consequently, the emulsion formed around the droplets.

The accumulation of asphaltenes at the interface results in the formation of a hard film. According to Ortiz (2010) when asphaltenes adsorb on the water/oil interface, they form an interfacial film with high elasticity.

The state of asphaltenes in crude oil also has an effect on its stabilizing properties of emulsions. The asphaltenes will stabilize emulsions when they are present in colloidal form. There is strong evidence that its properties are significantly increased when stabilizers are precipitated in the oil.

If emulsifying agents do not exist in crude oil, the instability of the system contributes to the coalescence, facilitating phase separation. If there is the presence of an emulsifying agent, there will be a greater stability of droplets hindering the natural separation of the phases.

The schematic diagram shown in Figure 15 represents a drop of water stabilized by asphaltenes and paraffin crystals. A region not stabilized is shown with the formation of an incomplete barrier.

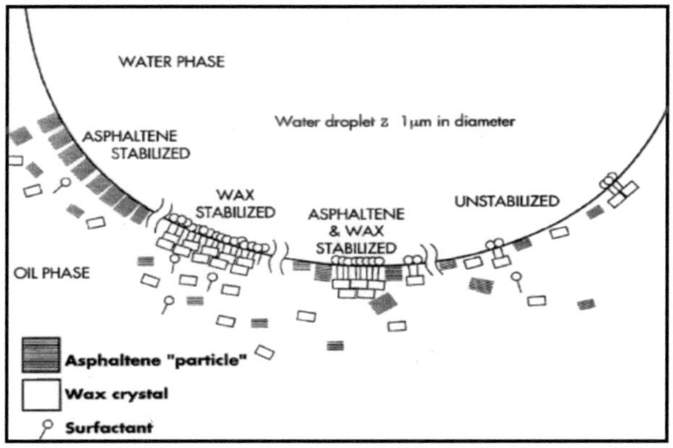

Fig. 15. Stabilization of a drop of water. Source: Lee, 1999.

Several studies demonstrate the importance of asphaltenes, resins and paraffins existing in crude oil in the promotion and stabilization of emulsions of water-in-oil.

9. Case study

Petrobras has about 65% of the area of its offshore exploration blocks in water depths greater than 400m, consequently it is increasing its activities in exploratory drilling in ever deeper waters, resulting in emulsions in almost all phases of production and processing of oil.

Therefore, there is a great need to understand the mechanism of stabilization of emulsions of oil in order to increase production rates and the efficiency of separation that can be accomplished by applying methods such as thermal, mechanical, electrical and/or chemical (Kokal, 2005; Nordgard et al, 2009).

Studies on the form of interaction between natural surfactants species present in oil and interfacial film can assist in developing more efficient methods of separation. From this point of view, asphaltenes can be studied by seeking a greater understanding of this complex fraction in terms of structure and composition, as a species that can contribute to the stabilization of these emulsions.

The laboratory DOPOLAB is currently studying the influence of asphaltenes and resins in the stability of Brazilian oil emulsions. For this, certain physico-chemical characteristics of oil are determined such as viscosity, density, °API, water content, chloride content, total acidity index; extraction of asphaltenes following the standard ASTM6560/00 and characterization of asphaltenes and resins through techniques of elemental analysis, IR and ^{1}H and ^{13}C NMR.

Interfacial tension tests, electrocoalescence and interfacial rheology studies using resins and asphaltenic fractions are in progress.

As preliminary tests to show surfactant properties of asphaltenes were performed interfacial tension measures in a Krüss Tensiometer DSA100 using the pendant drop method. The analysis time was 90min and a needle with a diameter of 1.463 mm and a 500μL syringe used.

The toluene + asphaltenes solutions were made solubilizing 0.5 g of asphaltene in 100 ml of toluene. First the interfacial tension was measured between toluene and water to have a default value and can then compare the results. The interfacial tension measurements of toluene+ asphaltenes solutions were determined in duplicate, so the values of the interfacial tension shown in Table 04 are the averages of duplicates determined.

Sample	Interfacial
Toluene	33,69 ±0,09
Toluene + AA	23,35 ± 1,14
Toluene + AB	30,73 ± 0,04
Toluene + AC	30,19 ± 0,38
Toluene + AD	21,43 ± 0,06
Toluene + AE	28,67 ± 0,14
Toluene + AF	30,93 ± 0,65

Table 4. Values of interfacial tension for asphaltenes. Source: DOPOLAB, 2011.

Analyzing the results, it is observed that the value of interfacial tension between water/toluene was 33.69 mN/m and that for all solutions containing asphaltenes the value was less than this. With these preliminary data we can see that asphaltenes have surfactant properties as the interfacial tension decreased.

Figure 16 shows graphically the behavior of the interfacial tension of each asphaltene studied.

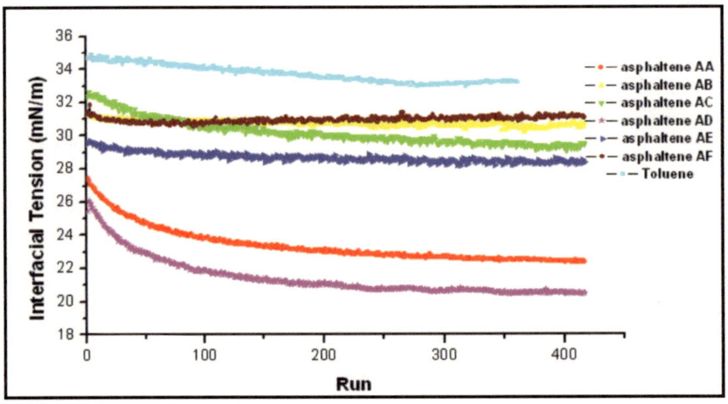

Fig. 16. Graph Interfacial Tension versus race of asphaltenes. Source: DOPOLAB, 2011.

Interfacial rheology tests are being carried out by varying the concentration of asphaltenes, resins and asphaltenes/resins to better understand the surfactant properties of these species in the oil.

10. Conclusion

In conclusion we mention that although the literature on asphaltenes is being quite extensive and current, much remains to understanding this complex fraction. The understanding of chemical structures present in this asphaltenic fraction seems to be really important and helpful in understanding their properties, and assist in proposing mechanisms to explain: their interactions with other species (resins), its precipitation;

participation in the stabilization of emulsions, and many other problems can be avoided and/or minimized since then.

11. Acknowledgment

The authors would like to thank Petrobras for providing the oil samples and give financial support.

12. References

Almehaideb, R.A.; Zekri, A.Y., (2001). Possible Use of Bacteria/Steam to treat Asphaltene Deposition in Carbonate Rocks. *European Formation Damage Conference, SPE* 11851, 1-11

Ancheyta, J.; Centeno, G.; Trejo, F., (2004). Precipitation, Fractionation and Characterization of Asphaltenes from Heavy and Light Crude Oils. *Fuel*, 83, 2169-2175

Bauget, F.; Langevin, D.; Lenormand, R.; (2001). Effects of Asphaltenes and Resins on Foamability of Heavy Oil. Paper SPE-2001 (71504) Presented at SPE *Annual Technical Conference and Exhibition in New Orleans, Lousiana*

Bernucci, L.B.; Motta, L.G.; Ceratti, J.A.P.; Soares, J. B. (2006). Pavimentação Asfáltica: Formação básica para engenheiros. *PETROBRAS*: Abeda,504p

Boussingault, J.B., (1837). *Ann. Chem. Phys.* (64), 141. Cited by Sheu (2002)

C. Ijogbemeye Oseghale and F. O. Ebhodaghe, (2001). Asphaltene Deposition and Remediation in Crude Oil Production: Solubility Technique. *Journal of Engineerind and Applied Scinces* 6 (4): 258-261, 2011

Caldas, J. N. (1997), Estudo Experimental e Modelagem Termodinâmica da Floculação dos Asfaltenos, *PhD Thesis*, COPPE/UFRJ, Rio de Janeiro

Calemma, V.; Iwanski, P.; Nali, M.; Scotti, R.; Montanari, L.; (1995). Structural Characterization of Asphaltenes of Different Origins. *Energy & Fuels*, 9, 225-230

Carnahan, N. F., (1989). Paraffin Deposition in Petroleum Production - *Journal of Petroleum Technology*, v 41, 1024-1025

Carvalho, C.C.V., (2003). Extração e Fracionamento da Asfaltenos de Petróleo. 100p. *Dissertation*, Escola de Química, Universidade Federal do Rio de Janeiro - UFRJ, Rio de Janeiro

Chrisman, E. C. A. N and Lima, V. S., (2009). The Influence of the Cut Temperature on Asphaltenics Fractions. *Chemical Engineering Transactions*, (17), 1455-1460, DOI :103303/CET0917243

Deo, M. D.; OH, K.; RING, T. A., (2004). Asphaltenes Aggregation in Organic Solvents. *Journal of Colloid and Interface Science*, 271, 212-219

Duda Y., Lira-Galeana C., (2006). Thermodynamics of asphaltene structure and aggregation. *Fluid Phase Equilibria* 241, 257–267

Faria, F.R.D., (2003). Caracterização de Petróleos: Influencia do Numero de Componentes no Cálculo de Processo. 55p. *Dissertation*, Escola de Química, Universidade Federal do Rio de Janeiro - UFRJ, Rio de Janeiro

Gauthier T., Danial-Fortain P., Merdrignac I., Guibard I., Anne-Agathe Quoineaud, (2008). Studies on the evolution of asphaltene structure during hydroconversion of petroleum residues. *Catalysis Today* 130, 429–438

Gonçalves M. L. A., Ribeiro D. A., Teixeira A. M. R. F., Teixeira M.G., (2007). Influence of asphaltenes on coke formation during the thermal cracking of different Brazilian distillation residues. *Fuel* 86 619–623

Asphaltene (Heptane Insolubles) in Petroleum Products, in Standards for Petroleum and its Products, *IP 143/90 ou ASTM 6560* pp. 143.1-143.7, Institute of Petroleum, London, UK

Kilpatrick, P. K.; Spiecker, P. M.; Gawrys, K. L.; (2003a). Aggregation and solubility Behavior of Asphaltenes and Their Subfractions. *Journal of Colloid and Interface Science, 267*, 178-193

Kilpatrick, P. K.; Spiecker, P. M.; Gawrys, K. L.; Trail, C. B.; (2003 b). Effects of Petroleum Resins on Asphaltenes Aggregation and Water-in-Oil Emulsion Formation. *Colloids and Surfaces*, 220, 9-27

Kokal, Sunil; (2005). Crude oil Emulsions: A State of The Art. *SPE, Production & Facilities*, Saudi Aramco

Lee Richard F.(1999). Agents Which Promote and Stabilize Water-in-Oil Emulsions. *Spill Science & Technology Bulletin*, Vol. 5, No. 2, pp 117-126

Leon, O., Rogel, E.; Espidel, J., (2000). Asphaltenes: Structural Characterization, Self Association and Stability Behavior. *Energy & Fuels*. 14, 6-10

León, O.; Rogel, E.; Contreras, E.; (2001). Amphiphile Adsorption on Asphaltenes Particles: Adsorption Isotherms and Asphaltene Stabilization. *Colloids and Surfaces*,189, 123-130

Lima, V. S., (2008). Avaliação da Influência da Temperatura de Corte sobre as Frações Asfaltênicas. *Dissertation*. Escola de Química/UFRJ

Marcusson, J., (1931). Die naturlichen and kunstlichen asphalte. Engelmann, Leipzig. Cited by Yen e Chinligarian (2000) e Sheu (2002)

Mariano, J.B., (2005). Impactos Ambientais do Refino de Petróleo. *Editora Interciência* - Rio de Janeiro

Menechini P. O., (2006). Extração, separação e caracterização de frações pesadas de petróleo. *Dissertation*. Escola de Química/UFRJ

Merdrignac I., Espinat D., (2007). Physicochemical Characterization of Petroleum Fractions:the State of the Art. *Oil & Gas Science and Technology* - Rev. IFP, Vol. 62, No. 1, pp. 7-32

Muller, Hendrik ; Pauchard, Vincent O., Hajji, Adnan A., (2009). Role Naphthenic Acids in Emulsion Tightness for a Low Total Acid Number (TAN)/High asphaltenes Oil: Characterization of the Interfacial Chemistry. *Energy & Fuels*, 23, 1280–1288

Mullins, O. C.; Groenzin, H.; Buch, L.; Gonzalez, E. B.; Andersen, S. I.; Galeana, C. L., (2003) Molecular Size of Asphaltene Fractions Obtained from Residuum Hydrotreatment. *Fuel*, 82, 1075-1084

Murgich, J.; Rodríguez, J. M.; Aray, Y.; (1999). Molecular Recognition and Molecular Mechanics of Micelles of Some Model Asphaltenes and Resins. *Energy & Fuels*, 10, 68-76

Murgich, J; (2002) Intermolecular Force in Aggregates of Asphaltenes and Resins. Petroleum *Science and Technology*, 20, 1029-1043

Mustafa Al-Sabawi; Deepyaman Seth;, Theo de Bruijn. (2011). Effect of modifiers in n-pentane on the supercritical extraction of Athabasca bitumen. *Fuel Processing Technology* 92 (2011) 1929–1938

Nellensteyn, F.J., (1933). The colloidal structure of bitumen, *In The Science of Petroleum*, Oxford Press, London, vol.4, 2760

Nordgard Erland L., Sørland Geir, And Sjoblom Johan, (2009). Behavior of Asphaltene Model Compounds at W/O Interfaces. *Langmuir Article* 2010, 26(4), 2352–2360

Normura, m.; Artok, l.; SU, Y.; Hirose, Y.; Hosokwa, M.; Murata, S., (1999). structure and reactivity of Petroleum-Derived asphaltene. *Energy & Fuels*, 13 92), 287-296

Ortiz, D. P.; Baydaka, E.N.; Yarranton H.W.(2010). Effect of surfactants on interfacial films and stability of water-in-oil emulsions stabilized by asphaltenes. *Journal off Colloid and Interface Science*,doi:10.1016/j.jcis.2010.08.032

Pryanto, S., Mansoori, G.A., Suwono, A., (2001). Measurement of property relationships of nano-structure micelles and coacervates of asphaltene in a pure solvent. *Chemical Engineering Science*, 56, 6933-6939

Rogel, Estrella; Ovalles, Cesar; Moir Michael (2010). Asphaltene Stability in Crude Oils and Petroleum Materials by Solubility Profile Analysis. *Energy Fuels* 2010, 24, 4369–4374

Rogel, E.J.; (2000). Simulation of Interactions in Asphaltenes Aggregates. *Energy & Fuels*, 14, 566-574

Sheu, E.Y., (2002). Petroleum asphaltene properties, characterization and issues. *Energy and Fuels*, 16, 74 - 82

Shkalikov, N. V.; Vasil'ev, S. G.; Skirda, V. D. (2010). Peculiarities of Asphaltene Precipitation in n-Alkane–Oil Systems Colloid. *Colloid Journal*, Vol. 72, No. 1, pp. 133–140

Siddiqui, M. N. (2003). Alkylation and oxidation Reactions of Arabian Asphaltenes. *Fuel* 82, 1323-1329

Silva, S.M.C., (2003). Estudo Experimental do Tamanho de Asfaltenos Dispersos em Meios Solventes e Petróleo, 123 p. *PhD Thesis*, Escola de Química, Universidade Federal do Rio de Janeiro - UFRJ, Rio de Janeiro

Silva M. G., (2008). Comportamento Reológico de Emulsões de água em óleo na Indústria Petrolífera. *Monograph*, Universidade Federal de Itajubá Programa de Recursos Humanos da Agência Nacional do Petróleo

Skoog, D.A.; Hooler, F.G.; Nieman, T.A., (2002). Princípios de Análise Instrumental. *Bookman*, 5a edição, Porto Alegre

Speight, J. G.; Long R.B.; Trowbridge, T.O., (1984). Factors influencing the separation of asphaltenes from heavy petroleum feedstock's. *Fuel*, 63, 616-621

Speight, J. G., (1992) Molecular Models for Petroleum Asphaltenes and Implications for Processing. Kentucky University e U.S. DOE EAST, oil Shale. *Symposium Proceedings*, 177

Speight, J. G.; Long R.B., (1995). The concept of asphaltene revisited. *AICHE: Spring National Meeting Preprint*, 78a

Speight, J. G; Yen, T.F.; Chinligarian, G.V., (1994). Chemical and Physical Studies of Petroleum Asphaltenes. Asphaltenes and Asphalts. Developments in Petroleum Science, v. 40, p. 7-61, *Elsevier Science*, Amsterdam

Speight, J. G.; Andersen, S. I., (1999). Thermodynamic Models for Asphaltene Solubility and Precipitation. *Journal of Petroleum Science and Engineering*, 22, 53-66

Speight, J. G., (1999 a). The Chemistry and Technology of Petroleum. *New York. 3rd Edition*, Marcel Dekker

Speight, J. G.; (1999 b). The Chemical and physical structure of petroleum: effects on recovery operations. *Journal of Petroleum Science and Engineering*, 22, 3-15;

Speight, J. G.; (2001). Handbook of Petroleum Analysis. *John Wiley & Sons*, Laramie, Wyoming

Thomas, J.E. (ORGANIZADOR), (2001). Fundamentos de engenharia de petróleo. - Rio de Janeiro: *Editora Interciência*: PETROBRAS

Tissot, B. P.; Welte, D. H.; (1978). Petroleum formation and occurrence: A New Approach to Oil and Gas Exploration. Editor: *Springer- Verlog*, Berlin, Heidelberg

Wang, J.; Fan, T.; Buckley, J. S.; (2002). Evaluating Crude Oils by SARA Analysis. *Paper SPE-*2002 (75228) Presented at SPE/DOE Improved Oil Recovery Symposium in Tulsa, Oklahoma

Wei Wang; Jing Gong; Panagiota Angeli (2011). Investigation on heavy crude-water two phase flow and related flow characteristics. *International Journal of Multiphase Flow* 37 (2011) 1156–1164

Trejo, F., Ancheyta J., (2007). Characterization of Asphaltene Fractions from Hydrotreated Maya Crude Oil. *Ind. Eng. Chem. Res.*, 46, 7571-7579

F. Trejo F.; Ancheyta J.; Morgan T. J.;, Herod A. A., Kandiyoti R., (2007). Characterization of Asphaltenes from Hydrotreated Products by SEC, LDMS, MALDI, NMR, and XRD. *Energy & Fuels* 2007, 21, 2121-2128

Yarranton, H. W. e Gafanova, O. V.; (2001). The Stabilization of Water-in-Hydrocarbon Emulsions by Asphaltenes and Resins. *Journal of Colloid and Interface Science*, 241,469-478

Yasar M., Akmaz S., Ali Gurkaynak M., (2007). Investigation of glass transition temperatures of Turkish asphaltenes. *Fuel* 86 1737–1748

Yarranton, H. W.; Alboudwarej, H.; Jakher, R.; (2000a). Investigation of Asphaltene; Association With Vapor Pressure Osmometry and Interfacial Tension Measurements. *Ind. Eng. Chem. Res.* 39, 2916-2924

Yarranton, H. W.; Hussein, H.; Masliyah, J. H.; (2000b). Water-in-Hydrocarbon Stabilized by Asphaltenes at Low Concentrations. *Journal of Colloid and Interface Science*, 228, 52-63

Yen, T.F.; Chinligarian, G.V., (2000). Introduction to asphaltenes and asphalts. In F.T. Yen e G.V. Chinlingarian (editors), Asphaltenes and Asphalts, 2. Developments in Petroleum Science, 40B, *Elsevier Science*, 1-15

www.autosuficiencia.com.br, visited in 05/2008

www.petrobras.com.br visited in 07/2010

www.monitormercantil.com.br visited in 09/2010

www.anp.gov.br visited in 09/2011

www.bp.com visited in 08/2011

http://pt.wikipedia.org/wiki/petrobras visied in 08/2011

www.dopolab.com.br., DOPOLAB - Laboratório de Desenvolvimento e Otimização de Processos Orgânicos, UFRJ, Brazil

2

Petroleum Asphaltenes

Lamia Goual
University of Wyoming
USA

1. Introduction

A crude oil at atmospheric pressure and ambient temperature has three main constituents: (i) oils (that is, saturates and aromatics), (ii) resins, and (iii) asphaltenes. Asphaltenes and resins are believed to be soluble, chemically altered fragments of kerogen, which migrated from the oil source rock during oil catagenesis. However, unlike resins, asphaltenes contain highly polar species that tend to associate. As a result, the interactions of asphaltenes with their environment are very complex. For example, asphaltene precipitation or deposition can occur in wellbores, pipelines, and surface facilities and is undesirable because it reduces well productivity and limits fluid flow (see Figure 1).

Fig. 1. Asphaltenes clogging a pipe. Courtesy of A. Pomerantz, Schlumberger.

The main factors that promote precipitation are pressure, temperature and composition variations due to gas injection, phase separation, mixing of fluid streams. Paradoxically, asphaltene precipitation is often observed in light crude oils that contain very low asphaltene content. This is because light oils contain large amounts of light alkanes in which asphaltenes have limited solubility. Heavy oils, which are usually rich in asphaltenes, contain large amounts of intermediate components that are good asphaltene solvents. However the refining of heavy oils in downstream operations is very challenging because it can lead to coking, fouling, and catalyst deactivation during processing or upgrading, as high temperatures and vacuum conditions are required. In subsurface formations, the adsorption of asphaltenes on mineral rocks can lead to wettability alteration and formation damage.

Asphaltenes impact virtually all aspects of utilization of crude oil and despite this importance; asphaltenes have not been well understood. Fortunately this situation has recently changed rather significantly. The ability to connect asphaltene science performed at different length scales into a single cohesive picture requires establishing structure–function relationships and is the heart of Petroleomics (Mullins et al., 2007). For example, asphaltene molecular weights are now known to be relatively small, ~800 g/mol (Boduszynski, 1981; Groenzin & Mullins, 1999). Compositional variations of the fluid containing asphaltenes directly affect their aggregation state. Under unfavorable conditions, asphaltene molecules tend to associate into small nanoaggregates that can grow into larger clusters and eventually flocculate and precipitate. The connection between molecular architecture and aggregation was not clarified until recently (Akbarzadeh et al., 2007; Mullins, 2010). The role of resins on asphaltene stability has long been controversial and a more unified view is gradually emerging (Goual et al., 2011). This chapter is based upon the author's experience in asphaltene research over the past decade. It focuses on asphaltene separation, characterization, structure, and role of resins.

2. Definition

The term "Asphaltene" originated in 1837 when Boussingault defined them as the distillation residue of bitumen: insoluble in alcohol and soluble in turpentine (Boussingault, 1937). Today, asphaltenes are defined as the heaviest components of petroleum fluids that are insoluble in light n-alkanes such as n-pentane (nC_5) or n-heptane (nC_7) but soluble in aromatics such as toluene. The solubility class definition of asphaltenes generates a broad distribution of molecular structures. These polydisperse molecules consist mostly of polynuclear aromatics (PNA) with different proportions of aliphatic and alicyclic moieties and small amounts of heteroatoms (such as oxygen, nitrogen, sulfur) and heavy metals (such as vanadium and nickel, which occur in porphyrin structures). Heavy resins and waxes can co-precipitate with asphaltenes and their amount is variable depending on the method of separation. Figure 2 shows the separated asphaltenes and resins from a crude oil according to ASTM-D2007 (Goual & Firoozabadi, 2002). Asphaltenes and resins differ in color and texture. Asphaltenes are black, shiny, and friable solids; while resins are dark brown, shiny, and gummy.

Fig. 2. Asphaltenes and resins separated from crude oils (Goual and Firoozabadi, 2002).

3. Separation

The standard procedures for asphaltene separation consist mainly of precipitation of asphaltenes by excess n-alkanes (typically 40 volumes of n-alkane to 1 volume of oil). In the Institute of Petroleum Standard IP 143 (IP 143/84, 1988), asphaltenes are separated from waxy crude oils with nC_7 then the precipitated phase is washed for 1 h with a reflux of hot heptane to remove waxes. In the ASTM D-3279 method (ASTM D3279-07, 2007), asphaltenes from petroleum residues are precipitated with nC_7 and filtered after 30 min of heating and stirring with a reflux system. In the ASTM D-893 method (ASTM D893-05a, 2010), asphaltenes are precipitated from lubricating oils by centrifugation in nC_7. In the Syncrude analytical method (Bulmer & Starr, 1979), heavy crude oils are mixed with benzene prior to asphaltene precipitation with nC_5; they are then filtered and washed after 2 h of settling in the dark. For each separation method, conditions such as n-alkane, contact time, temperature, filter size, and washing procedure need to be specified. Considering crude oils as a continuum of several thousands of molecules, it is very difficult to define a cut-off between asphaltenes, resins, and oils. The variables introduced in each method may generate different fractions of asphaltenes when using different methods. For example, if nC_7 is used as precipitant for asphaltenes instead of nC_5, then the C_5-C_7 fraction of C_5 asphaltenes will now be part of C_7 resins. A representation of the asphaltenes fractions using molecular weight and polarity/aromaticity is provided in Figure 3 (Long, 1981). In this Figure, the slope of the lines varies with the composition of asphaltene fractions. The concept emphasizes the lower molecular weight and increased polarity of the various asphaltene constituents (Speight, 2007).

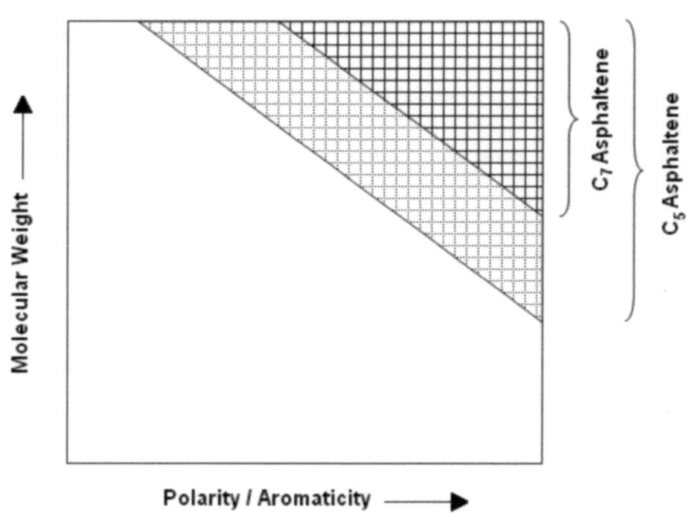

Fig. 3. Representation of n-alkane asphaltenes using molecular weight and polarity/ aromaticity.

Figure 4 depicts the main fractions that can be extracted from a solid-free petroleum fluid. The term "solid" here refers to mineral fines, clays, etc. Asphaltenes extracted from bitumen are very likely to contain solids. In this case, solids can be removed by centrifugation of a 5 wt% solution in toluene at 30,000 g for 3 h (Goual et al., 2006) or by filtration using 0.02 μm filter size. Resins are usually separated from maltenes by preparative liquid chromatography via adsorption on surface-active materials such as fuller's earths, attapulgus clay, alumina, or silica gel (ASTM-D2007; ASTM-D4124; Syncrude analytical method) then desorption with aromatic/polar solvents such as toluene/acetone mixtures. Other resin separation methods by precipitation in propane or ketones or alcohols also exist in the literature but remove only a fraction of resins, usually the heaviest and more polar. Like asphaltenes, resins are defined according to the solvents used for their separation (Goual & Firoozabadi, 2002). Discrepancies in the amount and chemical composition can be found if adsorbents are different or elution time is varied (Wallace et al., 1987).

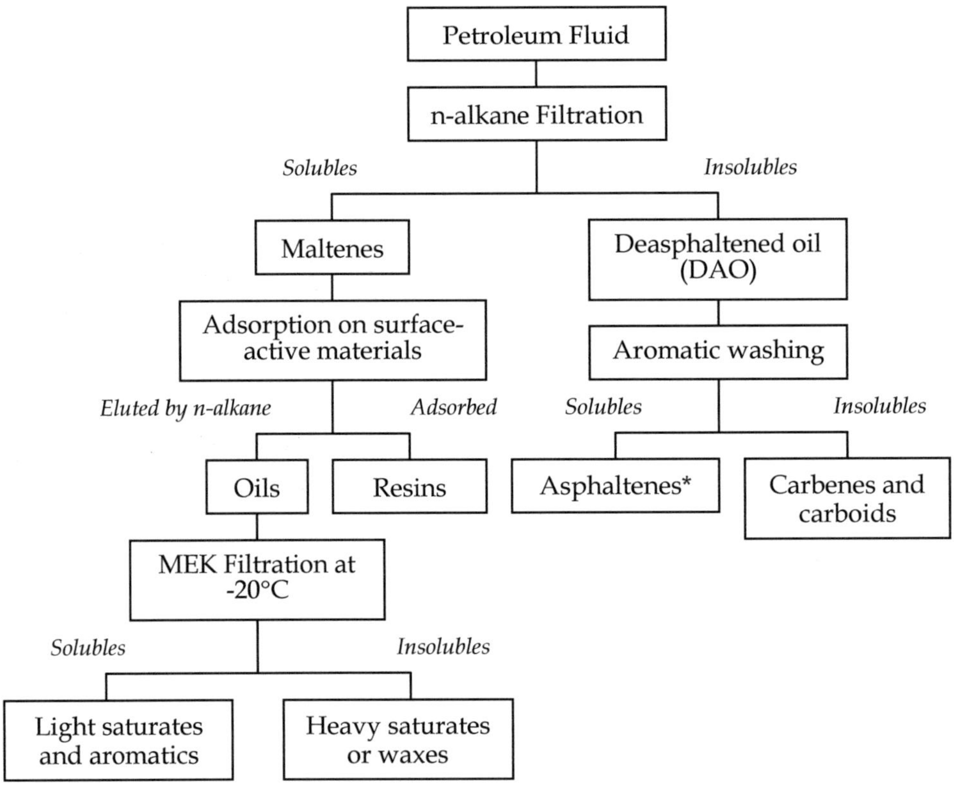

* For waxy oils, heavy waxes are removed from asphaltenes by hot-alkane washing (IP 143/84, 1988)

Fig. 4. General fractionation scheme for petroleum fluids.

Analytical methods such as thin layer chromatography with flame ionization detection (TLC-FID) (Karlsen & Larter, 1991) are widely used in the oil industry. These solubility based separation methods allow for the investigation of crude oil components based on polarity. However they can yield very different amounts of Saturates, Aromatics, Resins and Asphaltenes (SARA) depending on the nature of solvents used in the separation. At a panel discussion on standardization of petroleum fractions held at the 2009 Petrophase conference, a need to unify and improve the separation methods for asphaltenes and resins was expressed (Merino-Garcia et al., 2010). The diversity of operating definitions employed and measurement variability affect the ability of researchers to determine whether compound classes are present and to draw cross-comparisons among measurements from different laboratories. Moreover, the current separation methods create an artificial distinction between resins and asphaltenes yet provide no information on more meaningful distinctions, such as between associating and non-associating species or surface-active versus non-surface-active species. When compared to pressure drop asphaltenes from live oils, heptane-separated asphaltenes from dead oils contained a higher number of rings plus double bonds but exhibited a lower abundance of species containing sulfur (Klein et al., 2006). Thus, the solubility criterion for asphaltenes defines a significantly different chemical composition than the (more field-relevant) pressure-drop criterion.

A new generation of separation methods that permits the extraction of asphaltene nanoaggregates and clusters without solvent use has recently been proposed, namely, ultra- or nanofiltration (Zhao & Shaw, 2007, Marques et al., 2008). Nanofiltration experiments with 5-200 nm mesh show that the composition of filtered asphaltenes closely approximates C_5 asphaltenes over a broad temperature range (Zhao & Shaw, 2007). When used to fractionate small asphaltenic aggregates from larger ones, ultrafiltration tests revealed that small aggregates present lower aromaticity and higher aliphatic composition than larger ones. Their alkyl chains also appear to be shorter and more alkylated. Based on elemental analysis, that smaller aggregates contain a lower metal concentration and are preferentially enriched in vanadium than nickel when compared with larger ones (Marques et al., 2008). The advantage of nanofiltration is that asphaltenes are separated based on their size rather than their solubility, thus nanofiltered-fractions are significantly less polydispersed than alkane-asphaltenes. On the other hand, nanofiltration cannot substitute for solvent standards for routine measurements because of the time as well as the cost involved.

The presence of wax in petroleum fluids renders the separation of asphaltenes even more complex. Figure 4 shows that the heaviest components of wax tend to co-precipitate with asphaltenes and the alkane-soluble wax components can be extracted from oils through precipitation in methyl ethyl ketone (MEK) at -20 °C, then filtration (UOP 46-64). A major drawback of wax and asphaltene separation methods is the fact that they are often time-consuming and require large volumes of solvents. To overcome this limitation, an on-column separation of wax and asphaltenes in petroleum fluids has been proposed (Schabron & Rovani, 2008; Goual et al., 2008) and allows for the detection and separation of these fractions in minutes. The principle of the method is to first precipitate waxes and asphaltenes together on a ground polytetrafluoroethylene (PTFE)-packed column using MEK at -20°C and then re-dissolve the precipitate with solvents of increasing polarity at different temperatures. The development and demonstration of the on-column asphaltene precipitation and re-dissolution technique have been described in detail elsewhere (Schabron & Rovani, 2008). Figure 5 provides an example of separation profile for Dagang waxy crude oil from China using waxphaltene determinator (WAD). The material soluble in MEK elutes as a first peak and consists mostly of aliphatic light oils, including n-alkanes with carbon atoms less than C_{20} and highly branched alkanes. Other functional groups, such as naphthenic or aromatic rings, may be present. The precipitated material is re-dissolved in four steps using solvents of increasing polarity and different temperatures: heptane at -20 °C (for low polarity oils and moderately branched alkanes, possibly containing naphthenic components), heptane at 60 °C (for n-alkanes with carbon atoms higher than C_{20} and slightly branched alkanes), toluene at ~25 °C (for asphaltenes), and then methylene chloride at ~25 °C (for higher polarity asphaltene components).

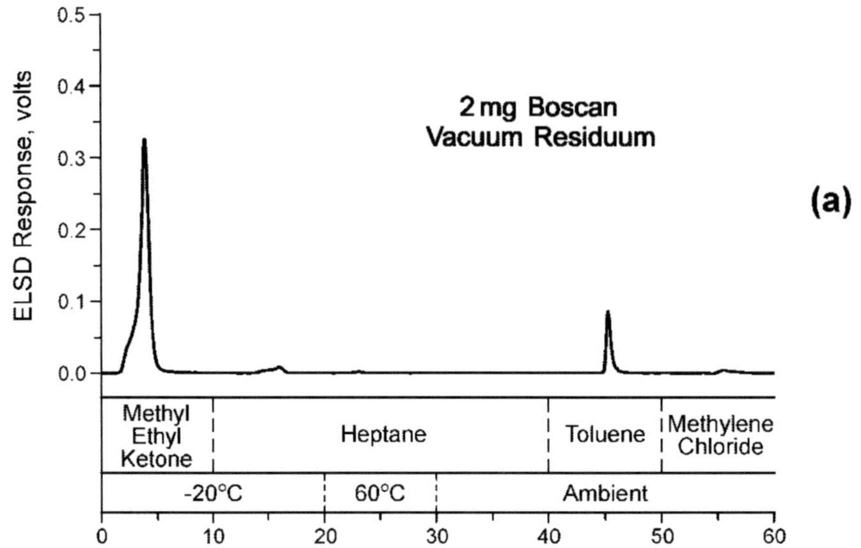

Fig. 5. Separation profile for Dagang crude oil using waxphaltene determinator (Goual et al., 2008).

Note that co-solvency effects have been observed with the WAD method. Co-solvency occurs when individual chemical components that are not soluble in a particular solvent can dissolve readily when they are part of a mixture with other species that impart co- solvency. For example, a significant portion of C_7 asphaltenes from petroleum residua is not soluble in cyclohexane although the whole residua that contain these components dissolve completely in cyclohexane (Schabron et al., 2001). Similar solubility behavior was observed with petroleum resins in *n*-alkanes (Goual and Firoozabadi, 2004). Co-solvency is also observed with wax components. The melting and freezing of *n*-alkane components can also occur in a manner that affects solubility (Goual et al., 2008). The combination of the co-solvency and freezing/melting effects in petroleum systems could be complex. These are also evident when individual chemicals or fractions are isolated from petroleum (Goual & Firoozabadi, 2004; Schabron & Rovani, 2008).

4. Characterization

Asphaltenes have a density between 1.1 and 1.20 g/mL (Speight, 2007), an atomic H/C ratio of 1.0-1.2 (Spieker et al., 2003), and a solubility parameter between 19 and 24 $MPa^{0.5}$ at ambient conditions (Hirschberg et al., 1984; Wiehe, 1996). They strongly affect the rheological behavior of petroleum fluids. Two concentrations regimes have been identified in crude oils: a diluted regime where viscosity increases linearly with asphaltene content, and a concentrated regime where viscosity depends more than exponentially on asphaltene content. Natural heavy oils correspond to the concentrated regime and the high viscosities may be due to the entanglement of solvated asphaltene particles (Argillier et al., 2001).

Studies on asphaltene molecular weight are more controversial. The tendency of asphaltenes to aggregates in toluene at concentrations as low as 50 mg/liter (Goncalves et al., 2004) has led to aggregate weights being misinterpreted as molecular weights with colligative methods such as vapor pressure osmometry (VPO) or gas exclusion chromatography (GPC). As a result, the molecular architecture of asphaltenes has been the subject of debate for several years. Results from advanced analytical techniques now agree that asphaltene molecular weight distributions are in the 400-1,500 Dalton (Da) range, with a mean mass between 700-800 Da. Examples of these methods are field-ionization mass spectrometry (FIMS) (Boduszynski, 1981), fluorescence correlation spectroscopy (FCS) (Schneider et al., 2007), time resolved florescence depolarization (TRFD) (Groenzin & Mullins, 1999, 2000), electrospray ionization, Fourier transform ion cyclotron resonance mass spectrometry (ESI FTICR MS) (Rodgers & Marshall, 2007; Hsu et al., 2011), atmospheric pressure photoionization mass spectrometry (APPI MS) (Merdrignac et al., 2004), field-desorption/field- ionization mass spectrometry (FDFI MS) (Qian et al., 2007), laser desorption ionization (LDI) (Hortal et al., 2006).

The polar character of asphaltenes impacts their adsorption at interfaces (Abudu & Goual, 2009; Saraji et al., 2010). Figure 6 depicts the adsorption amounts from 5 wt % crude oil in toluene and heptane on different substrates (Abudu & Goual, 2009). The substrates consist of hydrophilic surfaces (gold, silica, and stainless steel) and hydrophobic surfaces (polystyrene). The amount adsorbed from toluene on hydrophilic surfaces varies from 350 to 450 ng/cm^2, with the highest being on silica.

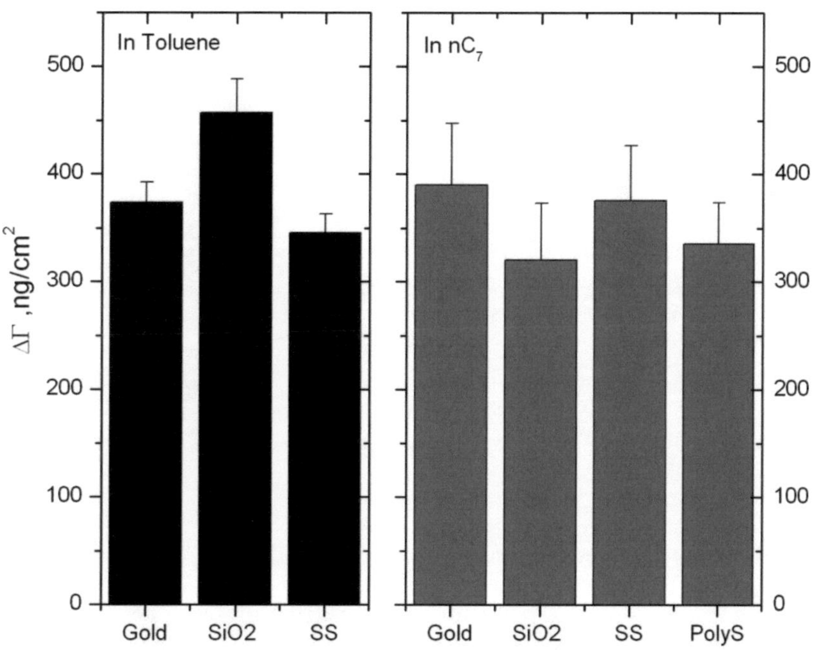

Fig. 6. Adsorption amounts from 5 wt% crude oil in toluene and heptane on different surfaces (gold, silica, alumina, polystyrene, and stainless steel).

Figure 6 also shows that asphaltenes are almost amorphous in heptane, evidenced by almost the same adsorption amounts on hydrophilic and hydrophobic surfaces (~350 ng/cm^2). XPS survey revealed the presence of relatively large percentages of surface heteroatoms (mainly oxygen and sulfur) in adsorbed films from toluene which suggests that asphaltenes are much more hydrophilic in toluene than in heptane. It is likely that asphaltene polar functions become caged inside clusters because of flocculation in heptane (Sheu & Acevedo, 2006).

What role polarity and/or aromaticity play in aggregation and molecular structure is still largely unknown. On the other hand, non-polar dispersive forces are believed to control the flocculation and precipitation of asphaltene clusters (Buckley et al., 1998). In a recent study of the interactions between asphaltene surfaces in organic solvents using an atomic force microscope (AFM), Wang and co-workers found that the ratio of toluene to heptane in the solvent could significantly change the nature and the magnitude of the interaction forces between asphaltene surfaces. In pure toluene, steric repulsion forces were measured whereas in pure heptane, van der Waals attraction forces were measured (Wang et al., 2010).

The polar character of asphaltenes originates from the presence of permanent dipoles (Goual & Firoozabadi, 2002) and electrical charge carriers (Goual et al., 2006). Measured dipole moments and electrical direct current (DC) conductivities of asphaltene monomers are in the range of 4-7 D (Goual & Firoozabadi, 2002) and 10^{-8} S/m (Goual, 2009), respectively. Electrodeposition studies of solid-free bitumen in organic solvents revealed that asphaltene charge carriers are mainly positively charged in organics solvents such as toluene (Goual et al., 2006). Figure 7 illustrates the contribution of these charge carriers to the DC conductivity.

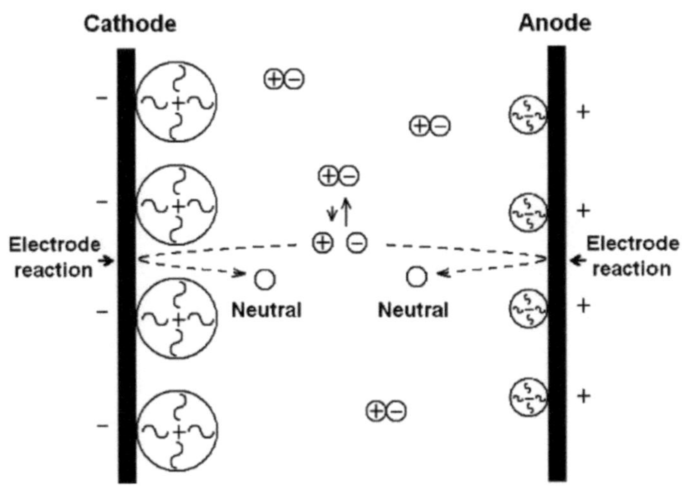

Fig. 7. Schematic of the contribution of charge carriers to the DC conductivity in crude oils

Lewis acids and bases form ion pairs and permanent dipoles. The dissociation of these species produces small-size charge carriers that are involved in charge-transfer reactions at the electrodes and are responsible for the DC conductivity. These acids and bases are also involved in asphaltene association processes leading to high aggregate mass. In these aggregates, the numbers of Lewis acids and bases are not balanced exactly. Electron exchange between the electrodes and these associated acids and bases is hindered (because of steric constrains) so they can only be electrocollected and not electrodeposited at the electrodes; they are quickly released from the electrode surface when the potential is turned off. These species do not contribute to the DC conductivity. The DC conductivity of asphaltenes increases with their concentration in toluene because of the increased ion mobility or reduced activation energy for charge transfer, as illustrated in Figure 8. Note that less than 10^{-4} mole fraction of asphaltene is charged in toluene. These charge carriers can act as tracers to monitor asphaltene dynamics.

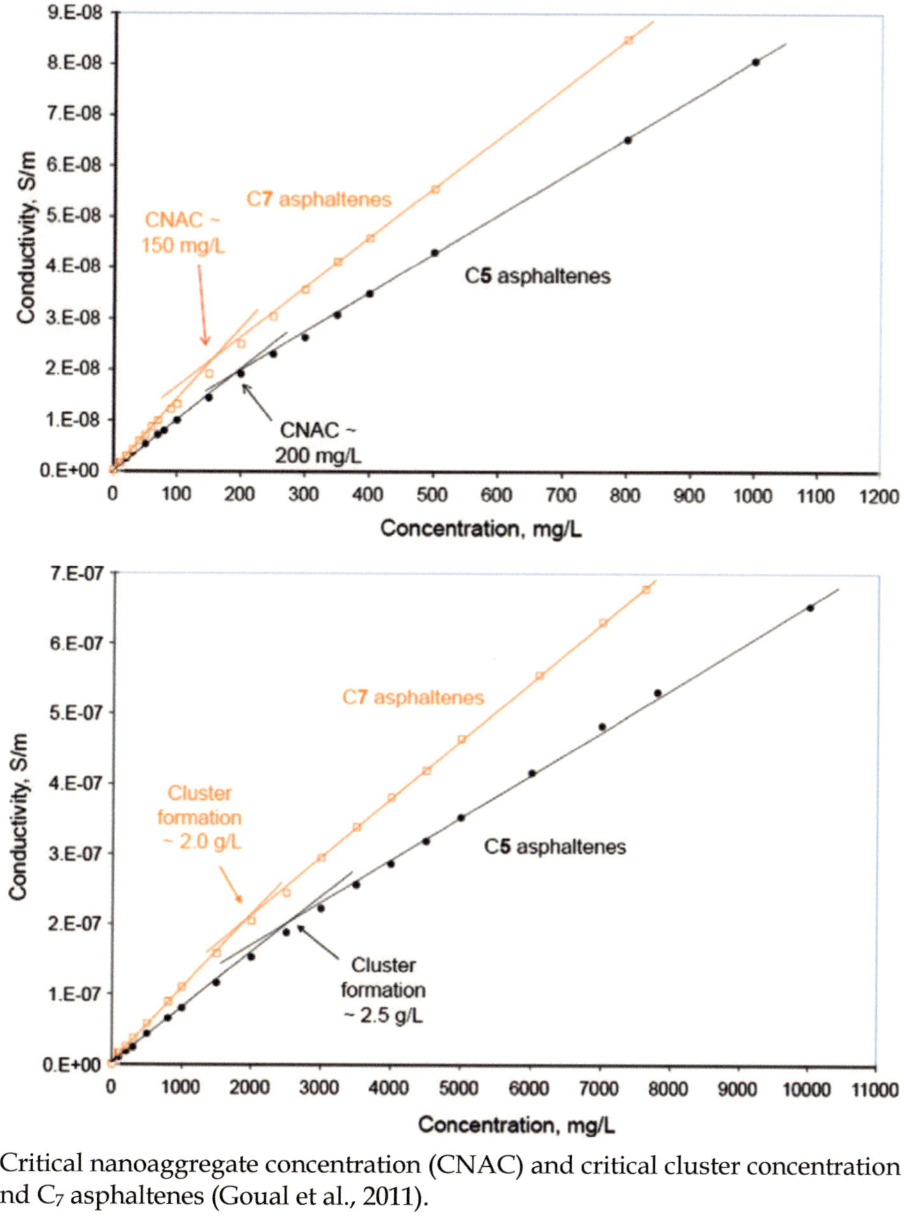

Fig. 8. Critical nanoaggregate concentration (CNAC) and critical cluster concentration (CCC) of C_5 and C_7 asphaltenes (Goual et al., 2011).

The critical nanoaggregate concentration (CNAC), above which nanoaggregates stop growing, can be seen in Figure 8 as a clear breakpoint in the variation of conductivity with concentration. The breakpoint occurs at around 100 mg/L for C_5 asphaltenes and 200 mg/L for C_7 asphaltenes, in agreement with previous work (Andreatta et al., 2005; Freed et al.,

2009; Zeng et al., 2009; Mostowfi et al., 2009). Similarly, the critical cluster concentration (CCC) can be seen at around 2,000 and 2,500 mg/L for C_7 and C_5 asphaltenes, respectively. The CNAC and CCC increase somewhat with the inclusion of the heaviest resins in the C_5 asphaltenes (Goual et al., 2011). A clear demonstration of the formation of asphaltene clusters in toluene was obtained by measurement of the kinetics of asphaltene flocs upon addition of n-heptane to asphaltene-toluene solutions using dynamic light scattering (Yudin & Anisimov, 2007). Below a concentration of ~3 g/liter, the kinetics of floc formation are diffusion limited aggregation; whereas above this concentration the kinetics are reaction limited aggregation. This is consistent with the DC conductivity data in Figure 8.

In low-frequency dielectric relaxation studies (Goual, 2009), DC conductivities were related to the effective diffusion coefficient of asphaltenes via the Nernst-Einstein equation, from which average sizes were calculated assuming asphaltenes are spherical in shape. It was shown that asphaltenes form viscoelastic films at solid interfaces when they are close to the flocculation threshold (Abudu & Goual, 2009) or when their size in toluene is smaller than ~3 nm (Goual & Abudu, 2010), as shown in Figure 9. The topographical features of rigid (d > 3 nm) and viscoelastic (d < 3 nm) asphaltene films obtained by AFM Tapping mode imaging in air are displayed in Figure 10. The viscoelastic films consist of soft multilayers (i.e., toluene-rich) of small aggregates that possibly interact with each other, whereas rigid films consist of a monolayer of particles with few large aggregates scattered on its surface.

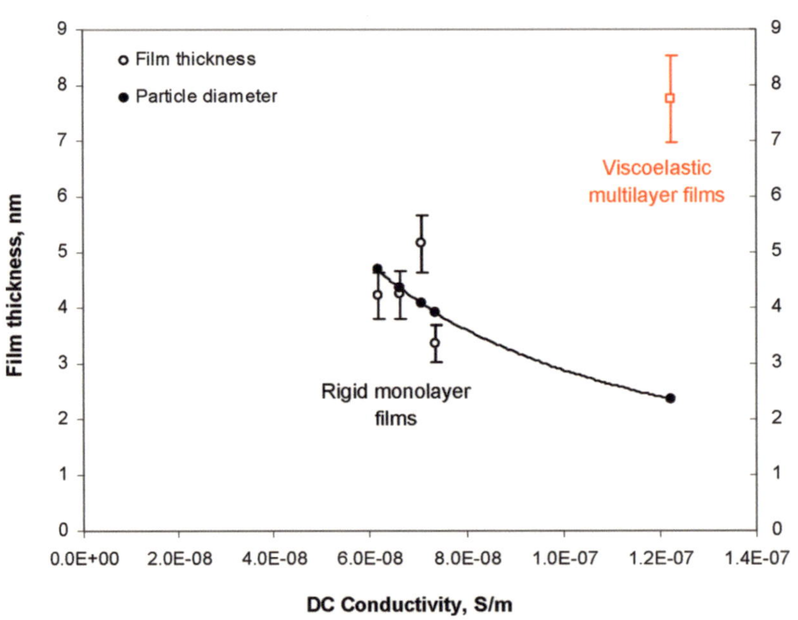

Fig. 9. Relation between film thickness, particle diameter, and DC conductivity of 0.1 wt % asphaltenes in toluene.

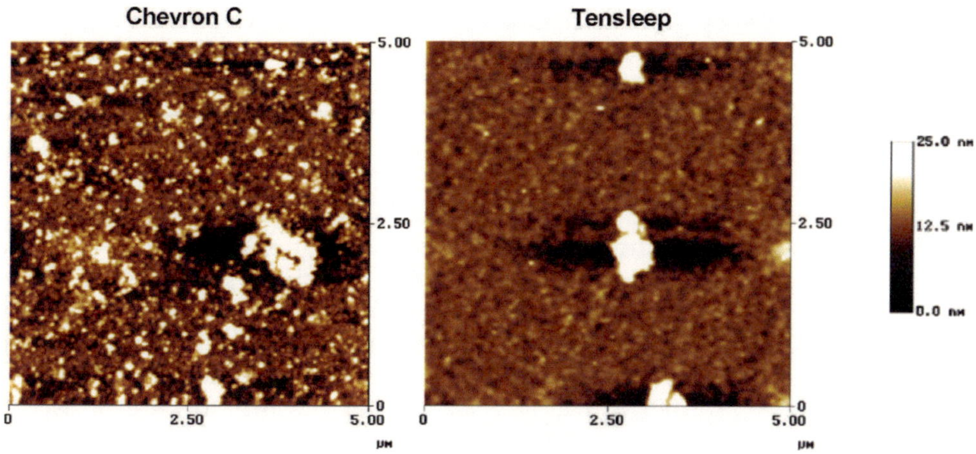

Viscoelastic layer (d < 3 nm) Rigid layer (d > 3 nm)

Fig. 10. AFM images of adsorbed particles on gold from 0.5 wt% crude oils in toluene.

Thus, data derived from low-frequency dielectric relaxation are important in predicting asphaltene-associated problems. Low frequency measurements can also produce sensor responses that are sensitive to interactions between asphaltenes and surfactants usually employed in chemical flooding.

5. Structure and role of resins

The chemical structure of asphaltenes is difficult to ascertain due to the complex nature of asphaltenes. Two models have been proposed to describe the molecular architecture in asphaltenes: (i) archipelago models, and (ii) island models. The Archipelago models consider that several aromatic moieties are bridged together via aliphatic chains (Strausz et al., 1992), whereas island models suggest that there is predominantly one fused polycyclic aromatic hydrocarbon (PAH) ring system per asphaltene molecule with pendant aliphatic chains (Dickie & Yen, 1967). New fragmentation studies by two-step laser desorption laser ionization mass spectrometry (L2MS) (Sabbah et al., 2011) and FTICR-MS (Hsu et al., 2011) support the contention that the dominant structural character of asphaltenes is island-like. These results are in accord with other methods such as TRFD (Groenzin & Mullins, 2000), high-Q ultrasonics (Andreatta et al., 2005), and NMR (Freed et al., 2009). Recently, DC conductivity studies showed that resins are unlikely to coat asphaltene nanoaggregates in anhydrous organic solvents (Sedghi & Goual, 2010). Thus, the long-time-standing Nellensteyn hypothetical model (Nellensteyn, 1938), where resins adsorb on asphaltenes to provide a steric stabilizing layer, is not valid. However when water is present in the solvent, the role of resins on asphaltene adsorption at the solvent/water interface becomes important. Resins and natural surfactants tend to diffuse first to the interface before being replaced by asphaltenes (Magual et al., 2006). The amount of adsorbed asphaltenes on water also depends on the resin to asphaltene ratio (Goual et al., 2005).

Based on the previous studies, a new asphaltene model has been codified in the "modified Yen model" and stipulates the dominant structure of asphaltene molecules, nanoaggregates and clusters of nanoaggregates (Mullins, 2010). This model was built upon the Yen Model, which has been in use for 40 years (Dickie & Yen, 1967). The Yen model has been very useful, particularly for considering bulk properties of phase-separated asphaltenes. Nevertheless, at the time the Yen model was proposed, there were many uncertainties in asphaltene molecular weight, architecture, and colloidal structure. Figure 11 shows the corresponding schematic of the modified Yen model, which incorporates the many substantial advances in asphaltene science particularly over the last ten years (Mullins, 2010). The island architecture exhibits attractive forces in the molecule interior (PAH) and steric repulsion from alkane peripheral groups. These structures were also observed in oil reservoirs with extensive vertical offset, where gravitational effects are evident (Creek et al., 2010; Mullins et al., 2011). Results from analytical methods such as time-resolved fluorescence depolarization (TRFD) (Groenzin & Mullins, 2000), Nuclear magnetic diffusion (Freed et al., 2007); Fluorescence correlation spectroscopy (FCS) (Andrews et al., 2006), DC conductivity (Zeng et al., 2009; Goual et al., 2011) indicate that the size of asphaltene molecules is ~ 1.5 nm.

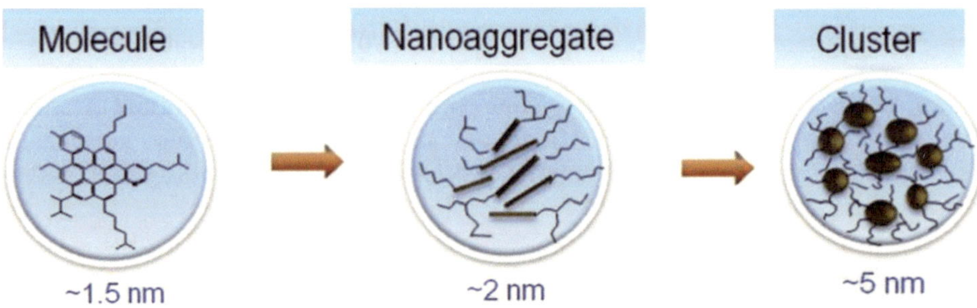

Fig. 11. The modified Yen model. Courtesy of O. Mullins, Schlumberger

With the broad scope and significant applications of the modified Yen model, it is important to validate all aspects of this model. In particular, there are wide ranging studies at many length scales involved in building the modified Yen model. Recently, the cohesiveness of two independent data streams was determined by centrifugation and DC conductivity measurements (Goual et al., 2011). The data indicate that asphaltene nanoaggregates have relatively small aggregation numbers of 4-6, in accord with previous work (Mullins, 2010), and that aggregation is entropically driven (Goual et al., 2011). The relatively small aggregation numbers are consistent with nanoaggregate sizes in the range of 2 nm measured by centrifugation (Mostowfi et al., 2009; Indo et al., 2009) and inferred by small angle X-ray scattering (SAXS) and Neutron scattering (SANS) studies (Eyssautier et al., 2011a). More recent x-ray and neutron scattering analyses confirmed the nanoaggregate architecture in the modified Yen model where PAHs in the aggregate interior are attractive while the alkane substituents act to sterically repel other asphaltene molecules and preclude further aggregate growth (Eyssautier et al., 2011a). The clusters form only at much higher concentration (see Figure 8) because the attractive forces of one nanoaggregate to another are much weaker. Nanofiltration studies indicate that the size of clusters is smaller than 30 nm (Ching et al., 2010). Very recent SAXS and SANS studies revealed that asphaltene clusters consist of ~12 nanoaggregates and, due to their fractal nature, they are very sensitive to temperature variations and solvent type (Eyssautier et al., 2011b). On the other hand, nanoaggregates are insensitive to temperature (Goual et al., 2011; Eyssautier et al., 2011b).

All the latest studies in asphaltene science demonstrate repeated consistency with the modified Yen model proposed in Figure 11. This model provides a foundation that has far reaching implications in many areas associated with the production of crude oil.

6. References

Abudu, A & Goual, L. (2009). Adsorption of Crude Oil on Surfaces Using Quartz Crystal Microbalance with Dissipation (QCM-D) under Flow Conditions. *Energy Fuels*, Vol.23, No.3, pp. 1237-1248

Akbarzadeh, K.; Hammami, A.; Kharrat, A.; Zhang, D.; Allenson, S.; Creek, J.; Kabir, S.; Jamaluddin, A.; Marshall, A.G.; Rodgers, R.P.; Mullins, O.C. & Solbakken, T. (2007). Asphaltenes-Problematic but Rich in Potential. *Oilfield Review*, Vol.19, No.2, pp. 22-43

Andreatta, G.; Bostrom, N. & Mullins, O.C. (2005). High-*Q* Ultrasonic Determination of the Critical Nanoaggregate Concentration of Asphaltenes and the Critical Micelle Concentration of Standard Surfactants. *Langmuir*, Vol.21, No.7, pp. 2728-2736

Andrews, A.B.; Guerra, R.E.; Mullins, O.C. & Sen, P.N. (2006). Diffusivity of Asphaltene Molecules by Fluorescence Correlation Spectroscopy. *Journal of Physical Chemistry A*, Vol.110, No.26, pp. 8093–8097

ASTM D3279-07. (2007). *Standard Test Method for n-Heptane Insolubles*, Philadelphia, PA, Available from http://www.astm.org/Standards/D3279.htm

ASTM D893-05a. (2010) *Standard Test Method for Insolubles In Used Lubricating Oils*, Philadelphia, PA, Available from http://www.astm.org/Standards/D893.htm

Argillier, J.F.; Barre, L.; Brucy, F.; Dournaux, J-L.; Henaut, I. & Bouchard, R. (2001). Influence of Asphaltene Content and Dilution on Heavy Oil Rheology, SPE paper 69711 presented at the SPE International Thermal Operations and Heavy Oil Symposium, Margarita Island, Venezuela, 12-14 March, 2001

Boduszynski, M.M. (1981). Asphaltenes in Petroleum Asphalts: Composition and Formation, In *Chemistry of Asphaltenes*, J.W. Bunger & N.C. Li (Eds.), 119–135, American Chemical Society, Washington DC, USA

Boussingault, M. (1937). Memoire sur la Composition des Bitumes. *Annales de Chimie et de Physique*, Vol.LXIV, pp. 141-151

Buckley, J.S.; Hirasaki, G.J.; Liu, Y.; Von Drasek, S.; Wang, J.X. & Gil, B.S. (1998). Asphaltene Precipitation and Solvent Properties of Crude Oils. *Pet. Sci. and Tech.*, Vol.16, No.3-4, pp. 251-285

Bulmer, J. T. and Starr, J. (1979). *Syncrude Analytical Method for Oil Sand and Bitumen Processing*. Syncrude Canada Ltd., Edmonton, Alberta, Canada

Ching, M.-J.T.M.;. Pomerantz, A.E.; Andrews, A.B.; Dryden, P.; Schroeder, R.; Mullins, O.C. & Harrison, C. (2010). On the Nanofiltration of Asphaltene Solutions, Crude Oils, and Emulsions. *Energy Fuels*, Vol.24, 5028–5037

Creek, J.; Cribbs, M.; Dong, C.; Mullins, O.C.; Elshahawi, H.; Hegeman, P.; O'Keefe, M.; Peters, K. & Zuo, J.Y. (2010). Downhole Fluids Laboratory. *Oilfield Review*, Vol.21, No.4, pp. 38-54

Dickie, J.P. & Yen, T.F. (1967). Macrostrucutres of Asphaltic Fractions by Various Instrumental Methods. *Anal Chem*, Vol.39, pp. 1847–1852

Eyssautier, J.; Levitz, P.; Espinat, D.; Jestin, J.; Gummel,J.; Grillo, I. & Barré, L. (2011a). Insight into Asphaltene Nanoaggregate Structure Inferred by Small Angle Neutron and X-ray Scattering. *J. Phys. Chem. B.*, Vol.115, No.21, pp. 6827–6837

Eyssautier,J.; Levitz, P.; Espinat, D.; Jestin, J.; Gummel,J.; Grillo, I. & Barré, L. (2011b). Petroleum residue in hydroprocessing conditions: What is the asphaltene signature? Presented at Petrophase XII Conference, London, UK, 10-14 July 2011

Freed, D.E.; Lisitza, N.V.; Sen, P.N. & Song, Y.Q. (2009). A Study of Asphaltene Nanoaggregation by NMR. *Energy Fuels*, Vol.23, No.3, pp. 1189-1193

Freed, D.E., Lisitza, N.V.; Sen, P.N. & Song, Y-Q. (2007). Molecular Composition and Dynamics of Oils from Diffusion Measurements, In *Asphaltenes, Heavy Oils and Petroleomics,* O.C. Mullins; E.Y. Sheu; A. Hammami & A.G. Marshall (Eds), 279-299, Springer, NY, USA

Goual, L. & Firoozabadi, A. (2002). Measuring Asphaltenes and Resins, and Dipole Moment in Petroleum Fluids, *AiChE Journal*, Vol.48, No.11, pp. 2646-2663

Goual, L. & Firoozabadi, A. (2004). Effect of Resins and DBSA on Asphaltene Precipitation from Petroleum Fluids. *AiChE Journal*, Vol.50, No.2, pp. 470-479

Goual, L.; Horváth-Szabó, G.; Masliyah, J.H. & Xu, Z. (2005). Adsorption of Bituminous Components at Oil/Water Interfaces Investigated by Quartz Crystal Microbalance: Implications to the Stability of Water-in-Oil Emulsions. *Langmuir*, Vol.21, No.18, pp. 8278-8289

Goual, L.; Horváth-Szabó, G.; Masliyah, J.H. & Xu, Z. (2006). Characterization of the charge carriers in bitumen. *Energy Fuels*, Vol.20, No.5, pp. 2099-2108

Goual, L.; Schabron, J.F.; Turner, T.F. & Towler, B.F. (2008). On-Column Separation of Wax and Asphaltenes in Petroleum Fluids. *Energy Fuels*, Vol.22, No.6, pp. 4019-4028

Goual L. (2009). Impedance Spectroscopy of Petroleum Fluids at Low Frequency. *Energy Fuels*, Vol.23, No.4, pp. 2090–2094

Goual, L. & Abudu, A. (2010). Predicting the Adsorption of Asphaltenes from Their Electrical Conductivity. *Energy Fuels*, Vol.24, No.1, pp. 469–74

Goual, L.; Sedghi, M.; Zeng, H., Mostowfi. F.; McFarlane, R. & Mullins, O.C. (2011). On the Formation and Properties of Asphaltene Nanoaggregates and Clusters by DC-Conductivity and Centrifugation. *Fuel*, Vol.90, No.7, pp. 2480–2490

Goncalves, S.; Castillo, J.; Fernandez, A. & Hung, J. (2004). Absorbance and Fluorescence Spectroscopy on the Aggregation of Asphaltene-Toluene Solutions. *Fuel*, Vol.83, No.13, pp. 1823-1828

Groenzin, H. & Mullins, O.C. (1999). Asphaltene Molecular Size and Structure. *J. Phys. Chem. A*, Vol.103, No.50, 11237–11245

Groenzin, H. & Mullins, O. C. (2000). Molecular Size and Structure of Asphaltenes from Various Sources. *Energy Fuels*, Vol.14, No.3, 677-684

Hirschberg, A.; de Jong, L.N.J.; Schipper, B.A. & Meijer, J.G. (1984). Influence of. Temperature and Pressure on Asphaltene Flocculation. *Soc. Pet. Eng. Journal*, Vol.24, No.3, pp. 283-293

Hortal, A.R.; Martínez-Haya, B.; Lobato, M.D.; Pedrosa, J.M. & Lago, S. (2006). On the Determination of Molecular Weight Distributions of Asphaltenes and Their Aggregates in Laser Desorption Ionization Experiments. Journal of Mass Spectrometry, Vol.41, No.7, pp. 960–968

Hsu, C.S.; Hendrickson, C.L.; Rodgers, R.P.; McKenna, A.M. & Marshall, A.G. (2011). Petroleomics: Advanced Molecular Probe for Petroleum Heavy Ends, Journal of Mass Spectrometry, Vol.46, pp. 337–343

Indo, K.; Ratulowski, J.; Dindoruk, B.; Gao, J.; Zuo, J. & Mullins, O.C. (2009). Asphaltene Nanoaggregates Measured in a Live Crude Oil by Centrifugation. *Energy Fuels*, Vol.23, pp. 4460-4469

IP 143/84. (1988). Asphaltene Precipitation with Normal Heptane. *Standard Methods for Analysis and Testing of Petroleum and Related Products*, Vol.1, Institute of Petroleum, London

Karlsen, D.A. & Larter, S.R. (1991). Analysis of Petroleum Fractions by TLC–FID: Applications to Petroleum Reservoir Description. *Organic Geochemistry*, Vol.17, No.5, pp. 603–617

Klein, G.C.; Kim, S.; Rodgers, R.P.; Marshall, A.G.; Yen, A. & Asomaning, S. (2006). Mass Spectral Analysis of Asphaltenes. I. Compositional Differences between Pressure-Drop and Solvent-Drop Asphaltenes Determined by Electrospray Ionization Fourier Transform Ion Cyclotron Resonance Mass Spectrometry. *Energy Fuels*, Vol.20, No.5, pp. 1965-1972

Long, R.B. (1981). The Concept of Asphaltenes, In *Chemistry of Asphaltenes*, J.W. Bunger & N.C. Li (Eds.), 17-27, American Chemical Society, Washington DC, USA

Magual, A.; Horváth-Szabó, G. & Masliyah, J.H. (2005). Acoustic and Electroacoustic Spectroscopy of Water-in-Diluted-Bitumen Emulsions. *Langmuir*, Vol.21, No.19, pp. 8649–8657

Marques, J.; Merdrignac, I.; Baudot, A.; Barré, L.; Guillaume, D.; Espinat, D. & Brunet, S. (2008). Asphaltenes Size Polydispersity Reduction by Nano- and Ultrafiltration Separation Methods: Comparison with the Flocculation Method. *Oil Gas Sci. Technol.*, Vol.63, pp. 139-149

Merdrignac, I.; Desmazières, B.; Terrier, P.; Delobel, A. & Laprévote, O. (2004). Analysis of Raw and Hydrotreated Asphaltenes Using Off-Line and On-Line SEC/MS Coupling. *Presented at the International Conference on Heavy Organics Deposition*, Los Cabos, Baja California, Mexico, November 14–19, 2004

Daniel Merino-Garcia, D.; Shaw, J.M.; Carrier, H.; Yarranton, H. & Goual, L. (2009). Petrophase 2009 Panel Discussion on Standardization of Petroleum Fractions. *Energy Fuels*, Vol.24, No.4, pp 2175–2177

Mullins, O.C.; Sheu, E.Y.; Hammami, A. & Marshall A.G. (2007). *Asphaltenes, Heavy Oils and Petroleomics*, Springer, NY, USA

Mostowfi, F.; Indo, K.; Mullins, O.C. & McFarlane, R. (2009). Asphaltene Nanoaggregates Studied by Centrifugation. *Energy Fuels*, Vol.23, pp. 1194-1200

Mullins, O.C. (2010). The Modified Yen Model. *Energy Fuels*, Vol.24, pp. 2179-2207

Mullins, O.C.; Andrews, A.B.; Pomerantz, A.E.; Dong, C.; Zuo, J.Y.; Pfeiffer, T.; Latifzai, A.S.; Elshahawi, H.; Barré, L. & Larter, S. (2011). Impact of Asphaltene Nanoscience on Understanding Oilfield Reservoirs. *Paper SPE 146649 presented at the SPE Annual Technical Conference and Exhibition*, Denver, 30 October-2 November, 2011

Nellensteyn, F. J. (1938). The Colloidal Structure of Bitumen, In *The Science of Petroleum;* Vol.4, 2760, Oxford University Press, London, UK

Qian, K.; Edwards, K.E.; Siskin, M.; Olmstead, W.N.; Mennito, A.S.; Dechert, G.J. & Hoosain, N.E. (2007). Desorption and Ionization of Heavy Petroleum Molecules and

Measurement of Molecular Weight Distributions. *Energy Fuels*, Vol.21, No.2, pp. 1042–1047

Rodgers, R.P. & Marshall, A.G. (2007). Petroleomics: Advanced Characterization of Petroleum-Derived Materials by Fourier-Transform Ion Cyclotron Resonance Mass Spectrometry (FT-ICR MS), In *Asphaltenes, Heavy Oils and Petroleomics,* O.C. Mullins; E.Y. Sheu; A. Hammami & A.G. Marshall (Eds), 63-93, Springer, NY, USA

Sabbah, S.; Morrow, A.L.; Pomerantz, A.D. & Zare, R.N. (2011). Evidence for Island Structures as the Dominant Architecture of Asphaltenes, Vol.25, 1597-1604

Saraji, S.; Goual, L. & Piri, M. (2010). Adsorption of Asphaltenes in Porous Media under Flow Conditions. *Energy Fuels*, Vol.24, pp. 6009-6017

Spiecker, P.W.; Gawrys, K.L. & Kilpatrick, P.K. (2003). Aggregation and Solubility Behavior of Asphaltenes and their Subfractions. *J. Colloid Interface Sci.*, Vol.267, pp. 178-193

Sedghi, M. & Goual, L. (2010). Role of Resins on Asphaltene Stability. *Energy Fuels*, Vol.24, pp. 2275–80

Schabron, J.F; Pauli, A.T. & Rovani, J.F., Jr. (2001). Molecular Weight/Polarity Map for Residua Pyrolysis. *Fuel*, Vol.80, No.4, 529–537

Schabron, J.F. & Rovani, J.F., Jr. (2008). On-Column Precipitation and re-Dissolution of Asphaltenes in Petroleum Residua. *Fuel,* Vol.87, pp. 165–176

Schneider, M.H.; Andrews, B.; Mitra-Kirtley, S. & Mullins, O.C. (2007). Asphaltene Molecular Size by Fluorescence Correlation Spectroscopy. *Energy Fuels*, Vol.21, pp. 2875-2882

Sheu, E. Y. & Acevedo, S. (2006). A Dielectric Relaxation Study of Precipitation and Curing of Furrial Crude oil. *Fuel*, Vol.85, No.14-15, 1953–1959

Speight, J.S. (2007). *The Chemistry and Technology of Petroleum*. Fourth Edition, CRC Press/Taylor & Francis, Boca Raton, FL, USA, ISBN 978-084-9390-67-8

Strausz, O.P.; Mojelsky, LT.W. & Lown, E.M. (1992). The Molecular Structure of Asphaltene: an Unfolding Story. *Fuel*, Vol.71, No.12, pp. 1355-1363

UOP 46-85. (1985). *Paraffin Wax Content of Petroleum Oils and Asphalts*. UOP Methods, UOP Inc., Des Plaines, IL, USA

Wallace, D., Henry, D.; Pongar, K. & Zimmerman, D. (1987). Evaluation of Some Open Column Chromatographic Methods for Separation of Bitumen Components. *Fuel*, Vol.66, No.1, pp. 44-50

Wang, S.; Liu, J.; Zhang, L.; Masliyah, J. & Xu, Z. (2010). Interaction Forces between Asphaltene Surfaces in Organic Solvents. *Langmuir*, Vol.26, pp. 183–190

Wiehe, I.A. (1996). Two-Dimensional Solubility Parameter Mapping of Heavy Oils, *Fuel Sci. & Tech. Int.*, Vol. 14, pp. 289-312

Yudin, I.K. & Anisimov, M.A. (2007). Dynamic Light Scattering Monitoring of Asphaltene Aggregation in Crude Oils and Hydrocarbon Solutions, In *Asphaltenes, Heavy Oils and Petroleomics,* O.C. Mullins; E.Y. Sheu; A. Hammami & A.G. Marshall (Eds), 439-468, Springer, NY, USA

Zhao, B. & Shaw, J.S. (2007). Composition and Size Distribution of Coherent Nanostructures in Athabasca Bitumen and Maya Crude Oil. *Energy Fuels*. Vol.21, pp. 2795–2804

Zeng, H.; Song, Y.Q.; Johnson, D.L. & Mullins, O.C. (2009). Critical Nanoaggregate Concentration of Asphaltenes by Low Frequency Conductivity. *Energy Fuels*, Vol.23, pp. 1201-1208

3

Adsorption and Aggregation of Asphaltenes in Petroleum Dispersed Systems

Jamilia O. Safieva[1], Kristofer G. Paso[1],
Ravilya Z. Safieva[2] and Rustem Z. Syunyaev[2]
[1]*Ugelstad Laboratory, Norwegian University of Science and Technology (NTNU), Trondheim*
[2]*Gubkin Russian State University of Oil and Gas, Moscow,*
[1]*Norway*
[2]*Russia*

1. Introduction

In current usage, the word "petroleum" has attained an almost magical meaning due to its importance to several aspects of modern life, influencing economic, policy, social factors. The price of petroleum is an important factor which will affect the character, rate and terms of future societal development.

The task of understanding actual physical and chemical structures in petroleum systems is non-trivial. Petroleum systems are highly complex and contain a potentially rich palette of possible structures and components. In technological processes, petroleum systems typically evolve in open systems of energy flows and matter flows.

2. Crude oil is a nanostructured "soft" system

«Petroleum is a complex mixture of hydrocarbons which is located in the Earth's crust in liquid, gaseous, and solid forms. Natural gas, heavy viscous oil and bitumens are forms of petroleum». This definition provided by a British encyclopedia is somewhat limited. Current scientific knowledge affords a more correct description of petroleum as a system containing numerous hydrocarbon components with different chemical natures. Such a correct definition allows a variety of physical properties and states related to internal structuring.

Classical physics defines three states of matter: solid, liquid, and gas. It provides adequate models of gaseous and solid states. The liquid state is somewhat more difficult to characterize, due to several critical obstacles. In addition, little attention has previously been paid to boundary states (coexistence of any two or even all three states at certain thermodynamic conditions). Different sciencific disciplines created separate terminologies such as metamaterials, which properties derive from artificially created periodic microstructure. Concepts such as multiphase heterogeneous or particularly ordered media or complex materials have appeared. Finally the Nobel winner in physics J.-P. de Gennes (de Gennes, 1992) united all terminologies under a common term «soft materials» and founded the concept of a new field in condensed matter physics which received the conditional name "Physics of a Soft Matter", covering physics of polymers, liquid crystals, critical phenomena, biological systems and colloids. Properties of Soft Materials (SM) strongly correlate at micro-, meso-, and macro-levels under external influences.

Open systems in technology provide a means for formation of space-time structures. The synergetic approach was developed by a brilliant chemist, Nobel winner I.Prigogine (Prigogine, 1977). The condition of "soft" objects is determined by the inclination of the systems to undergo ordering through the action of intermolecular attractive forces (enthalpy factor) and disordering factor - leading any system to chaos or disintegration (entropy factor). The competition between order and chaos comprises an important condition of "soft" objects.

Various structural elements and the range of intermolecular interactions cause morphological diversity of supermolecular structures at all levels. These processes are called self-assembly or self-organization. These key phenomena serve as the basis for a new physical chemistry approach - the physical chemistry of intermolecular interaction or nanochemistry, which was founded by Nobel winner J.M. Lehn (Lehn, 1987).

Petroleum systems undoubtedly refer to objects of "Soft Matter". Today, the "Petroleomics concept" (Mullins et al., 2007) follows the slogan "Characterization of all chemical constituents of petroleum, their interactions and their reactivity". Petroleum component structures and their interaction potentials define the reactivity of petroleum system.

A universal criterion is that petroleum systems are mostly multiphase and heterogeneous with highly developed interfaces. The degree of dispersity is inversely proportional to a characteristic linear scale of inclusions. The degree of dispersity is a kernel of classification of disperse systems and should be accounted for as an additional variable in all equations describing the thermodynamic state of a system. At nano-scale ranges, this fact becomes especially important (Anisimov, 2004).

3. Petroleum dispersed systems

Approaching petroleum systems as homogeneous fluids with averaged characteristics is justifiable for the study of hydrodynamic flow processes in porous media at macroscopic scales. Upon recovery, the viscosity of filtered fluids varies considerably with the presence of additives (polymers, surfactants). In micellar solutions, the surface tension is reduced by several orders of magnitude. Not all methods of recovery are equally successful in various oilfields. Representations of an "average molecule" in petroleum components are incorrect. Phase behavior and dispersion conditions/structure of petroleum systems are determining factors in processing technology at all stages of the petroleum value chain, including recovery, transport, storage, processing and applications. A majority of these processes are accompanied by phase transitions, where the activity of the disperse structures is crucial.

Petroleum systems contain complex matter which is the subject of new field of condensed matter physics. Petroleum systems are typical oleo-dispersed, lyophilic systems, or systems of low polarity dispersive media. The medium is in dynamic balance with elements of a dispersed phase. Traditional tools for studying petroleum systems are limited only by the definition of fractional, grouping, chemical content, and etc. However, observed behaviour of physical and chemical properties in the molecular or dispersed state can differ significantly even with identical chemical contents. Regulation of micro-structural parameters at production, transportation, refinery and application of oil and oil fractions by practically accessible methods comprises the basis of new effective physical and chemical technologies (Syunyaev et al., 1991). Phase transformations from one physical condition to another are accompanied by nucleation, dispersion and formation of micro-heterogeneous systems. The degree of dispersity is considered a necessary additional parameter of state. This data contains initial information which is crucial for proper selection of new technological strategies (Syunyaev et al., 1991).

Recent exhaustion of easily accessible petroleum and gas resources has resulted in more focused attention directed towards extraction and processing of heavy, high viscosity oils and natural bitumens comprising a significant portion of remaining world energy reserves. These resources exist in highly concentrated dispersed systems with significant contents of resin-asphaltene substances (RAS). These petroleum systems demand investigation of microstructure caused by the presence of RAS. For the proper study of Petroleum Dispersed Systems, it is not sufficient to only determine fractional, group, and chemical compounds, elemental analysis, and etc. Of no less importance is to define the interval of the molecular or dispersed state for the tested system at given external conditions. A uniform integrated approach based on analysis and management of microstructure of petroleum systems opens opportunities to regulate properties of intermediate and final products in all technological sequences in the petroleum industry.

The Integrated Theory of Adsorption and Aggregation Equilibrium (ITAAE) is postulated. Separately, there are few classical theories of adsorption. Theories exist for aggregation of complex molecules, such as surfactants, polymers, dyes, and biomolecules. The evolution of such "soft" systems is defined by the hierarchy of InterMolecular Potentials (IMP). There is no doubt that petroleum macromolecules (asphaltenes, paraffines, adamantanes, tethrameric acids etc.) or structural units (nanoaggregates and nanoclusters) belong to that family. According to approaches of supramolecular chemistry or nanochemistry, a variety of these molecules results in a wide morphological spectrum of supermolecular structures in petroleum at nano and micro-scales. These are heavy petroleum, bitumens, emulsions, pitches, tars, coke, foams and gas hydrates.

ITAAE is necessary to understand situations where the contribution of collective IMP in the bulk phase is comparable with the influence of the interface. This situation occurs in porous media. Petroleum disperse systems (PDS) are multicomponent systems with a hierarchy of intermolecular interactions (IMP). Various selective contributions to IMP determine appropriate terms to establish solubility parameters. This theory is a key to understand structuring of PDS on nano- and microscales. Special attention should be drawn to the new and still not fully understood effect of superficial aggregation when surface forces manage formation of superficial units during and after Langmuire and Brunauer-Emmett-Teller (BET) adsorption. This kinetic effect is closely related to phase stability in open systems.

4. Adsorption of petroleum asphaltenes onto mineral surfaces

Asphaltenes and resins comprise the most polar petroleum macromolecules. The polarity causes the relatively high interfacial activity. Strong asphaltene and resin adsorption tendencies lead these substances to contribute to several undesirable phenomena in petroleum industry: well bore plugging and pipeline deposition; stabilization of water/oil emulsions; sedimentation and plugging during crude oil storage; adsorption on refining equipment and coke formation. Knowledge of kinetic and thermodynamic adsorption parameters opens a possibility to regulation of capillary number and wettability. Actually it affords a method for physical and chemical engineering of liquid-solid interfaces in the oil industry.

Adsorption is the redistribution of the components of substance between bulk phase and surface layer. Dissolved molecules displace molecules of solvent during the process of adsorption from solution. Different mechanisms of surface layers formation are possible due to different nature of adsorbed components, their concentrations, morphology of adsorbent's surface, its wettability. At low concentrations Henry low is performed. In this case adsorption linearly depends on concentration of dissolved substance. The most physically proved model for monomolecular adsorption is a Langmuir model. BET theory is used in more complicated cases of polymolecular adsorption (Hunter, 2001). Asphaltene adsorption on reservoir's rock surface is of special interest in practice of oil production. Asphaltenes are the most high-molecular polar oil components and they have high surface activity (Sayyouh et al., 1991; Syunyaev&Balabin, 2007). Adsorption precedes asphaltenic, resinous and paraffinic formation of macroscopic deposits on reservoir's pores and on surfaces of production equipment. Early colmatage of reservoir may lead to premature well shutdown (Ekholm et al., 2002). Surface layers of asphaltenes and resins can change rock characteristics. This influences filtrational characteristics of the processes of oil and water migration in porous media (Basniev et al., 1993). Adsorption of these substances determines surface hydrophobization. Some researchers consider that gel-like films form at quartz surface during the adsorption process. These films greatly reduce oil penetrability in porous media (TatNIPI, 1988). Adsorption and following deposit formation are significant problems in refinery (Syunyaev et al., 1991; Speight, 1999). Filtration of fluids is based on Darcy Law:

$$Q = -\frac{k}{\eta} \times grad P ,$$

where Q - filtration rate, k - permeability of petroleum collector, η - viscosity of fluid, $grad P$ - pressure gradient. Coefficient k is connected with porosity of rock m. In simple case when porous media is as a set of similar channels of diameter D the next equation combines all these parameters

$$D = 4\sqrt{\frac{2k}{m}}$$

As a sequence it follows that adsorption of petroleum macromolecules: asphaltenes, resins and paraffines modifies inner pore surface and sequently influences to permeability **k**. Porous media is classified to next classes (Table 1).

Movement in supercapillary pores is obeyed on macroscopic hydrodynamic laws. On the contrary in micropores fluids are immobile practically. Movement in mesopores is defined by a balance between hydrodynamical and intermolecular forces. Interactions of molecules with walls of such pores are rather high. The most widespread oil reservoirs are quartz sands and dolomite rocks and clays with bulky block-structured packing.

Asphaltene and resin adsorption was studied by a great number of research groups (Ekholm et al., 2002; TatNIPI, 1988; Balabin&Syunyaev, 2008; Syunyaev et al., 2009; Acevedo et al., 2000; Lopez-Linares et al., 2006; Gonzalez&Middea, 1987; Drummond&Israelachvili, 2004; Toulhoat et al., 1994; Akhlaq et al., 1997; Dudasova et al., 2007; Alboudwarej et al., 2005; Batina et al., 2003; Batina et al., 2005; Castillo et al., 1998; Acevedo et al., 1998; Acevedo et al., 2000; Acevedo et al, 2003; Ekholm et al., 2002; Xie&Karan, 2005; Dudasova et al., 2008; Rudrake et al., 2009; Abdallah&Taylor, 2007; Labrador et al., 2007; Turgman-Cohen et al.; 2009). Adsorption of asphaltenes on solid surfaces has been characterized by several experimental techniques, including measurement of contact angles (Drummond & Israelachvili, 2004; Toulhoat et al., 1994; Akhlaq et al., 1997), UV and NIR spectroscopy (Dudasova et al., 2007; Alboudwarej et al., 2005), atomic force microscopy (Toulhoat et al., 1994; Batina et al., 2003; Batina et al., 2005), Fourier transform infrared microscopy (Batina et al., 2005), photothermal surface deformation (Castillo et al., 1998; Acevedo et al., 1998; Acevedo et al., 2000; Acevedo et al., 2003), quartz microbalance (Ekholm et al., 2002; Xie&Karan, 2005; Dudasova et al., 2008; Rudrake et al., 2009), X-ray photoelectron spectroscopy (Dudasova et al., 2007; Abdallah&Taylor, 2007) and ellipsometry (Labrador et al., 2007; Turgman-Cohen et al., 2009). In this study we studied influence of porosity of the rock on adsorption parameters of the asphaltenes.

Macropores	Supercapillary	D>100 μm	Free motion on hydrodynamic forces	Large and medium size Sands, Carbonates
Mesopores	Capillary	D: 0,1-100 μm	Influence of capillary forces	Cemented Sands, Limestones and Dolomites
	Subcapillary	D: 2nm - 100 nm	No movement. Fluids are fixed by intermolecular forces	Clays, Fine crystalline Limestones and Dolomites
Micropores		D < 2 nm	No movement. Fluids are fixed by intermolecular forces	Clays, Ceolites

Table 1. Classification of Porous materials.

Asphaltenes were extracted from West Syberian crude by standard method described in previous work (Balabin&Syunyaev, 2008). The value 750 g/mol was chosen as average molecular mass of asphaltenes (Mullins et al., 2007). Adsorption of asphaltenes was studied at concentration 1 g/l in benzene (density 0,88 g/cm^3, molar mass 78 g/mol). Four fractions of quartz sand, three fractions of dolomite and two fractions of mica were used as adsorbents. Quartz and dolomite particles are considered to be quasispherical particles. Mica models crumbling rock with plate-like particles. Adsorbents were provided by company "Batolit". Its parameters are shown in Table 2. The particle size distribution was evaluated using optical microscope (OPTITECH SME-F2).

Mineral adsorbents					
Adsorbent	Material Density, d_0, g/cm^3		Powder apparent density, d_p, g/sm^3	Medium size, mcm	Porosity m, 1- d_p/d_0
Dolomite	D-03	3,00	0,65	3	0,75
	D-10	3,00	1,05	10	0,58
	D-30	3,00	1,15	30	0,55
Mica	M-05	2,77	0,35	5	0,84
	M-30	2,77	0,55	30	0,77
Quartz	Q-5	2,65	1,085	5	0,59
	Q-10	2,65	1,085	10	0,59
	Q-100	2,65	1,110	100	0,58
	Q-200	2,65	1,180	200	0,55
Metal adsorbent (steel shot)					
	Density, d_0, g/cm^3		Medium diameter, mm		
DSL - 0,4	7,2		0,5		
DSL - 1,4	7,2		1,4		
DSL - 3,6	7,2		3,6		

Table 2. Adsorbent properties.

Near-InfraRed-spectroscopy (NIR) became more and more popular technique for study of petroleum systems (Balabin&Syunyaev, 2008; Oh&Deo, 2002). NIR device - Near-IR FT Spectrometer InfraLUM FT-10 (LUMEX, Russia) was used for measurements. The spectra have been registered at the temperature from 22 to 25 °C. No cell thermostating was used. Background spectrum was taken before and after each measurement; then, the averaged background spectrum was subtracted from the sample spectrum. This technique allowed obtaining an analytical signal with satisfactory accuracy and precision. The instrument calibration was performed using four pure hydrocarbons (toluene, hexane, benzene, and isooctane). Figure 1 represents the scheme of NIR experiment for adsorption observation.

Kinetic experiments were carried out by continuously measuring the light absorption in a NIR range (8,500-13,000 cm^{-1}) in benzene asphaltenes solution in contact with powders.

Quantity of adsorbed asphaltenes was determined by registration of transmittance spectra of bulk phase above the adsorbent top. Knowing the time dependent bulk concentration of asphaltenes $C(t)$ and the initial concentration we are able to evaluate the adsorbate mass $m(t)$ and the adsorbed mass density $\Gamma(t)$. Fundamental Buger-Lambert-Beer law was used for concentration measurements. Final formula for determination of adsorption (Syunyaev et al., 2009) is presented below.

$$\Gamma(t) = \frac{m(t)}{S_{ads}} = \frac{V_0}{S_{ads}} \cdot C_{SS} = \frac{V_0}{S_{ads}}[C_0 - C_{AS}(t)] \quad (1)$$

where C_0 is assigned bulk concentration at initial moment of time , $\mathbf{C_{AS}}$ is time –dependent measured concentration, $S_{ads} = s_{SP} \cdot m_{ads}$ is the total surface area of adsorbent; s_{SP} is the specific surface area of adsorbent (e.g., per gram); m_{ads} is the mass of adsorbent (e.g., in

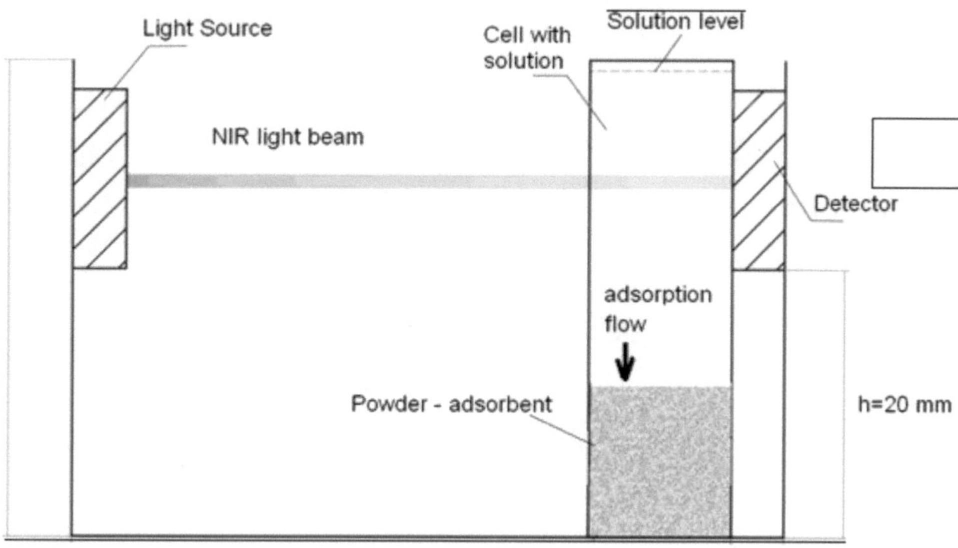

Fig. 1. Experimental setup for measuring adsorption on powders.

grams); V_0 is the volume of solution in the cell (5 ml in our experiments). Accuracy of the $\Gamma(\mathrm{t})$ evaluation depends on the accuracy of $C(t)$ evaluation.

The first order equation of reversible Langmuir adsorption kinetics (Hunter, 2001; Hiemenz&Rajagopalan, 1997) was used for data approximation:

$$\Gamma(\mathrm{t}) = \Gamma_{\max}\frac{KX_0}{KX_0+1}[1-\exp(-(k_a+k_d)t] = \Gamma_{\max}\cdot \mathrm{a}\cdot[1-\exp(-(k_a+k_d)t] \quad (2)$$

where $\Gamma_{\max}$ is maximal adsorbed mass density and coefficient a defines relative fraction of adsorbent surface occupied by asphaltene molecules. Equation (2) needs asphaltene concentration in mole fractions. The mole fraction for our concentration is placed below

$$C_0 = 0{,}1\,\mathrm{g/l} \quad \mathrm{X}_0 = 1{,}18\cdot 10^{-5}$$

In (2) $K = \frac{k_a}{k_d}$ is equilibrium constant of adsorption/desorption at concentration X_0. k_a and k_d are the rate constants of adsorption and desorption, respectively. Next step is to determine the equilibrium adsorption parameter Γ_{eq}, which can be achieved at assigned concentration. $\Gamma_{\max}$ determines maximum possible adsorbent capacity when surface is covered by monomolecular layer.

$$\Gamma_{eq} = a\cdot\Gamma_{\max}$$

Boundary spectra area contains noise. For more accuracy removal of noise area was done. Truncated spectra were integrated for integral transmittance A calculation. Example of asphaltene solution spectra is presented at Figure 2.

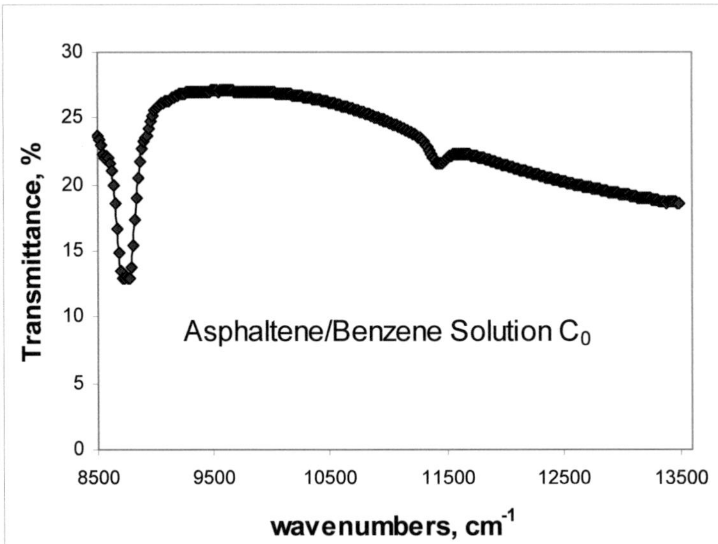

Fig. 2. Transmittance Spectra of Initial Asphaltene solution in Benzene (C_0=1 g/l) in Near Infrared Region.

Kinetic dependences of integral spectra were received for all samples of powders. All spectra were treated identically. Data processing steps for different samples are presented in Figures 3 and 4. All spectra were normalized to values at start time-point.

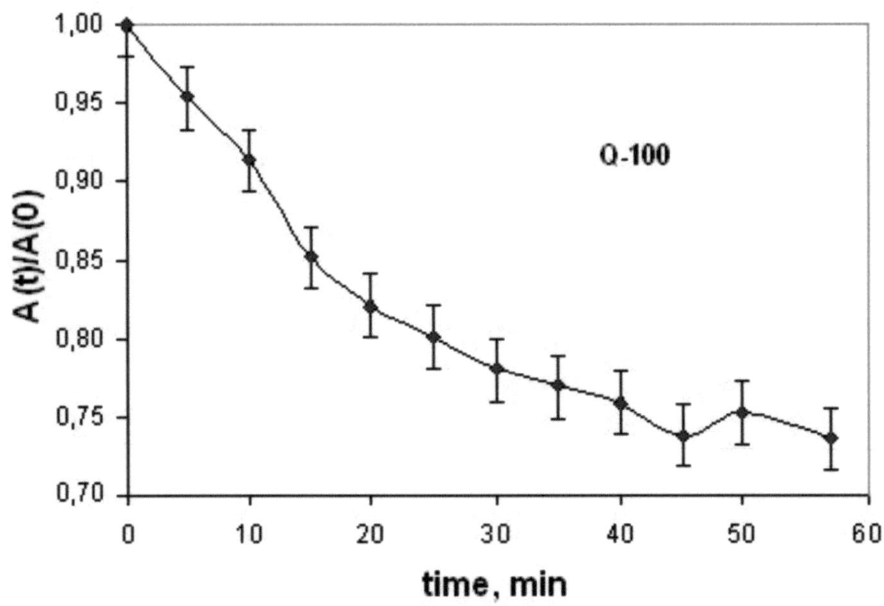

Fig. 3. Time change of integral transmittance of asphaltene solution (normalized to a value in initial time point). Sample Q-100.

Figure 3 represents example of normalized solution bulk concentration C_{AS} spectra for Q-100 quartz sand fraction. Figure 4 and Figure 5 show next steps of dependencies interpretation for adsorption kinetics analysis. Actually it demonstrates surface concentration C_{SS} changes by example of dolomite D-10 powder.

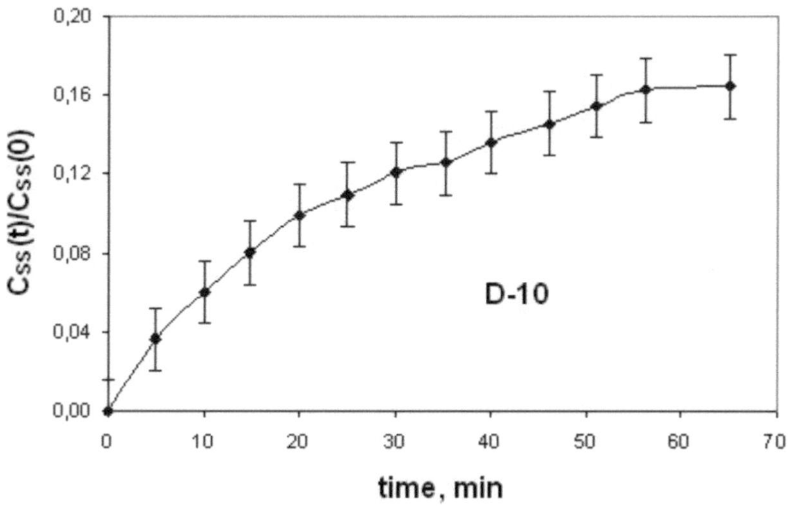

Fig. 4. Kinetics of change of surface concentration (normalized to a value in initial time point). Sample D-10.

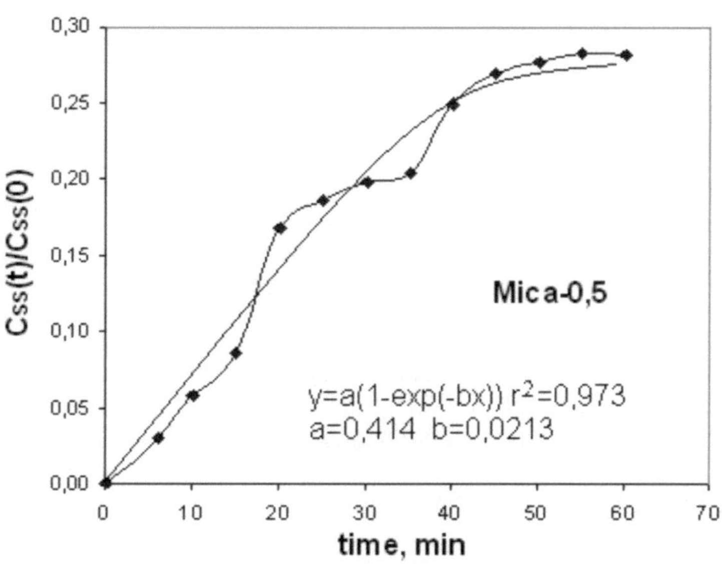

Fig. 5. Fitting of kinetic curve of surface concentration. Adsorption is proportional to concentration (normalized to a value in initial time point). Sample M-05.

5. Kinetic parameters of adsorption

Combined usage of Langmuir adsorption kinetics equation (2) and equation (1) allows us to calculate characteristics of adsorption-desorption process of asphaltenes onto surface of searched surfaces: relative fraction of occupied surface by molecules (a), the maximal adsorbed mass density (Γ_{max}), the equilibrium constant of adsorption (K), the rate constants of adsorption (k_A) and desorption (k_D). These parameters are resulted in Table 3.

Sample	a	k_a+k_d, min^{-1}	$K=\frac{k_A}{k_D}$	k_D, min^{-1}	k_A, min^{-1}	ΔG, kJ/mol
Q-5	0,23	0,04	2 570	0,000017	0,044	19,1
Q-10	0,29	0,03	3 423	0,000009	0,032	19,8
Q-100	0,29	0,04	3 490	0,000013	0,044	19,9
Q-200	0,27	0,03	3 098	0,000008	0,026	19,6
Mica-05	0,41	0,02	5 978	0,000004	0,021	21,2
Mica-30	0,28	0,04	3 210	0,000012	0,038	19,7
D-03	0,17	0,03	1 745	0,000019	0,034	18,2
D-10	0,18	0,04	1 795	0,000022	0,040	18,2
D-30	0,19	0,05	1 985	0,000025	0,050	18,5

Table 3. Kinetic and thermodynamic parameters of asphaltene adsorption onto rock sands.

The rate of asphaltene adsorption is greater than desorption rate for studied concentration. Gibbs adsorption energy values define character of adsorption as physical, not chemical. Rate of asphaltene adsorption is less then resins (in two orders) reported earlier (Balabin&Syunyaev, 2008). Parameters of resin adsorption from benzene solutions on quartz are submitted in Table 4.

Sample	k_A+k_D, min^{-1}	$K=\frac{k_A}{k_D}$	k_D, min^{-1}	k_A, min^{-1}	ΔG, kJ/mol
Resin	0.180	6.0	0.0259	0.15	4.4

Table 4. Adsorption parameters of resins on quartz sand.

The adsorption of asphaltenes is practically irreversible. Significantly larger masses and molecule sizes of asphaltenes appear to be the reason. Diffusion of such molecules to solid surface is embarrassing. The mechanism of diffusion limited adsorption is realized (Syunyaev et al., 2009; Diamant & Andelman, 1996). Gibbs energy values are more or less the same for surfaces of all investigated materials: quartz, dolomite, and mica. It is known that quartz and dolomite are the main components of oil reservoir framework rocks. The porosity has no influence on kinetic parameters of adsorption. Asphaltenes adsorption at the surfaces of quartz and dolomite is the most active.

As a whole, the designed values of asphaltene adsorption parameters on metal surfaces are close to values obtained for mineral powders. Parameters of adsorption-desorption processes on metal surface are listed in Table 5.

Sample	a	$k_A + k_D$, min^{-1}	$K = \frac{k_A}{k_D}$	k_D, min^{-1}	k_A, min^{-1}	ΔG, kJ/mol
DSL-0,5	0.039	0.00022	2.88E+04	7.64E-09	2.20E-04	25
DSL-1,4	0.0756	0.00052	9.34E+03	5.52E-09	5.16E-04	27.9
DSL-3,6	0.044	0.00192	7.53E+05	2.55E-09	1.92E-03	32.9

Table 5. Parameters of asphaltene adsorption on metal surfaces.

According to designed Gibbs adsorption energies minerals can be distributed in the following order: mica > quartz > dolomite for fine-grained mineral powders, mica = dolomite > quartz for coarse-grained mineral powders. Among investigated mineral adsorbents, mica is the most active in relation to asphaltenes.

In our experiments adsorption leads to reduction of bulk asphaltene concentration. Inside porous media asphaltene molecules occupy accessible sites on surface. Mineral adsorbents model the porous media of a petroleum collector. Structural parameters were appreciated using model of "nonideal soil" when the porous media is described as the spatial volume consisting of constant radius spheres package. The alternative description of the porous media as «an ideal ground», represent system of incorporating cylindrical capillaries with the characteristic length close to radius of a grain. Calculated values of the specific area of the adsorbents according to model of "nonideal soil" are presented in Table 6. After estimation of the specific area the maximal capacity of Langmuire asphaltene monolayer was calculated from common relation for molar volume **(V_m)**

$$V_M = \frac{M}{N_A \rho},$$

where M - molecular mass (750 g/mol), N_A - Avogadro constant (mol^{-1}), ρ-asphaltene density (ρ = 1.1 g/cm^3). As the cross section of a "spherical" asphaltene molecule makes 1, 2 nm^2 so surface capacity of adsorbed layer makes 1 mg/m^2.

Meaning of this parameter is boundary criterion of concentration changes in a bulk phase (C_0=0.1g/l) and in a space between grains of the porous media. Calculated masses of adsorbed asphaltenes and the appreciated pore volumes allow assuming concentration of the asphaltene solution inside porous space (Table 6). Langmuire model yields diminished values of adsorbed mass. It may be a sequence of well-known effect of concentration polarization (Bacchin et al., 2002) when in the domain near membranes concentration exceeds medium bulk concentration. On internal structure the membranes are similar to porous media. Average calculated concentrations exceeding bulk are presented in Table 6.

Mica is chosen as the substrate in further investigations of asphaltene deposits morphology by atomic force microscopy (AFM). Preliminary data shows that the mechanism of monomolecular adsorption is not realized. The kinetic models can be based on BET theoretical approach (Hunter, 2001; Hiemenz&Rajagopalan, 1997) which describes polymolecular adsorption.

Regular deviations from exponential curve form are observed in all obtained dependencies (Figure 5). These deviations are small and are inside the limits of experimental errors. But these deviations have regular character and are observed earlier also for resins (Balabin

Sample	Γ_{eq}, mg	Adsorbent specific area, m²/g	Γ_{max}, mg (calculated Langmuire monolayer capacity)	Average asphaltene concentration in pores , g/l
M-05	4.07	0.22	0.87	0.32
M-30	1.41	0.04	0.14	0.22
D-03	1.89	0.33	1.33	0.12
D-30	1.2	0.03	0.13	0.50
Q-10	2.81	0.11	0.45	1.08
Q-200	0.38	0.01	0.02	0.19

Table 6. Average concentration of asphaltene solution in pores of mineral model adsorbents.

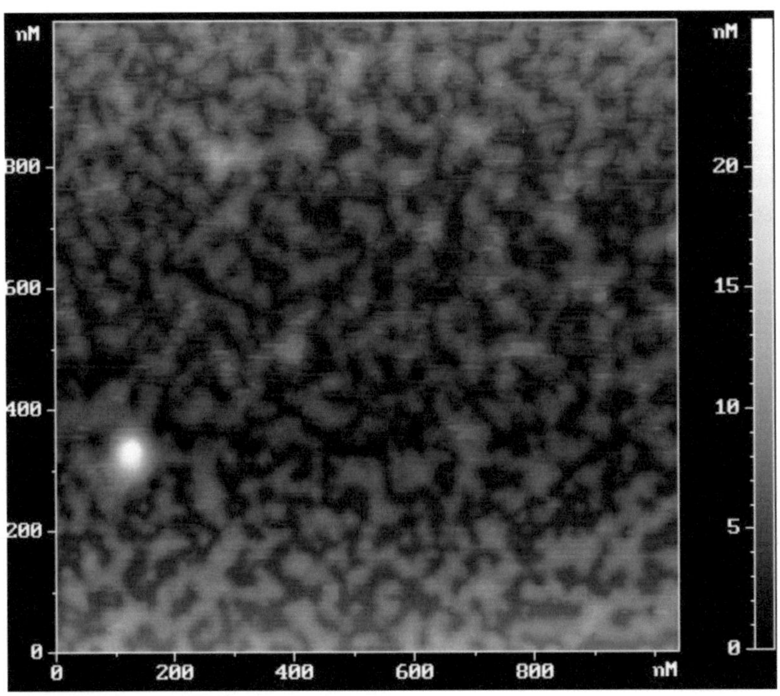

Fig. 5. AFM scans of adsorbed asphaltenes at mica surface.

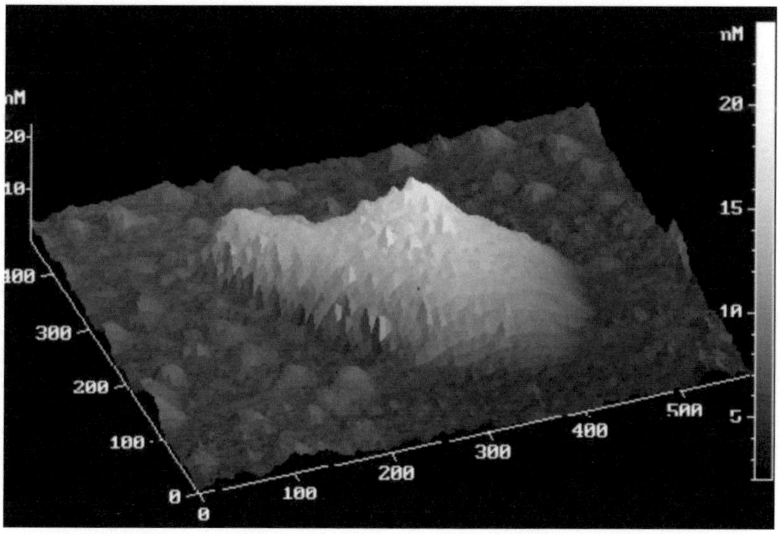

Fig. 6. Polymolecular multilayered adsorption of asphaltenes at mica surface.

& Syunyaev, 2008). Concentration oscillations in water solutions of dye were also observed by laser refractometry. It is possible to suppose that adsorption is kinetically limited by diffusion. The mechanism of diffusion relaxation suggested by I. Akhatov is realized (Akhatov, 1988, as cited in Syunyaev at al., 2009). The classic Fick's law can be generalized by introduction of additional relaxation item

$$\tau\frac{\partial j}{\partial t}+j=-D\frac{\partial C}{\partial x},$$

where j is diffusion flow, D is the coefficient of translational diffusion, C is concentration of substance, τ is the characteristic time of relaxation.

The equation is transformed to equation in partial derivatives of second order.

$$\tau\frac{\partial^2 c}{\partial t^2}+\frac{\partial c}{\partial t}=D\frac{\partial^2 c}{\partial x^2}.$$

Solution for equation appears to be superposition of two concentration time-dependent functions. One of these functions is the derivation for damped oscillations equation.

For porous media neighboring with bulk phase and at condition when adsorption time is much less than diffusion time

$$\tau_{\text{ads}} << \tau_{dif},$$

mechanism of 'adsorption pump" is actualized. While filling porous media by solution the adsorption process occurs faster. This results in decreasing of bulk equilibrium concentration inside porous media. As diffusion process goes more slowly, discontinuous jump of concentrations on border with porous media is appeared. So the region with decreased concentration in border with bulk phase arises. As a consequence the diffusion flow emerges in this border. Than concentration jump spreads to bulk phase with gradually decreasing amplitude due to diffusive spreading (Figure 7). This is damping concentration wave in bulk phase, which we observe in experiment (Figure. 4).

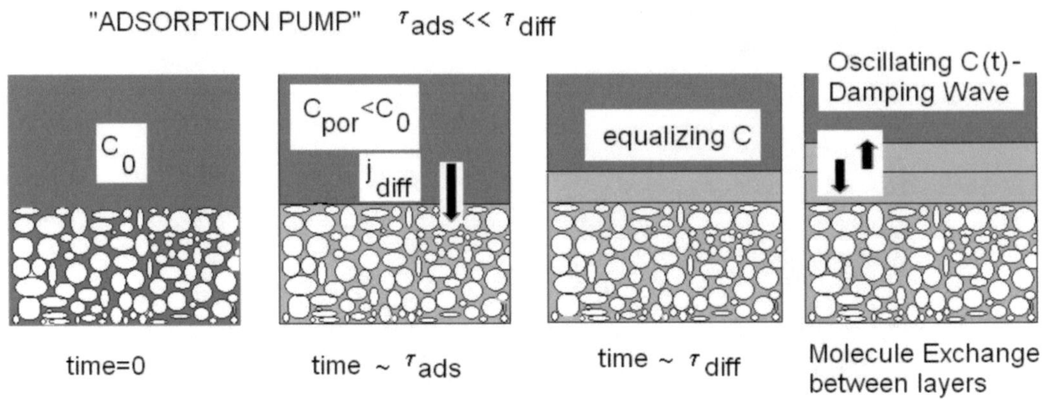

Fig. 7. "Adsorption Pump". Comments are in the text.

6. Conclusion

Asphaltene adsorption at the surface of mineral powders which are a framework of oil reservoir was studied by Near Infrared (NIR) spectroscopy. Experimental procedure of surface asphaltenes concentration measurement is proposed. Adsorption parameters are evaluated in a Langmuir approximation of monomolecular layer occupation. Adsorption characteristics are important for permeability coefficient estimation in Darcy law. That is necessary for further development of filtration theory. Increasing of asphaltene concentration in intergrain space of porous media in comparison with bulk concentration is similar to concentration polarization in membrane technology. The mechanism of initiation of concentration waves in bulk phase neighboring with porous media is offered. Integrated investigation of polymolecular multilayered adsorption using abilities of AFM-analysis is required in future.

7. References

Abdallah W. A., Taylor S. D. (2007). *Nucl. Instrum. Methods Phys. Res.*, Sect. B., 258, p. 213–217.

Acevedo S., Castillo J., Fernandez A., Goncalves S., Ranaudo M. A. (1998). *Energy&Fuels*, 12., p. 386–390.

Acevedo S., Ranaudo M. A., Garcia C., Castillo J., Fernandez A. (2003). *Energy&Fuels*, 17, p. 257–261.

Acevedo S., Ranaudo M., Garcia C., Castillo J., Fernandez A., Caetano M., Goncalvez S. (2000). *Colloids Surf.*, 166, p. 145-152

Akhlaq M. S., Gotze P., Kessel D., Dornow W. (1997). *Colloids Surf. A*, 126., p. 25–32.

Alboudwarej H., Pole D., Svrcek W. Y., Yarranton H. W. (2005). *Ind. Eng. Chem.*, Res., 44, p. 5585–5592.

Anisimov M.A. et al. (2004). Thermodynamics at the Meso- and Nanoscale, *in Dekker Encyclopedia of Nanoscience and Nanotechnology*, pp. 3893-3904, Marcel Dekker, New York.

Bacchin P., Si-Hassen D., Starov V., Clifton M.J., Aimar P. (2002). *Chemical Engineering Science*, 57, p. 77-91

Balabin R.M., Syunyaev R.Z. (2008). Petroleum resins adsorption onto quartz sand: Near infrared (NIR) spectroscopy study. *J. of Colloid and Interface Science*. 318, p. 167-174

Basniev K.S., Kochina I.N., Maksimov V.M. (1993). *Underground Hydromechanics*, Nedra, Moscow.

Batina N., Manzano-Martinez J. C., Andersen S. I., Lira-Galeana C. (2003). *Energy&Fuels*, 17, p. 532–542.

Batina N., Reyna-Cordova A., Trinidad-Reyes Y., Quintana-Garcia M., Buenrostro-Gonzalez E., Lira-Galeana C., Andersen S. I. (2005). *Energy&Fuels*, 19, p. 2001–2005.

Castillo J., Goncalves S., Fernandez A., Mujica V. (1998). *Opt. Commun.*, 145, p. 69–75.

de Gennes J.-P. (1992). Soft Matter (Nobel Lecture). *Angewandte Chemie*. International Edition in English, 31: p. 842-845.

Diamant H., Andelman D. (1996). *J. Phys. Chem.*, 100 (32), p. 13732-13742.

Drummond C., Israelachvili J. (2004). *J. Pet. Sci. Eng.*, 45. p. 61–81.

Dudasova D., Simon S., Hemmingsen P. V., Sjoblom J. (2007). *Colloids Surf. A.*, 317, p. 1–9.

Dudasova D., Silset A., Sjoblom J. (2008). *J. Dispersion Sci. Technol.*, 29, p. 139–146.

Ekholm P., Blomberg E., Claesson P., Auflem I.H., Sjöblom J. and Kornfeldt A. (2002). A Quartz Crystal Microbalance Study of the Adsorption of Asphaltenes and Resins onto a Hydrophilic Surface. *Journal of Colloid and Interface Science*, 247(2), p. 342-350.

Gonzalez G., Middea (1987). *A. J. Dispersion Sci. Technol.*, 8 (5-6), p. 525-548.

Hiemenz P.C., Rajagopalan R. (1997). *Principles of Colloid and Surface Chemistry 3rd Ed.* Marcel Dekker, New York.

Hunter R.J. (2001). *Foundations of Colloid Science.* Second Edition, Oxford, New York.

Labrador H., Fernandez Y., Tovar J., Munoz R., Pereira J.C., (2007). *Energy&Fuels*, 21, p. 1226–1230.

Lehn J.-M. (1987). *Supramolecular Chemistry - Scope and Perspectives Molecules - Supermolecules – Molecular Devices*. Nobel lecture, Dec.08.

Lopez-Linares F., Carbognani L., Gonzalez M.-F., Sosa-Stull C., Figueras M., Pereira-Almao P. (2006). *Energy & Fuels*, 20, p. 2748-2750.

Mullins O.C., Sheu E.Y., Hammami A., Marshall A.G (Eds). (2007). *Asphaltenes, Heavy Oils, and Petroleomics*. Springer.

Oh K., Deo M.D., (2002). *Energy&Fuels*, 16, p. 694.

Prigogine, I. (1977). *Time, Structure and Fluctuations*. Nobel Lecture, Dec. 08.

Rudrake A., Karan K., Horton J.Hugh., (2009). *J.of Colloid Interface Science*, 332, p. 22-31.

Safieva R.Z., Syunyaev R.Z. (2007). *Physical and Chemical Properties of Petroleum Dispersed Systems and Oil-and Gas Technologies.* NIC "Regular and Chaotic Dynamics", Moscow-Izhevsk.

Sayyouh M.H., Hemeida A.M., A1-Blehed M.S. and Desouky S.M. (1991). Role of polar compounds in crude oils on rock wettability, *J. of Petroleum Science and Engineering*, 6, p. 225-233.

Speight J.G. (1999). *The Chemistry and Technology of Petroleum*. Dekker, New York.

Syunyaev R.Z., Balabin R.M. (2007). Frequency dependence of oil conductivity at high pressure. *J. Disp. Sci. Technol.*, 28, p. 419.

Syunyaev R. Z., Balabin R. M., Akhatov I. S., and Safieva J. O. (2009). Adsorption of Petroleum Asphaltenes onto Reservoir Rock Sands Studied by Near-Infrared (NIR) Spectroscopy. *Energy & Fuels*, 23, p. 1230-1236.

Syunyaev Z.I., Safieva R.Z., Syunyaev R.Z. (1991). *Petroleum Dispersed Systems.* Khimiya, Moscow.

TatNIPI. (1988). Scientific report. Adsorption of Asphaltenes onto Quartz and it's Influence to Wettability.

Toulhoat H., Prayer C., Rouquet G. (1994). *Colloids Surf. A.*, 91., p. 267–283.

Turgman-Cohen S., Smith M. B., Fischer D. A., Kilpatrick P. K., Genzer J., (2009). *Langmuir.* 25 (11), p. 6260-6269.

Xie K., Karan K. (2005). *Energy&Fuels.* 19, p. 1252–1260.

4

Natural Surfactants from Venezuelan Extra Heavy Crude Oil - Study of Interfacial and Structural Properties

B. Borges
Departamento de Química, Universidad Simón Bolívar, Caracas, Venezuela

1. Introduction

Studies of the composition and physicochemical properties of heavy and extra heavy crude have been of great importance and interest to academia and the petroleum industry, due to the ability of these oils to form water in oil emulsions. Water in oil emulsions are known to form during crude oil production, oil sands extraction processes, and oil spills in aquatic environments. Often, these water-in-oil emulsions are undesirable since they can cause several problems including: production of an off-specifications crude oil (high solids and water content, >0.5%); corrosion and catalyst poisoning in pipes and equipment for water settling; and environmental issues when oil spills occur in rivers and oceans. Treatment of these emulsions is still a challenge in the petroleum industry due to their high stability versus coalescence. These emulsions are very stable, this stability is attributed largely to the adsorption of compounds with interfacial activity, such as asphaltenes, resins and carboxylic acids present in the oil, at the water - crude interface. Asphaltenes are a complex mixture of high-molecular weight compounds, where 90% or more of the mass is composed of carbon and hydrogen; the sulfur, oxygen, and nitrogen contents in asphaltenes are approximately 5%, 2%, and 1%, respectively, and trace quantities of other heteroatoms are present the high molecular weight components of crude oil. Reported data has shown that asphaltenes, are adsorbed at the crude oil-water interface. For instance, the high stability of the w/o emulsions could be due to strong interfacial films formed by asphaltenes. These films would be very resistant to coalescence (Yarranton et al, 2000; Ortiz et al, 2010; Spiecker, & Kilpatrick, 2004; Pauchard et al, 2009; Chaverot et al, 2010; Sjöblom et al 1992; McLean et al 1997)

The isolation of natural surfactants, using various methods such as separation by emulsification (Acevedo et al, 1992), chromatographic methods (Ramljak et al, 1977, Acevedo et al, 1999; Borges, 2009) has been reported. In this research, a modified chromatographic procedure, based on the proposed by Ramljak in 1977, has been used to isolate natural surfactants in crude oil and thus adapt to the properties of Venezuelan extraheavy crude oil from the Orinoco oil belt, followed by structural characterization of these surfactants, especially those derived from the acid. As mentioned above these natural surfactants play an important role in the stability of emulsions, there is great interest in knowing how they are structurally. Also, the study of interfacial properties, specifically the interfacial tension is an essential tool for determining the ability of surfactants to reduce the tension to adsorb at water-oil interface. This thermodynamic property is used in this research to try to elucidate the action of natural surfactants, specifically the nature aliphatic carboxylic acids and asphaltenes in the formation of the interfacial film and consequently its close relationship with the stability of emulsions of w/o. This study consisted in measured interfacial tension versus concentration for the different natural surfactants. Interfacial tension measurements was doing by the hanging drop method, a bitumen droplet of known volume is formed at the tip of a hypodermic needle immersed in an aqueous solution.

Is important to know that the Orinoco Oil Belt is located along the southern margin of the Eastern Venezuela Basin, (south of the Guárico, Anzoátegui, Monagas, and Delta Amacuro states) parallel to the Orinoco River, covering a geographic area on the order of 55,000 sq km. Within it lies one of the largest oil deposits in the world, roughly 1.3 trillion barrels of "oil in place". Petróleos de Venezuela S.A. has estimated that the producible reserves of the Orinoco Belt are up to 235 billion barrels which would make it the largest petroleum reserve in the world, before Saudi Arabia. The area is divided, from West to East, into four distinct production zones: Boyaca, Junin, Ayacucho and Carabobo.

2. Experimental methods

2.1 Crude oil fractionation

2.1.1. Junín Asphaltenes (AJ) and Junín Maltenes (MJ): The crude oil studied was Junin, which is found at the Orinoco oil belt and has an API gravity of 8°. Asphaltenes were precipitated from the crude oil by the addition of 60 volumes of n-heptane. Previously, the crude oil was diluted in 1:1 toluene - crude oil. Said mixture was stirred mechanically for 6 hours and was then left standing for 24 hours. After this time, the solid was filtered, the solvent of the supernatant liquid was evaporated, leaving a resin - the maltenes - which was dried later and quantified. The solid was washed in a Soxhlet extractor with n-heptane until the solvent turned clear.

2.1.2. Chromatographic system with solvent recirculation: to extracted acid compounds, where the stationary phase was silica gel modified by potassium hydroxide (Ramljak *et al.*, 1977). The experimental procedure was as follows:

Stationary Phase: KOH was dissolved in isopropanol in a 1:25 w/v ratio at high temperature and it was mixed with a silica gel suspension (70-230 mesh) in chloroform ($CHCl_3$) in a 1:2 w/v ratio. It was stirred for 15 minutes. The KOH : silica gel ratio was 1:10. The mixture was transferred to a chromatographic column with solvent recirculation, the column was washed with 600 mL of moderately heated (50°C) $CHCl_3$ and it was refluxed for 15 minutes to remove excess KOH.

Sample: Junin crude oil was dissolved in the lowest amount of $CHCl_3$, keeping a 1:10 ratio of crude oil to silica gel . Once prepared, the sample was placed on the top of the column.

2.1.2.1. Extraction of the Basic - Neutral Fraction (FBN): 600 mL of $CHCl_3$ were placed in the flask and were moderately heated to start reflux until the solvent eluting from the column became colorless. The solvent in the flask was evaporated to dryness and was attached to a vacuum pump to be quantified later.

2.1.2.2. Extraction of the Acid Fraction (AcJ): The residue which was adsorbed on the column was treated with 600 mL of a 20% solution of formic acid (HCOOH) in $CHCl_3$, repeating the previous procedure. Once the acid fraction was obtained, 90% of the solvent was evaporated; the organic residue was washed several times with water until the pH of the aqueous layer was close to 5 in order to remove excess HCOOH. Then, it was dried with magnesium sulfate ($MgSO_4$), filtered, and the rest of the solvent was evaporated. Finally, it was placed under high vacuum for 6 hours and it was quantified.

2.1.3. Acid-Free Asphaltenes (ASA): The basic - neutral fraction was passed another time through the chromatographic system using $CHCl_3$ as the elution solvent. Once the solvent was evaporated and the sample dried, the acid-free asphaltenes were precipitated using the previously described method.

2.2 Structural characterization

2.2.1. Molecular Weights by Vapor Pressure Osmometry (VPO): A Jupiter Instrument Co. model 833 vapor pressure osmometer was used to determine the molecular weights by VPO, using nitrobenzene as the solvent at a working temperature of 100 °C. The constant of the equipment was calculated using pyrene (202 g/mol) as the standard. The solutions analyzed were prepared in concentrations ranging from 1 to 6 g/L.

2.2.2. Elemental Analysis: A Herrman Moritz Macanal 10 micro analyzer was used to apply the high combustion method according to the AFNOR M03-032 standard, for C and H, and the AFNOR M03-08 and AFNOR M-025 standards for S.

2.2.3. Fourier Transform Spectroscopy Infrared (FTIR): Infrared spectra were obtained using a Bruker Optik GmbH Tensor 27 FT-IR, controlled by an Opus/IR software, which uses Fourier transform to process data. Samples were analyzed as liquid films in KBr cells and dry KBr pellets.

2.2.4. Nuclear Magnetic Resonance Spectroscopy (^{1}H NMR and ^{13}C NMR): Proton and carbon nuclear magnetic resonance spectra (^{1}H NMR and ^{13}C NMR, respectively) were obtained in a polynuclear JEOL Eclipse Plus 400 spectrometer (400 MHz), using tetramethylsilane as the reference and deuterated chloroform and carbon tetrachloride as the solvent for ^{13}C NMR and ^{1}H NMR, respectively. ^{13}C NMR spectrum were accumulated during 24 hours.

2.3 Interfacial characterization

Interfacial tension measurements by the Hanging Drop Method. A bitumen droplet of known volume is formed at the tip of a hypodermic needle immersed in an aqueous solution. The shape of a drop of liquid hanging from a syringe tip is determined from the balance of forces which include the surface tension of that liquid. The surface or interfacial tension at the liquid interface can be related to the drop shape through the Young-Laplace equation. (Hiemmenz, 1988)

This device consists of a visualization cell filled with double distilled water, where a drop of the solution is formed; a plastic syringe provided with a U-shaped needle was used to create the emerging drop; a lamp is used for illumination; a video system, to obtain the images; a computer, to obtain and digitalize the images for further treatment. The interfacial tension is calculated using a computer program called DROP (Lopez de Ramos *et al.*, 1993). A battery of solutions of AJ, MJ, AcJ, and ASA samples in toluene was prepared with different concentrations ranging from 100 mg/L to 90000 mg/L. In order to test the tension values, a drop of solution and the double distilled water pre-saturated with toluene at neutral pH were put in contact in the visualization cell for 5 minutes until apparent equilibrium was reached. Then, the measurement was made. The method was validated by comparing the different values reported for pure substances and showed errors below 2%. Experiments were carried out at room temperature.

3. Results and discussion

The analysis presented in this work suggests that the chromatographic method used for separation of acids from extraveavy crude oil is efficient, effective and relatively fast. Table 1 shows the yields of the different isolated fractions. AcJ acids represent 1.6% of the crude oil, *i.e.* half of the acids present in the Carabobo crude oil (Acevedo et al., 2005). However, these crude oil components showed a good interfacial activity, as will be shown later. Table 2 shows the H/C values. AcJ acids, which are amber colored resins, are a highly aliphatic fraction according to this ratio (H/C =1.6), contrasting with AJ and ASA, which are blackish brown and show an H/C value of about 1.12, which was expected for asphaltenes.

Samples	Yield (%p/p)
MJ	87
AJ	11
ASA	8.5
AcJ	1.6

Table 1. Yield of the different samples compared to the crude oil.

Samples	%C	%H	%S	H/C
MJ	85.32	11.8	3.43	1.671
AJ	84.33	8.43	5.30	1.199
ASA	84.72	8.39	4.83	1.118
AcJ	85.21	11.62	5.30	1.636

Table 2. Elementary analysis.

The molecular weights determined by VPO are reported in Table 3. The acid components (AcJ) showed a molecular weight of 474 g/mol, which is low and coincides with the values reported by Stanford et al., who using other techniques reported values from 225 to 1000 for polar acid species which stabilize water in crude oil emulsions. Values reported for AJ and ASA are not significantly different as the error of this technique (VPO) is about 10% and they are within the expected range.

Samples	Molecular Mass (Da)
AJ	1010
ASA	950
AcJ	474

Table 3. Molecular weights obtained by Vapor Pressure Osmometry.

Figure 1 shows the FTIR spectra for the AcJ and AJ fractions. The spectra show a band at 3320 cm-1 attributed to $\upsilon_{O\text{-}H}$; the carbonyl signal, $\upsilon_{C=O}$, at 1709 cm-1; the $\upsilon_{CH3+CH2}$ signal at 2922 for the acids, as opposed to the spectrum of asphaltenes, where no carbonyl signal and no signal due to OH stretching can be seen; other signals typical for these kind of compounds can be observed (Khanna et al., 2006).

Figures 2 and 3 show the ^{13}C NMR and ^{1}H NMR spectra of the acid fraction. The signals corresponding to carboxylic groups (178 and 9.8 ppm, respectively) and the bands corresponding to the aromatic zone (about 126.6 and 7, respectively) and the aliphatic zone (from 14 to 52 ppm and from 0.5 to 3 ppm, respectively) can be clearly seen. The latter are the most intense signals. Figures 4 and 5 show the ^{13}C NMR and ^{1}H NMR spectra for the asphaltenes (AJ), which are similar to those previously reported by other researchers (Acevedo et al., 2005b).

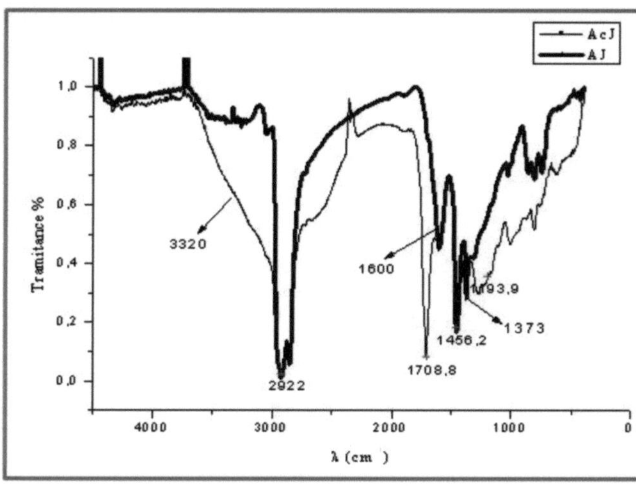

Fig. 1. Comparison of AcJ and AJ FTIR spectra.

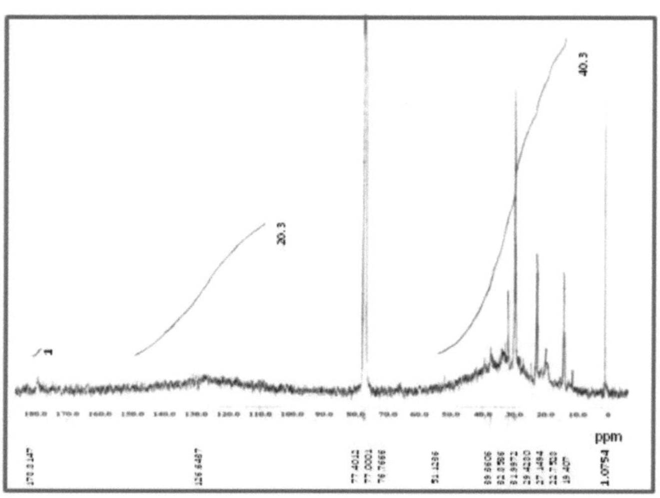

Fig. 2. ^{13}C NMR spectrum for AcJ.

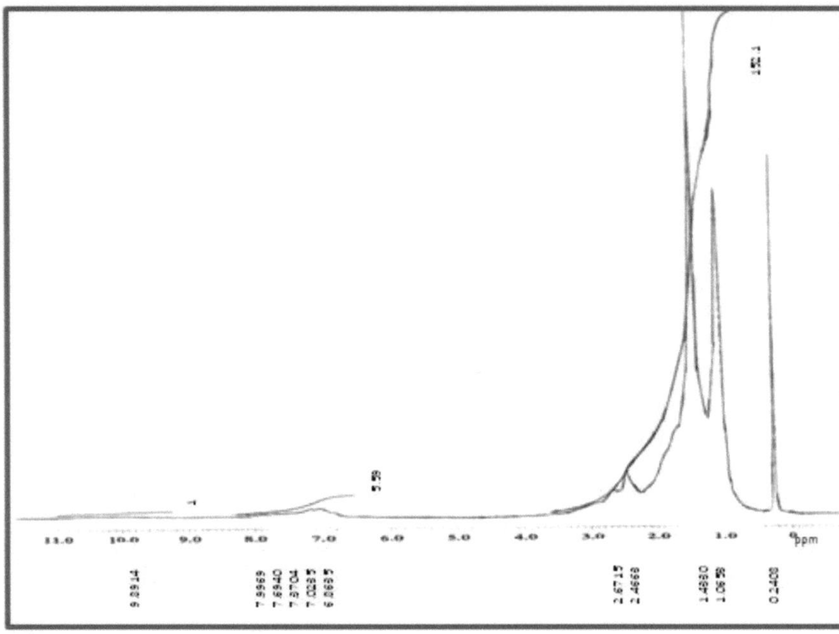

Fig. 3. ^{1}H NMR spectrum for AcJ.

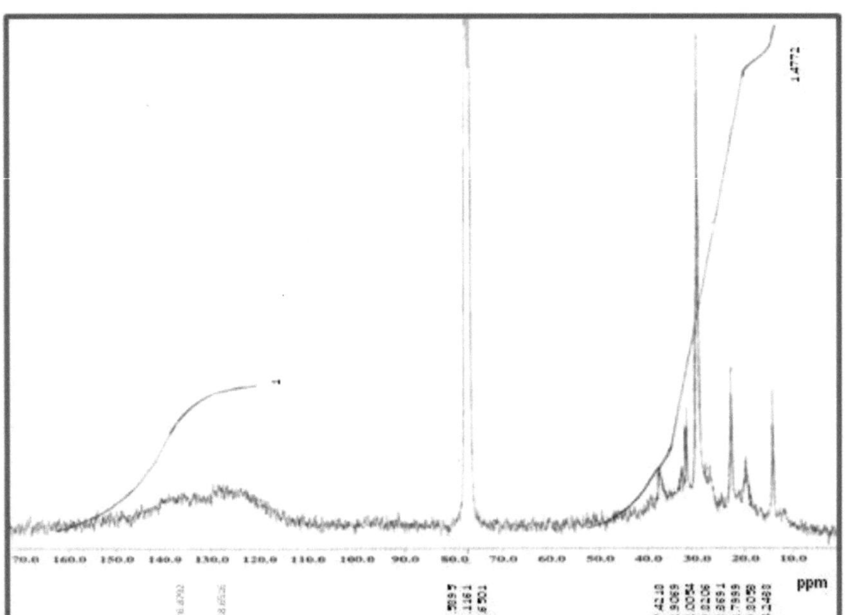

Fig. 4. ^{13}C NMR spectrum for AJ.

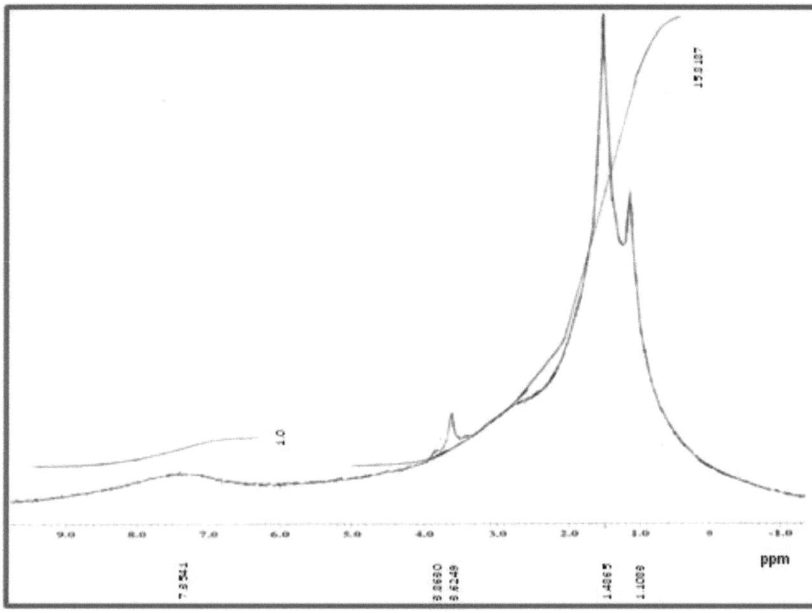

Fig. 5. ^{1}H NMR spectrum for AJ.

The adsorption of asphaltenes and other natural surfactants at the water/toluene interface has been studied here. The adsorption isotherms for the different systems studied (see Figure 5) show that there is a systematic decrease of the interfacial tension with the increase in the sample concentration until a concentration is reached after which the tension remains unchanged. This value corresponds to the saturation of the aggregates in the solution which coincides with the results reported by Acevedo et al. in 2005. The evidence presented in this work suggests that asphaltenes adsorb to the oil-water interface.

Maltenes show the highest tension values. The concentration of surface active species increases from MJ to AcJ, which was expected, because the polarity of the sample increases in the aforementioned direction. The acids fraction has low molecular weights, which favors the mobility of this species at the interface. As the tension curves are parallel, the concentration of each of the species to reach a specific interfacial tension was also calculated. The value of the AcJ concentration so obtained must correspond to the amount of acids present in each of the samples. Therefore, Table 5 shows the percentage of AcJ present in maltenes, asphaltenes, and acid-free asphaltenes. As can be seen, asphaltenes show the highest concentration of acids, suggesting that the AcJ are trapped by the asphaltene molecules which act as molecular traps, because these compounds are linked by hydrogen bonds, among other kind of interactions (π-π interactions) (Acevedo et al. 2007). On the other hand, acid-free asphaltenes (ASA) show a higher interfacial tension than AJ indicating that the acids were removed at least almost completely from the asphaltenes. It can also be seen that the saturation concentration for ASA is about 6000 mg/L, and it remains unchanged regardless of the concentration of asphaltenes in the solution, *i.e.* it shows $\delta\gamma/\delta(\ln C) \approx 0$ across a broad range of concentrations. This behavior may be related to the ability of asphaltenes to form aggregates and adsorb at the water - crude oil interface and the posterior flocculation of these aggregates to originate solid interfacial films.

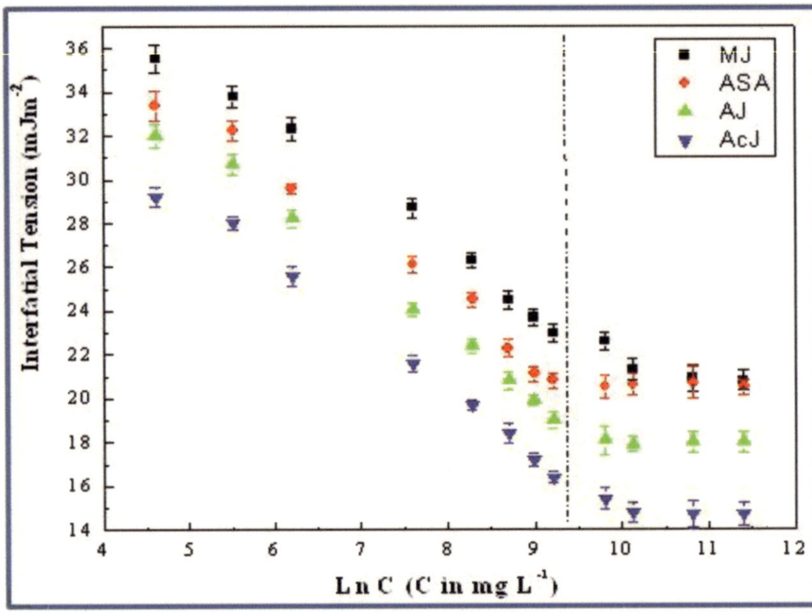

Fig. 6. Adsorption isotherm for maltenes (MJ), acid-free asphaltenes (ASA), asphaltenes (AJ) and acids (AcJ) at a water / toluene interface. The vertical line represents the maximum values used for the calculation of the slopes for each of the isotherms.

Sample	Slope (mJ/m^2)	Apparent Area per Molecule ($Å^2/mol$)	Molecular Weight (Da)	Percentage of Acids (AcJ)
MJ	-2.86±0.11	144±6	347±14	11±6
AJ	-2.91±0.09	141±4	340±10	41±4
ASA	-2.93±0.09	140±4	337±10	23±4
AcJ	-2.94±0.08	140±4	337±10	100±4

Table 4. Apparent area per molecule adsorbed at the interface and its respective molecular weight.

Another interesting aspect to discuss is the calculation of the apparent area of the species adsorbed at the interface and its relationship with the molecular weight of said species. As is well known, Gibbs adsorption equation allows us calculate the adsorption Γ (the concentration of the species that is adsorbed at the surface) from measurements of tension as a function of the concentration of the surfactant in the solution (Hiemmenz, 1998).

$$\Gamma = -\frac{1}{RT}\frac{d\gamma}{d\ln C} \tag{1}$$

If Γ is expressed in mol/cm^2, the average area per molecule at the interface in $Å^2$/molecule is:

$$A = \frac{1}{\Gamma} = \frac{RT}{d\gamma / d\ln C} = \frac{411.6}{d\gamma / d\ln C} \tag{2}$$

where R is the gas constant, T is temperature and $\frac{d\gamma}{d\ln C}$ is the slope of the linear region of each of the adsorption isotherms. This equation allows us deduce some information regarding the orientation and conformation of the molecules at the interface.

On the other hand, we can establish a relationship between the average area per molecule of a species and its molecular weight by using an expression including both parameters:

$$M = \rho\, N_A\, A \in \quad (3)$$

where: M: molecular weight (g/mol), ρ: density (1 g/cm^3), N_A: Avogadro's number (6.023*10^{23} molecules/mol), A: area (Å^2/molecule) and ∈:thickness (about. 4Å) (Acevedo et al., 1994).

This expression was used to carry out the respective calculations. The results for apparent area and molecular weights for the species adsorbed at the interface are shown in Table 4. It can be seen that the calculated values are the same since they are within the error range, obviously because the slopes are practically the same. This suggests that the species that is in direct contact with both phases is the same kind of compound, which has a molecular weight of about 350 Dalton. This leads us to suppose that these compounds are the acids present in the crude oil which had a VPO molecular weight of 470 Dalton. These values of molecular weights are similar to those reported when using other techniques such as ESI FTICR MS and FTICR MS for acid compounds adsorbed at the interfacial film of oil in water emulsions (Stanford et al., 2007).

4. Conclusions

Acids present in the crude, which have molecular weights of about 400 Da, are present in higher concentrations in asphaltenes, and contribute significantly to the reduction of interfacial tension. In other words, asphaltenes act as molecular traps for acids. Also the results suggest the formation of a mixed interface composed of asphaltenes and carboxylic acids of low molecular weight. This film acts as a barrier separating the two phases and prevents coalescence process occurs. Furthermore, it was possible to confirm with simpler methods, that substances adsorbed at the water - crude oil interface have low molecular weights. This is using of Gibbs adsorption equation allows us calculate the adsorption Γ (the concentration of the species that is adsorbed at the surface) from measurements of tension as a function of the concentration of the surfactant in the solution. Once the parameter Γ is obtained, calculate the apparent area of the species adsorbed at the interface and its relationship with the molecular weight of said species.

5. References

Acevedo, S., Escobar, G., Gutierrez, L. & Rivas, H. (1992). Isolation and characterization of natural surfactants from extra heavy crude oils, asphaltenes and maltenes. Interpretation of their interfacial tension-pH behaviour in terms of ion pair formation, *Fuel* Vol.71: 619-623.

Acevedo, S., Escobar, G., Ranaudo, M. & Gutierrez, L. (1994). Discotic shape of asphaltenes obtained from g.p.c. data, *Fuel* Vol. 73(No.11):1807-1809.

Acevedo, S., Escobar, G., Ranaudo, M., Khazen, J., Borges, B., Pereira, J. & Méndez, B. (1999). Isolation and Characterization of Low and High Molecular Weigh Acids Compounds fron Cerro Negro Extraheavy Crude Oil. Role of These Acids in the Interfacial Properties of the Crude Oil Emulsions, *Energy & Fuels* Vol. 13(No.2):333-335.

Acevedo, S., Borges, B., Quintero, F., Piscitelly, V. and Gutierrez, L. (2005a). Asphaltenes and Other Natural Surfactants from Cerro Negro Crude Oil, Stepwise Adsortion at the Water/Toluene Interface: Film Formation and Hydrophobic Effects, *Energy & Fuels* Vol.19:1948-1953.

Acevedo, S., Gutierrez, L., Negrin, g., Pereira, J., Méndez, B., Demolme, F., Dessalces, G. and Broseta, D. (2005b). Molecular Weight of Petroleum asphaltenes: A Comparison between Mass spectrometry and vapor pressure osmometry, *Energy&Fuels* Vol.19:1545-1560.

Acevedo, S., Castro, A., Negrin, J., Fernández, A., Escobar, G. and Piscitelly, V. (2007). Ralations between Asphaltene Structures and Their Physical and Chemical Properties: The Rosary-Type Structure, *Energy & Fuels* Vol. 21:2165-2175.

Borges, B. (2009). Interfacial and structural properties of different natural surfactants extracted from Venezuelan Junin extraheavy crude oil, *Petroleum Science and technology* Vol. 24(No.18):2212-2222.

Chaverot, P., Cagna, A., Glita, S., & Rondelez, F. (2008). Interfacial Tension of Bitumen-Water Interfaces. Part 1: Influence of Endogenous Surfactants at Acidic pH, *Energy & Fuels* Vol.22:790–798.

Hiemmenz, P. 1988. Principles and Surface Chemistry. New York: Marcel Dekker.

Khanna, S., Khan, H., Nautiyal, S., Agarwal, K., Aloopwan, M., Tyagi, O. & Sawhney, S. (2006). IR and HNMR Analysis of Aphaltic materials Present in Some Indian Crude oils of Gujarat Region, *Petroleum Science and technology* Vol. 24:23-30.

López, A., Redner, R., & Cerro R. (1993). Surface Tension from Pendant Drop Curvature, *Langmuir* Vol.9:3691-3694.

McLean, J.D., Kilpatrick, P.K., 1997. Effects of asphaltene aggregation in model heptane–toluene mixtures on stability of water-in-oil emulsions, *Journal of Colloid and Interface Science* Vol.196: 23–34.

Ortiz, D., Baydak, E., Yarranton, H. (2010). Effect of surfactants on interfacial films and stability of water-in-oil emulsions stabilized by asphaltenes, *Journal of Colloid and Interface Science* Vol.351:542–555

Pauchard, V., Sjöblom, J., Kokal, S., Bouriat, P., Dicharry, C., Hendrik Müller, H., & Adnan al-Hajji. (2009). Role of Naphthenic Acids in Emulsion Tightness for a Low-Total-Acid-Number (TAN)/High-Asphaltenes Oil, *Energy & Fuels* Vol. 23: 1269–1279.

Ramljak, Z., Solc, A., Arpino, P., Schmitter, J. & Guiochom, G. (1977). Separation of acids from asphaltenes, Analytical Chemistry Vol.49(No.8):1222-1225.

Yarranton, H.W., Hussein, H., & Masliyah, J.H., (2000). Water-inhydrocarbon emulsions stabilized by asphaltenes at low concentrations, *Journal of Colloid and Interface Science* Vol.228: 52–63.

Stanford, L., Rodgers, R., Marshall, A., Czarnecki, J. and Wu, X. (2007). Compositional Characterization of Bitumen/Water Emulsion Films by Negative-and positive-Ion Electrospray Ionization Ion Cyclotron Resonance Mass Spectrometry, Energy&Fuels Vol. 21(No.2):963-972.

Part 2

Characterization of Crude Oil

5

Thermodynamic Models for the Prediction of Petroleum-Fluid Phase Behaviour

Romain Privat and Jean-Noël Jaubert
Ecole Nationale Supérieure des Industries Chimiques, Université de Lorraine
France

1. Introduction

Petroleum fluids, which include natural gases, gas condensates, crude oils and heavy oils are in the category of complex mixtures. A complex mixture is defined as one in which various families of compounds with diverse molecular properties are present. In petroleum fluids, various families of hydrocarbons such as paraffins, naphthenes and aromatics exist. Such mixtures also contain some heavy organic compounds such as resins and asphalthenes and some impurities such as mercaptans or other sulphur compounds. Non hydrocarbons are typically nitrogen (N_2), carbon dioxide (CO_2), hydrogen sulphide (H_2S), hydrogen (H_2) among few others. Water is another fluid that is typically found co-existing with naturally occurring hydrocarbon mixtures.

Not only do oil and gas engineering systems handle very complex mixtures, but they also operate within exceptionally wide ranges of pressure and temperature conditions. Extremely low temperatures are required in liquefied natural gas (LNG) applications, while very high temperatures are needed for thermal cracking of heavy hydrocarbon molecules. Between these two extremes, hydrocarbon fluids are found underground at temperatures that can reach 90 °C or more, while surface conditions can hover around 20 °C. Pressure can vary from its atmospheric value (or lower in the case of vacuum distillation) to a number in the hundreds of MPa. Due to their complexity, multi-family nature and ample range of conditions, petroleum fluids undergo severe transformations and various phase transitions which include, but not limited to, liquid–vapour, liquid–liquid and liquid–liquid–vapour.

The major obstacle in the efficient design of processes dealing with hydrocarbon mixtures, e.g. primary production, enhanced oil recovery, pipeline transportation or petroleum processing and refining is the difficulty in the accurate and efficient prediction of their phase behaviour and other properties such as mixing enthalpies, heat capacities and volumetric properties. There is thus a great deal of interest to develop accurate models and computational packages to predict the phase behaviour and thermodynamic properties of such mixtures with the least amount of input data necessary. To reach this objective, equations of state have been widely used during the last decades.

The goal of this chapter is to present and analyse the development of Van der Waals-type cubic equations of state (EoS) and their application to the correlation and prediction of phase equilibrium and volumetric properties. A chronological critical walk through the most important contributions during the first part of the 1900s is made, to arrive at the equation proposed by Redlich and Kwong in 1949. The contributions after Redlich and Kwong to the modern development of equations of state and the most recent equations proposed in the literature like the PPR78 and PR2SRK models are analyzed. The application of cubic equations of state to petroleum fluids and the development of mixing rules is put in a proper perspective. The main applications of cubic equations of state to petroleum mixtures including high-pressure phase equilibria and supercritical fluids are summarized. Finally, recommendations on which equations of state and which mixing rules to use for given applications in the petroleum industry are presented.

2. Some words about cubic equations of state History

Our current industrialized world transports and produces chemicals on an unprecedented scale. Natural gas and oil are today key-raw materials from which are derived the gaseous and liquid fuels energizing factories, electric power plants and most modes of transportation as well. Processes of evaporation and condensation, of mixing and separation, underlie almost any production method in the chemical industry. These processes can be grandly complex, especially when they occur at high pressures. Interest for them has mainly started during the industrial revolution in the 19th century and has unceasingly grown up since then. A huge leap in understanding of the phase behaviour of fluids was accomplished during the second half of this century by the scientists Van der Waals and Kamerlingh Onnes - and more generally by the Dutch School. It is undisputable that their discoveries were built on many talented anterior works as for instance:

- the successive attempts by Boyle (in 1662), Mariotte (more or less in the same time) and Gay-Lussac (in 1801) to derive the perfect-gas equation,
- the first observation of critical points by Cagniard de la Tour in 1822,
- the experimental determination of critical points of many substances by Faraday and Mendeleev throughout the 19th century,
- the measurement of experimental isotherms by Thomas Andrews (1869) showing the behaviour of a pure fluid around its critical point.

As the major result of the Dutch School, the Van der Waals equation of state (1873) - connecting variables P (pressure), v (molar volume) and T (temperature) of a fluid - was the first mathematical model incorporating both the gas-liquid transition and fluid criticality. In addition, its foundation on – more or less rigorous – molecular concepts (Van der Waals' theory assumes that molecules are subject to attractive and repulsive forces) affirmed the reality of molecules at a crucial time in history. Before Van der Waals, some attempts to model the real behaviour of gases were made. The main drawback of the P-v-T relationships presented before was that they did not consider the finite volume occupied by the molecules, similarly to the perfect-gas equation. Yet, the idea of including the volume of the molecules into the repulsive term was suggested by Bernoulli at the end of the 18th century and was then ignored for a long time. Following this idea, Dupré and Hirn (in 1863 - 1864) proposed to replace the molar volume v by $(v-b)$, where b is the molar volume that molecules exclude by their mutual repulsions. This quantity is proportional to the temperature-independent molecular volume v_0, and named *covolume* by Dupré (sometimes also called *excluded volume*). However, none of these contributions were of general use and none was able to answer the many questions related to fluid behaviour remaining at that time. It was Van der Waals with his celebrated doctoral thesis on "*The Continuity of the Liquid and Gaseous States*" (1873) and his famous equation of state who proposed for the first time a physically-coherent description of fluid behaviour from low to high pressures. To derive his equation, he considered the perfect-gas law (i.e. $Pv = RT$) and took into account the fact that molecules occupy space by replacing v by $(v-b)$, and the fact that they exert an attraction on each other by replacing P by $P + a/v^2$ (cohesion effect). Therefore, due to mutual repulsion of molecules the actual molar volume v has to be greater than b while molecular attraction forces are incorporated in the model by the coefficient a. Note also that in Van der Waals' theory, the molecules are assumed to have a core in the form of an impenetrable sphere. Physicists and more particularly thermodynamicists rapidly understood that Van der Waals' theory was a revolution upsetting classical conceptions and modernizing approaches used until then to describe fluids. As a consecration, Van der Waals was awarded the Nobel Prize of physics the 12th December 1910; he can be seen as the father of modern fluid thermodynamics. The equation that Van der Waals proposed in his thesis (Van der Waals, 1873) writes:

$$\left(P + \frac{a}{v^2}\right)(v-b) = R(1+\alpha t) \tag{1}$$

where P and v are the externally measured pressure and molar volume. $R = 8.314472\ J \cdot mol^{-1} \cdot K^{-1}$ is the gas constant and α is a measure of the kinetic energy of the molecule. This equation was rewritten later as follows (see for example the lecture given by Van der Waals when he received the Nobel Prize):

$$P(T,v) = \frac{RT}{v-b} - \frac{a}{v^2} \tag{2}$$

It appears that the pressure results from the addition of a repulsive term $P_{\text{rep}} = RT/(v-b)$ and an attractive term $P_{\text{att}} = -a/v^2$. Writing the critical constraints (i.e. that the critical isotherm has a horizontal inflection point at the critical point in the P-v plane), it becomes possible to express the a and b parameters with respect to the experimental values of T_c and P_c, resp. the critical temperature and pressure:

$$a = \Omega_a \frac{R^2 T_c^2}{P_c} \quad \text{and} \quad b = \Omega_b \frac{RT_c}{P_c} \quad \text{with:} \begin{cases} \Omega_a = 27/64 \\ \Omega_b = 1/8 \end{cases} \tag{3}$$

The Van der Waals equation is an example of *cubic equation*. It can be written as a third-degree polynomial in the volume, with coefficients depending on temperature and pressure:

$$v^3 - v^2\left(b + \frac{RT}{P}\right) + \frac{a}{P}v - \frac{ab}{P} = 0 \tag{4}$$

The cubic form of the Van der Waals equation has the advantage that three real roots are found at the most for the volume at a given temperature and pressure. EoS including higher powers of the volume comes at the expense of the appearance of multiple roots, thus complicating numerical calculations or leading to non-physical phenomena. In the numerical calculation of phase equilibrium with cubic equations, one simply discards the middle root, for which the compressibility is negative (this root is associated to an unstable state). Van der Waals' EoS and his ideas on intermolecular forces have been the subjects of many studies and development all through the years. (Clausius, 1880) proposed an EoS closely similar to the Van der Waals equation in which (i) an additional constant parameter (noted *c*) was added to the volume in the attractive term, (ii) the attractive term is made temperature-dependent. Containing one more adjustable parameter than Van der Waals' equation, Clausius' equation showed a possible way for increasing the model accuracy.

$$\text{Clausius: } P(T,v)=\frac{RT}{v-b}-\frac{a/T}{(v+c)^2} \tag{5}$$

In the middle of the 20th century, (Redlich & Kwong, 1949) published a new model derived from Van der Waals' equation, in which the attractive term was modified in order to obtain better fluid phase behaviour at low and high densities:

$$\text{Redlich-Kwong: } P(T,v)=\frac{RT}{v-b}-\frac{a(T)}{v(v+b)} \text{ with: } \begin{cases} a(T)=a_c\cdot\alpha(T)\ ;\ b=\Omega_b RT_c/P_c \\ a_c=\Omega_a R^2T_c^2/P_c \text{ and } \alpha(T)=\sqrt{T_c/T} \\ \Omega_a=1/\left[9\left(\sqrt[3]{2}-1\right)\right]\approx 0.4274 \\ \Omega_b=\left(\sqrt[3]{2}-1\right)/3\approx 0.08664 \end{cases} \tag{6}$$

Similarly to Clausius' equation, a temperature dependency is introduced in the attractive term. The *a* parameter is expressed as the product of the constant coefficient $a_c=a(T_c)$ and $\alpha(T)$ (named alpha function) which is unity at the critical point. Note that only the two pure-component critical parameters T_c and P_c are required to evaluate *a* and *b*. The modifications of the attractive term proposed by Redlich and Kwong - although not based on strong theoretical background - showed the way to many contributors on how to improve Van der Waals' equation. For a long time, this model remained one of the most popular cubic equations, performing relatively well for simple fluids with acentric factors close to zero (like Ar, Kr or Xe) but representing with much less accuracy complex fluids with nonzero acentric factors. Let us recall that the acentric factor ω_i, defined by Pitzer as:

$$\omega_i=-\log\left[P_i^{sat}(T=0.7T_{c,i})/P_{c,i}\right]-1 \tag{7}$$

where P_i^{sat} is the vaporization pressure of pure component *i*, is a measure of the acentricity (i.e. the non-central nature of the intermolecular forces) of molecule *i*. The success of the Redlich-Kwong equation has been the impetus for many further empirical improvements. In 1972, the Italian engineer Soave suggested to replace in Eq. (6) the α function by a more general temperature-dependent term. Considering the variation in behaviour of different fluids at the same reduced pressure (P/P_c) and temperature (T/T_c), he proposed to turn from a two-parameter EoS (the two parameters are T_c and P_c) to a three-parameter EoS by introducing the acentric factor as a third parameter in the definition of $\alpha(T)$. The acentric factor is used to take into account molecular size and shape effects since it varies with the chain length and the spatial arrangement of the molecules (small globular molecules have a nearly zero acentric factor). The resulting model was named the Soave-Redlich-Kwong equation or the SRK equation (Soave, 1972), and writes:

$$\text{SRK:}\ P(T,v)=\frac{RT}{v-b}-\frac{a(T)}{v(v+b)}\ \text{with:}\begin{cases}a(T)=a_c\cdot\alpha(T)\ ;\ b=\Omega_b RT_c/P_c\\ a_c=\Omega_a R^2T_c^2/P_c\ \text{and}\ \alpha(T)=\left[1+m\left(1-\sqrt{T/T_c}\right)\right]^2\\ m=0.480+1.574\omega-0.176\omega^2\\ \Omega_a=1/\left[9\left(\sqrt[3]{2}-1\right)\right]\approx 0.4274\\ \Omega_b=\left(\sqrt[3]{2}-1\right)/3\approx 0.08664\end{cases}\quad(8)$$

The alpha function $\alpha(T)$ used in the SRK equation is often named *Soave's alpha function*. The accuracy of this model was tested by comparing vapour pressures of a number of hydrocarbons calculated with the SRK equation to experimental data. Contrary to the Redlich-Kwong equation, the SRK equation was able to fit well the experimental trend. After Soave's proposal, many modifications were presented in the literature for improving the prediction of one or another property. One of the most popular ones is certainly the modification proposed by (Peng & Robinson, 1976) (their equation is named PR76 in this chapter). They considered the same alpha function as Soave but coefficients of the m function were recalculated. In addition, they also modified the volume dependency of the attractive term:

$$\text{PR76:}\ P(T,v)=\frac{RT}{v-b}-\frac{a(T)}{v(v+b)+b(v-b)}\ \text{with:}\begin{cases}a(T)=a_c\cdot\alpha(T)\ ;\ b=\Omega_b RT_c/P_c\\ a_c=\Omega_a R^2T_c^2/P_c\ \text{and}\ \alpha(T)=\left[1+m\left(1-\sqrt{T/T_c}\right)\right]^2\\ m=0.37464+1.54226\omega-0.26992\omega^2\\ X=\frac{1}{3}\left(-1+\sqrt[3]{6\sqrt{2}+8}-\sqrt[3]{6\sqrt{2}-8}\right)\approx 0.2531\\ \Omega_a=8(5X+1)/(49-37X)\approx 0.45723\\ \Omega_b=X/(X+3)\approx 0.077796\end{cases}\quad(9)$$

The accuracy of the PR76 equation is comparable to the one of the SRK equation. Both these models are quite popular in the hydrocarbon industry and offer generally a good representation of the fluid phase behaviour of few polar and few associated molecules (paraffins, naphthenes, aromatics, permanent gases and so on). (Robinson & Peng, 1978) proposed to slightly modify the expression of the m function in order to improve the representation of heavy molecules i such that $\omega_i > \omega_{n-decane}=0.491$. This model is named PR78 in this chapter.

$$\text{PR78:}\begin{cases}m=0.37464+1.54226\omega-0.26992\omega^2 & \text{if}\ \omega\le 0.491\\ m=0.379642+1.487503\omega-0.164423\omega^2+0.016666\omega^3 & \text{if}\ \omega>0.491\end{cases}\quad(10)$$

In order to improve volume predictions (Péneloux et al., 1982) proposed a consistent correction for cubic EoS which does not change the vapour-liquid equilibrium conditions. Their method consists in translating the EoS by replacing v by $v+c$ and b by $b+c$. The volume translation c can be estimated from the following correlation which involves the Rackett compressibility factor, z_{RA}:

$$\text{Volume-translated equations: } \begin{cases} c = \dfrac{RT_c}{P_c}\left[0.1156 - 0.4077 z_{RA}\right], \text{ for the SRK equation} \\ c = \dfrac{RT_c}{P_c}\left[0.1154 - 0.4406 z_{RA}\right], \text{ for the PR76 equation} \end{cases} \quad (11)$$

3. General presentation of cubic equations of state

All the cubic equations can be written under the general following form:

$$P(T,v) = \frac{RT}{v-b} - \frac{a(T)}{(v - r_1 b)(v - r_2 b)} \quad (12)$$

wherein r_1 and r_2 are two universal constants (i.e. they keep the same value whatever the pure component). As shown with Van der Waals' equation (see Eq. (4)), cubic EoS can be written as third-degree polynomials in v at a fixed temperature T and pressure P:

$$v^3 - v^2\left[b(r_1 + r_2 + 1) + RT/P\right] + v\left[b^2(r_1 r_2 + r_1 + r_2) + RTb(r_1 + r_2)/P + a/P\right] - b\left(r_1 r_2 b^2 + r_1 r_2 bRT/P + a/P\right) = 0 \quad (13)$$

As a drawback of cubic equations, the predicted critical molar compressibility factor, $z_c = P_c v_c/(RT_c)$, is found to be a universal constant whereas experimentally, it is specific to each pure substance. In a homologous chemical series, the z_c coefficient diminishes as the molecular size increases. For normal substances, z_c is found around 0.27. Table 1 here below provides values of r_1, r_2 and z_c for all the equations presented above:

Equation of state	r_1	r_2	z_c
Van der Waals or Clausius	0	0	$3/8 = 0.375$
Redlich-Kwong or SRK	0	-1	$1/3 \approx 0.333$
PR76 or PR78	$-1-\sqrt{2}$	$-1+\sqrt{2}$	≈ 0.3074

Table 1. values of r_1, r_2 and z_c for some popular cubic equations of state.

Experimental values of T_c and P_c can be used to determine the expressions of $a_c = a(T_c)$ and b. Indeed, by applying the critical constraints, one obtains the following results:

$$\begin{cases} X = \left[1 + \sqrt[3]{(1-r_1)(1-r_2)^2} + \sqrt[3]{(1-r_2)(1-r_1)^2}\right]^{-1} \\ a_c = \Omega_a R^2 T_c^2 / P_c \quad \text{and} \quad b = \Omega_b RT_c / P_c \\ \Omega_a = (1 - r_1 X)(1 - r_2 X)\left[2 - (r_1 + r_2)X\right] / \left[(1 - X)\left[3 - X(1 + r_1 + r_2)\right]\right]^2 \\ \Omega_b = X / \left[3 - X(1 + r_1 + r_2)\right] \end{cases} \quad (14)$$

Note that a cubic equation of state applied to a given pure component is completely defined by the universal-constant values r_1 and r_2, by the critical temperature and pressure of the substance (allowing to calculate b and a_c) but also by the considered alpha function (allowing to evaluate the attractive parameter $a(T) = a_c \cdot \alpha(T)$). The next subsection is dedicated to present some $\alpha(T)$ functions frequently used with cubic equations.

4. Presentation of some alpha functions usable with cubic equations of state

Modifications of the temperature-dependent function $\alpha(T)$ in the attractive term of the SRK and PR equations have been mainly proposed to improve correlations and predictions of vapour pressure for polar fluids. Some of the most used are presented in this section.

- The most popular alpha function is certainly that of (Soave, 1972):

$$\alpha(T) = \left[1 + m\left(1 - \sqrt{T / T_c}\right)\right]^2 \tag{15}$$

wherein the m parameter is a function of the acentric factor (expressions of m for the SRK, PR76 and PR78 models are resp. given in Eqs. (8), (9) and (10)).

- (Mathias & Copeman, 1983) developed an expression for the alpha function aimed at extending the application range of the PR equation to highly polar components:

$$\begin{cases} \alpha(T) = \left[1 + C_1\left(1 - \sqrt{T / T_c}\right) + C_2\left(1 - \sqrt{T / T_c}\right)^2 + C_3\left(1 - \sqrt{T / T_c}\right)^3\right]^2 & \text{if } T \le T_c \\ \alpha(T) = \left[1 + C_1\left(1 - \sqrt{T / T_c}\right)\right]^2 & \text{if } T > T_c \end{cases} \tag{16}$$

Parameters C_1, C_2 and C_3 are specific to each component. They have to be fitted on vapour-pressure data.

- (Stryjek & Vera, 1986) proposed an alpha function for improving the modelling capacity of the PR equation at low reduced temperatures:

$$\begin{cases} \alpha(T) = \left[1 + m\left(1 - \sqrt{T / T_c}\right)\right]^2 \\ \text{with } m = m_0 + m_1\left(1 + \sqrt{T / T_c}\right)(0.7 - T / T_c) \\ \text{and } m_0 = 0.378893 + 1.4897153\omega - 0.17131848\omega^2 + 0.0196554\omega^3 \end{cases} \tag{17}$$

m_1 has to be fitted on experimental vapour-pressure data. Compared to the original PR EoS, this alpha function uses a higher order polynomial function in the acentric factor for the m parameter that allows a better modelling of heavy-hydrocarbon phase behaviour.

- (Twu et al., 1991, 1995a, 1995b) proposed two different alpha functions. The first one requires - for each pure component - to fit the model parameters (N, M and L) on experimental pure-component VLE data:

$$\alpha(T) = (T / T_c)^{N(M-1)} \exp\left[L\left[1 - (T / T_c)^{NM}\right]\right] \tag{18}$$

The second one is a predictive version of the first one, only requiring the knowledge of the acentric factor. Following Pitzer's corresponding-states principle, Twu et al. proposed:

$$\alpha(T) = \alpha^0(T) + \omega\left[\alpha^1(T) - \alpha^0(T)\right] \tag{19}$$

wherein the expressions of α^0 and α^1 are given by Eq. (18). Parameters *L*, *M* and *N* involved in these two functions are provided in Table 2.

alpha function parameters	$T \le T_c$		$T > T_c$	
	α^0	α^1	α^0	α^1
		PR:		
L	0.125283	0.511614	0.401219	0.024955
M	0.911807	0.784054	4.963070	1.248089
N	1.948150	2.812520	–0.200000	–8.000000
		SRK:		
L	0.141599	0.500315	0.441411	0.032580
M	0.919422	0.799457	6.500018	1.289098
N	2.496441	3.291790	–0.200000	–8.000000

Table 2. Generalized parameters of the predictive version of the Twu et al. alpha function.

5. Petroleum-fluid phase behaviour modelling with cubic equations of state

The SRK and the PR EoS are the primary choice of models in the petroleum and gas processing industries where *high-pressure* models are required. For a pure component, the three required parameters (T_c, P_c and ω) were determined from experiments for thousands of pure components. When no experimental data are available, various estimation methods, applicable to any kind of molecule, can be used. Extension to mixtures requires mixing rules for the energy parameter and the covolume. A widely employed way to extend the cubic EoS to a mixture containing p components, the mole fractions of which are x_i, is via the so-called Van der Waals one-fluid mixing rules [quadratic composition dependency for both parameters - see Eqs. (20)and (21)] and the classical combining rules, i.e. the geometric mean rule for the cross-energy [Eq. (22)] and the arithmetic mean rule for the cross covolume parameter [Eq. (23)]:

$$a = \sum_{i=1}^{p}\sum_{j=1}^{p} x_i x_j a_{ij} \tag{20}$$

$$b = \sum_{i=1}^{p}\sum_{j=1}^{p} x_i x_j b_{ij} \tag{21}$$

$$a_{ij} = \sqrt{a_i a_j}\,(1 - k_{ij}) \tag{22}$$

$$b_{ij} = \tfrac{1}{2}\left(b_i + b_j\right)(1 - l_{ij}) \tag{23}$$

Doing so, two new parameters, the so-called *binary interaction parameters* (k_{ij} and l_{ij}) appear in the combining rules. One of them, namely k_{ij} is by far the most important one. Indeed, a non null l_{ij} is only necessary for complex polar systems and special cases. This is the reason why, dealing with petroleum fluids, phase equilibrium calculations are generally performed with $l_{ij} = 0$ and the mixing rule for the co-volume parameter simplifies to:

$$b = \sum_{i=1}^{p} x_i b_i \tag{24}$$

We know by experience that the k_{ij} value has a huge influence on fluid-phase equilibrium calculation. To illustrate this point, it was decided to plot - using the PR EoS - the isothermal phase diagram for the system 2,2,4 trimethyl pentane (1) + toluene (2) at $T = 333.15\ K$ giving to k_{12} different values (see Figure 1). The obtained curves speak for themselves:

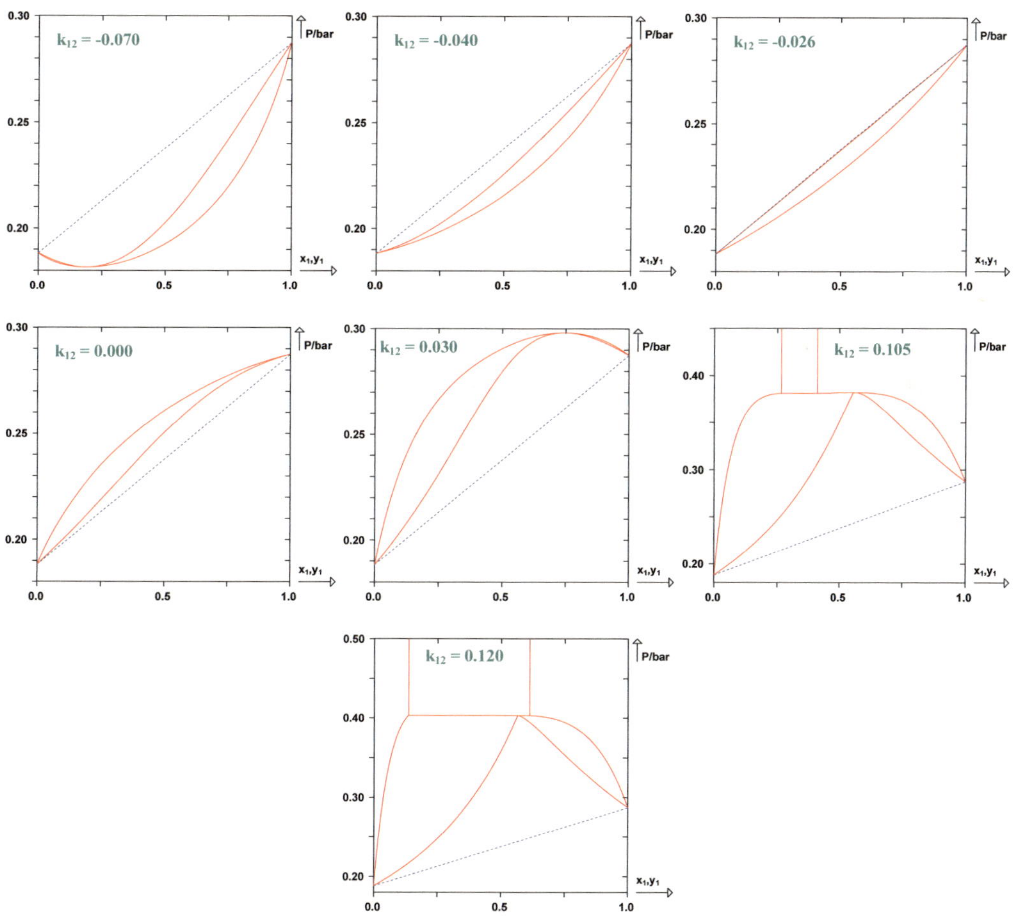

Fig. 1. Influence of k_{12} on the calculated isothermal phase diagram using the PR EoS for the system 2,2,4 trimethyl pentane (1) + toluene (2) at $T/K = 333.15$. The dashed line is Raoult's line.

- for $k_{12} = -0.07$, the system holds a negative homogeneous azeotrope
- for $k_{12} = -0.04$, the system still shows negative deviations from ideality but the azeotrope does not exist anymore
- for $k_{12} = -0.026$, the bubble curve is a straight line and the liquid phase behaves as an ideal solution
- for $k_{12} = 0.0$, the system shows positive deviations from ideality
- for $k_{12} = 0.03$, the system holds a positive homogeneous azeotrope
- for $k_{12} = 0.105$, the system simultaneously holds a positive homogeneous azeotrope and a three-phase line
- for $k_{12} = 0.12$, the system holds a heterogeneous azeotrope.

In front of such a big influence, the safest practice is to fit the k_{ij} value to phase equilibrium data. Such an approach however requires the knowledge of experimental data for all the binary systems it is possible to define in a multi-component system. Unfortunately such data are not always available inciting many researchers to develop correlations or group-contribution methods to estimate the k_{ij}.

5.1 Correlations to estimate the binary interaction parameters

Most of the proposed correlations are purely empirical and thus often unsuitable for extrapolation. Moreover, they often use additional properties besides those (the critical properties and the acentric factor) required by the cubic EoS itself. They are however very useful to help solve a phase equilibrium calculation problem.

Following London's theory, (Chueh & Prausnitz, 1967) proposed a correlation suitable for mixtures of paraffins which only requires knowledge of the critical volume of the two pure substances i and j (see Eq. (25) in Table 3). In 1990, a correlation allowing the estimation of the binary interaction parameters for a modified version of the SRK equation of state was developed by (Stryjek, 1990). It is applicable to mixtures of n-alkanes and is temperature dependent. The author pointed out that although the temperature dependence of the k_{ij} is moderate for mixtures of paraffins, the use of a temperature-dependent k_{ij} significantly improves the modelling of such systems. The proposed correlation has the form given by Eq. (26) in Table 3. (Gao et al., 1992) proposed a correlation suitable for mixtures containing various light hydrocarbons (paraffins, naphthenes, aromatics, alkynes). The theoretical approach adopted by these authors is a continuation of that of Chueh and Prausnitz previously mentioned. Their correlation requires the knowledge of the critical temperature and the critical compressibility factor. Such a parameter is unfortunately not well-known for many hydrocarbons, especially heavy ones. They proposed Eq. (27) shown in Table 3. (Kordas et al., 1995) developed two mathematical expressions to estimate the k_{ij} in mixtures containing methane and alkanes. The first one is suitable for alkanes lighter than the n-eicosane and the second one for heavy alkanes. In such an equation (see Eq. (28) in Table 3), ω is the acentric factor of the molecule mixed with methane. Unfortunately, Kordas et al. failed to generalize their correlations to mixtures containing methane and aromatic molecules. In the open literature, we also can find many correlations to estimate the k_{ij} for systems containing CO_2 and various hydrocarbons. Such equations are of the highest importance because we know by experience that for such systems the k_{ij} is far from zero. As an example, (Graboski & Daubert, 1978) proposed a correlation suitable for mixtures containing CO_2 and paraffins. The use of this correlation (see Eq. (29) in Table 3), developed for a modified version of the SRK EoS, needs the knowledge of the solubility parameters δ. In Eq. (29), subscript i stands for the hydrocarbon and subscript j for CO_2 (or N_2 or H_2S). Another well-known correlation is the one developed by (Kato et al., 1981) which has the great advantage to be temperature-dependent. It can be applied to the PR EoS and to mixtures containing CO_2 and n-alkanes. It is given in Table 3 (see Eq. (30)). The three coefficients (a, b and c) only depend on the acentric factor of the n-alkane. A similar approach was followed by (Moysan et al., 1986) who developed a correlation in which the temperature-dependent k_{ij} can be estimated knowing the acentric factor of the hydrocarbon mixed with CO_2. A key point of Moysan's work is its applicability to systems containing H_2, N_2 and CO. Another temperature-dependent correlation for systems containing CO_2 and n-alkanes was developed by (Kordas et al., 1994). In their approach, the k_{ij} depends on the CO_2 reduced temperature ($T_{r,i} = T / T_{c,i}$) and on the alkane acentric factor (ω_j). It can be found in Table 3, Eq. (31). Kordas et al. explain that their correlation can also be used with other hydrocarbons (branched alkanes, aromatics or naphthenes) under the condition to substitute in Eq. (31) the acentric factor (ω_j) of the studied hydrocarbon by an *effective acentric factor*, the value of which has been correlated to the molar weight and to the density at 15 °C. The paper by (Bartle et al., 1992) also contains a correlation suitable for many systems containing CO_2. Various correlations for nitrogen-containing systems were also developed. Indeed, as shown by (Privat et al., 2008a), the phase behaviour of such systems is particularly difficult to correlate with a cubic EoS even with temperature-dependent k_{ij}. We can cite the work by (Valderrama et al., 1990) who proposed a correlation principally applicable to systems containing nitrogen and light alkanes. The mathematical shape of this correlation was inspired by the previous work of Kordas et al. (Eq. (31)). It is given in Table 3, Eq. (32). Such a correlation can be applied to various cubic EoS and to mixtures not only containing alkanes and N_2 but also CO_2 and H_2S. The same authors (Valderrama et al., 1999) recently improved their previous work. Although only applicable to the PR EoS, a similar work was realized by (Avlonitis et al. 1994). Once again, the k_{ij} depends on temperature and on the acentric factor (see Eq. (33) in Table 3). We cannot close this section before saying a few words about the correlation proposed by (Nishiumi et al., 1988). As shown by Eq. (34) in Table 3, it is probably the most general correlation never developed since it can be used to predict the k_{ij} of the PR EoS for any mixture containing paraffins, naphthenes, aromatics, alkynes, CO_2, N_2 and H_2S. The positive aspect of this correlation is its possible application to many mixtures. Its negative aspect is the required knowledge of the critical volumes.

Although very useful these many correlations only apply to specific mixtures (e.g. mixtures containing hydrocarbons and methane, mixtures containing hydrocarbons and nitrogen, mixtures containing hydrocarbons and carbon dioxide, mixtures containing light hydrocarbons, etc.) and to a specific cubic EoS. Moreover, they are often empirical and thus unsuitable for extrapolation. In addition, they may need additional properties besides those (the critical properties and the acentric factor) required by the cubic EoS itself. Finally, they do not always lead to temperature-dependent k_{ij} whereas we know by experience that the temperature has a huge influence on these interaction parameters.

Mathematical expression of the correlation	
$k_{ij} = 1 - \left(\dfrac{2\sqrt{V_{c,i}^{1/3} V_{c,j}^{1/3}}}{V_{c,i}^{1/3} + V_{c,j}^{1/3}} \right)^n$	(25)
$k_{ij} = k_{ij}^0 + k_{ij}^T \left[(T/K) - 273.15 \right]$	(26)
$1 - k_{ij} = \left(\dfrac{2\sqrt{T_{c,i} T_{c,j}}}{T_{c,i} + T_{c,j}} \right)^{Z_{c,ij}}$ with $Z_{c,ij} = \dfrac{Z_{c,i} + Z_{c,j}}{2}$	(27)
$\begin{cases} k_{ij} = -0{,}13409\omega + 2{,}28543\omega^2 - 7{,}61455\omega^3 + 10{,}46565\omega^4 - 5{,}2351\omega^5 \\ k_{ij} = -0{,}04633 - 0{,}04367 \ln \omega \end{cases}$	(28)
$k_{ij} = A + B\left\|\delta_i - \delta_j\right\| + C\left\|\delta_i - \delta_j\right\|^2$	(29)
$k_{ij} = a(T-b)^2 + c$	(30)
$k_{ij} = a(\omega_j) + b(\omega_j) \times T_{r,i} + c(\omega_j) \times T_{r,i}^3$	(31)
$k_{ij} = A(\omega_j) + B(\omega_j) / T_{r,j}$	(32)
$k_{ij} = Q(\omega_j) - \dfrac{T_{r,i}^2 + A(\omega_j)}{T_{r,i}^3 + C(\omega_j)}$	(33)
$1 - k_{ij} = C + D\dfrac{V_{c,i}}{V_{c,j}} + E\left(\dfrac{V_{c,i}}{V_{c,j}} \right)^2$ *with* $\begin{cases} C = c_1 + c_2 \left\|\omega_i - \omega_j\right\| \\ D = d_1 + d_2 \left\|\omega_i - \omega_j\right\| \end{cases}$	(34)

Table 3. Correlations to estimate the binary interaction parameters.

A way to avoid all these drawbacks would be (i) to develop a group-contribution method capable of estimating temperature-dependent k_{ij} and (ii) to develop a method allowing switching from a cubic EoS to another one. These issues are developed in the next subsections.

5.2 Group contribution methods to estimate the binary interaction parameters

In any group-contribution (GC) method, the basic idea is that whereas there are thousands of chemical compounds of interest in chemical technology, the number of functional groups that constitute these compounds is much smaller. Assuming that a physical property of a fluid is the sum of contributions made by the molecule's functional groups, GC methods allow for correlating the properties of a very large number of fluids using a much smaller number of parameters. These GC parameters characterize the contributions of individual groups in the properties.

5.2.1 The ARP model

The first GC method designed for estimating the k_{ij} of a cubic EoS was developed by (Abdoul et al., 1991) and is often called the ARP model (ARP for Abdoul-Rauzy-Péneloux, the names of the three creators of the model). Even though accurate, such a model has some disadvantages. Indeed, Abdoul et al. did not use the original PR EoS but instead a non-conventional *translated PR type EoS*. Moreover, in order to estimate the attractive parameter *a(T)* of their EoS, these authors defined two classes of pure compounds. For components which are likely to be encountered at very low pressure, the Carrier-Rogalski-Péneloux (CRP) *a(T)* correlation (Carrier et al., 1988) which requires the knowledge of the normal boiling point was used. For other compounds, they used a Soave-like expression (Rauzy, 1982) which is different from the one developed by Soave for the SRK EoS and different from the one developed by Peng and Robinson for their own equation. Moreover, the decomposition into groups of the molecules is not straightforward and is sometimes difficult to understand. For example, propane is classically decomposed into two CH_3 groups and one CH_2 group but 2-methyl propane {CH_3-CH(CH_3)-CH_3} is decomposed into four CH groups and not into three CH_3 groups and one CH group. Isopentane is formed by one CH_3 group, half a CH_2 group and three and a half CH groups. Lastly, because this model was developed more than twenty years ago, the experimental database used by Abdoul et al. to fit the parameters of their model (roughly 40,000 experimental data points) is small in comparison to databases today available. The group parameters obtained from this too small data base may lead to unrealistic phase equilibrium calculations at low or at high temperatures. For all these reasons, this model has never been extensively used and never appeared in commercial process simulators.

5.2.2 The PPR78 model

Being aware of the drawbacks of the ARP model, Jaubert, Privat and co-workers decided to develop the PPR78 model which is a GC model designed for estimating, as a function of temperature, the k_{ij} for the widely used PR78 EoS (Jaubert & Mutelet, 2004; Jaubert et al., 2005, 2010; Vitu et al. 2006, 2008; Privat et al. 2008a, 2008b, 2008c, 2008d). Such an equation of sate was selected because it is used in most of the petroleum companies but above all because it is available in any computational package.

5.2.2.1 Presentation

Following the previous work of Abdoul et al., $k_{ij}(T)$ is expressed in terms of group contributions, through the following expression:

$$k_{ij}(T)=\frac{-\dfrac{1}{2}\displaystyle\sum_{k=1}^{N_g}\sum_{l=1}^{N_g}(\alpha_{ik}-\alpha_{jk})(\alpha_{il}-\alpha_{jl})A_{kl}\cdot\left(\frac{298.15}{T}\right)^{\left(\frac{B_{kl}}{A_{kl}}-1\right)}-\left(\dfrac{\sqrt{a_i(T)}}{b_i}-\dfrac{\sqrt{a_j(T)}}{b_j}\right)^2}{2\dfrac{\sqrt{a_i(T)\cdot a_j(T)}}{b_i\cdot b_j}} \tag{35}$$

In Eq. (35), T is the temperature. a_i and b_i are the attractive parameter and the covolume of pure i. N_g is the number of different groups defined by the method (for the time being, twenty-one groups are defined and $N_g = 21$). α_{ik} is the fraction of molecule i occupied by group k (occurrence of group k in molecule i divided by the total number of groups present in molecule i). $A_{kl} = A_{lk}$ and $B_{kl} = B_{lk}$ (where k and l are two different groups) are constant parameters determined during the development of the model ($A_{kk} = B_{kk} = 0$). As can be seen, to calculate the k_{ij} parameter between two molecules i and j at a selected temperature, it is only necessary to know: the critical temperature of both components ($T_{c,i}$, $T_{c,j}$), the critical pressure of both components ($P_{c,i}$, $P_{c,j}$), the acentric factor of each component (ω_i, ω_j) and the decomposition of each molecule into elementary groups (α_{ik}, α_{jk}). It means that no additional input data besides those required by the EoS itself is necessary. Such a model relies on the Peng-Robinson EoS as published by Peng and Robinson in 1978 (Eq. (10)). The addition of a GC method to estimate the temperature-dependent k_{ij} makes it predictive; it was thus decided to call it PPR78 (predictive 1978, Peng Robinson EoS). The twenty-one groups which are defined until now are summarized below.

For alkanes: group1 = CH_3, group2 = CH_2, group3 = CH, group4 = C, group5 = CH_4 i.e. methane, group6 = C_2H_6 i.e. ethane

For aromatic compounds: group7 = CH_{aro}, group8 = C_{aro}, group9 = $C_{\text{fused aromatic rings}}$

For naphthenic compounds: group10 = $CH_{2,cyclic}$, group11 = CH_{cyclic} = C_{cyclic}

For permanent gases: group12 = CO_2, group13 = N_2, group14 = H_2S, group21 = H_2.

For water-containing systems : group16 = H_2O

For mercaptans: group15 = -SH,

For alkenes: group17 = $CH_2{=}CH_2$ i.e. ethylene, group18 = $CH_{2,alkenic}$ = $CH_{alkenic}$, group19 = $C_{alkenic}$, group20 = $CH_{2,cycloalkenic}$ = $CH_{cycloalkenic}$

The decomposition into groups of the hydrocarbons (linear, branched or cyclic) is very easy, that is as simple as possible. No substitution effects are considered. No exceptions are defined. For these 21 groups, we had to estimate 420 parameters (210A_{kl} and 210B_{kl} values) the values of which are summarized in Table 4. These parameters have been determined in order to minimize the deviations between calculated and experimental vapour-liquid equilibrium data from an extended data base containing roughly 100,000 experimental data points (56,000 bubble points + 42,000 dew points + 2,000 mixture critical points).

The following objective function was minimized:

$$F_{obj} = \frac{F_{obj,bubble} + F_{obj,dew} + F_{obj,crit.\ comp} + F_{obj,crit.\ pressure}}{n_{bubble} + n_{dew} + n_{crit} + n_{crit}} \quad (36)$$

	Group 1	Group 2	Group 3	Group 4	Group 5	Group 6	Group 7	Group 8	Group 9	Group 10	Group 11	Group 12	Group 13	Group 14	Group 15	Group 16	Group 17	Group 18	Group 19	Group 20
Group 1	0	-	-	-	-	-	-	-	-	-	-	-	-	-	-	-	-	-	-	-
Group 2	$A_{12}=74.81$ $B_{12}=165.7$	0	-	-	-	-	-	-	-	-	-	-	-	-	-	-	-	-	-	-
Group 3	$A_{13}=261.5$ $B_{13}=388.8$	$A_{23}=51.47$ $B_{23}=79.61$	0	-	-	-	-	-	-	-	-	-	-	-	-	-	-	-	-	-
Group 4	$A_{14}=396.7$ $B_{14}=804.3$	$A_{24}=88.53$ $B_{24}=315.0$	$A_{34}=-305.7$ $B_{34}=-250.8$	0	-	-	-	-	-	-	-	-	-	-	-	-	-	-	-	-
Group 5	$A_{15}=32.94$ $B_{15}=-35.00$	$A_{25}=36.72$ $B_{25}=108.4$	$A_{35}=145.2$ $B_{35}=301.6$	$A_{45}=263.9$ $B_{45}=531.5$	0	-	-	-	-	-	-	-	-	-	-	-	-	-	-	-
Group 6	$A_{16}=8.579$ $B_{16}=-29.51$	$A_{26}=31.23$ $B_{26}=84.76$	$A_{36}=174.3$ $B_{36}=352.1$	$A_{46}=333.2$ $B_{46}=203.8$	$A_{56}=13.04$ $B_{56}=6.863$	0	-	-	-	-	-	-	-	-	-	-	-	-	-	-
Group 7	$A_{17}=90.25$ $B_{17}=146.1$	$A_{27}=29.78$ $B_{27}=58.17$	$A_{37}=103.3$ $B_{37}=191.8$	$A_{47}=158.9$ $B_{47}=613.2$	$A_{57}=67.26$ $B_{57}=167.5$	$A_{67}=41.18$ $B_{67}=50.79$	0	-	-	-	-	-	-	-	-	-	-	-	-	-
Group 8	$A_{18}=62.80$ $B_{18}=41.86$	$A_{28}=3.775$ $B_{28}=144.8$	$A_{38}=6.177$ $B_{38}=-33.97$	$A_{48}=79.61$ $B_{48}=-326.0$	$A_{58}=139.3$ $B_{58}=464.3$	$A_{68}=-3.088$ $B_{68}=13.04$	$A_{78}=-13.38$ $B_{78}=20.25$	0	-	-	-	-	-	-	-	-	-	-	-	-
Group 9	$A_{19}=62.80$ $B_{19}=41.86$	$A_{29}=3.775$ $B_{29}=144.8$	$A_{39}=6.177$ $B_{39}=-33.97$	$A_{49}=79.61$ $B_{49}=-326.0$	$A_{59}=139.3$ $B_{59}=464.3$	$A_{69}=-3.088$ $B_{69}=13.04$	$A_{79}=-13.38$ $B_{79}=20.25$	$A_{89}=0.000$ $B_{89}=0.000$	0	-	-	-	-	-	-	-	-	-	-	-
Group 10	$A_{1-10}=40.38$ $B_{1-10}=95.90$	$A_{2-10}=12.78$ $B_{2-10}=28.37$	$A_{3-10}=101.9$ $B_{3-10}=-90.93$	$A_{4-10}=177.1$ $B_{4-10}=601.9$	$A_{5-10}=36.37$ $B_{5-10}=26.42$	$A_{6-10}=8.579$ $B_{6-10}=76.86$	$A_{7-10}=29.17$ $B_{7-10}=69.32$	$A_{8-10}=34.31$ $B_{8-10}=95.39$	$A_{9-10}=34.31$ $B_{9-10}=95.39$	0	-	-	-	-	-	-	-	-	-	-
Group 11	$A_{1-11}=98.48$ $B_{1-11}=231.6$	$A_{2-11}=-54.90$ $B_{2-11}=-319.5$	$A_{3-11}=-226.5$ $B_{3-11}=-51.47$	$A_{4-11}=17.84$ $B_{4-11}=-109.5$	$A_{5-11}=40.15$ $B_{5-11}=255.3$	$A_{6-11}=10.29$ $B_{6-11}=-52.84$	$A_{7-11}=-26.42$ $B_{7-11}=-789.2$	$A_{8-11}=-105.7$ $B_{8-11}=-286.5$	$A_{9-11}=-105.7$ $B_{9-11}=-286.5$	$A_{10-11}=-50.10$ $B_{10-11}=-891.1$	0	-	-	-	-	-	-	-	-	-
Group 12	$A_{1-12}=164.0$ $B_{1-12}=269.0$	$A_{2-12}=136.9$ $B_{2-12}=254.6$	$A_{3-12}=184.3$ $B_{3-12}=762.1$	$A_{4-12}=287.9$ $B_{4-12}=346.2$	$A_{5-12}=137.3$ $B_{5-12}=194.2$	$A_{6-12}=135.5$ $B_{6-12}=239.5$	$A_{7-12}=102.6$ $B_{7-12}=161.3$	$A_{8-12}=110.1$ $B_{8-12}=637.6$	$A_{9-12}=267.3$ $B_{9-12}=444.4$	$A_{10-12}=130.1$ $B_{10-12}=225.8$	$A_{11-12}=91.28$ $B_{11-12}=82.01$	0	-	-	-	-	-	-	-	-
Group 13	$A_{1-13}=52.74$ $B_{1-13}=87.19$	$A_{2-13}=82.28$ $B_{2-13}=202.8$	$A_{3-13}=365.4$ $B_{3-13}=521.9$	$A_{4-13}=263.9$ $B_{4-13}=772.6$	$A_{5-13}=37.90$ $B_{5-13}=37.20$	$A_{6-13}=61.59$ $B_{6-13}=84.92$	$A_{7-13}=185.2$ $B_{7-13}=490.6$	$A_{8-13}=284.0$ $B_{8-13}=1892$	$A_{9-13}=718.1$ $B_{9-13}=1892$	$A_{10-13}=179.5$ $B_{10-13}=546.6$	$A_{11-13}=100.9$ $B_{11-13}=249.8$	$A_{12-13}=98.42$ $B_{12-13}=221.4$	0	-	-	-	-	-	-	-
Group 14	$A_{1-14}=158.4$ $B_{1-14}=241.2$	$A_{2-14}=134.6$ $B_{2-14}=138.3$	$A_{3-14}=193.9$ $B_{3-14}=307.8$	$A_{4-14}=305.1$ $B_{4-14}=-143.1$	$A_{5-14}=181.2$ $B_{5-14}=288.9$	$A_{6-14}=157.2$ $B_{6-14}=217.1$	$A_{7-14}=21.96$ $B_{7-14}=13.04$	$A_{8-14}=1.029$ $B_{8-14}=-8.579$	$A_{9-14}=1.029$ $B_{9-14}=-8.579$	$A_{10-14}=120.8$ $B_{10-14}=163.0$	$A_{11-14}=-16.13$ $B_{11-14}=-147.6$	$A_{12-14}=134.9$ $B_{12-14}=201.4$	$A_{13-14}=319.5$ $B_{13-14}=550.1$	0	-	-	-	-	-	-
Group 15	$A_{1-15}=799.9$ $B_{1-15}=2109$	$A_{2-15}=459.5$ $B_{2-15}=627.3$	$A_{3-15}=425.5$ $B_{3-15}=514.7$	$A_{4-15}=682.9$ $B_{4-15}=1544$	$A_{5-15}=706.0$ $B_{5-15}=1483$	NOT AVAILABLE	$A_{7-15}=285.5$ $B_{7-15}=392.0$	$A_{8-15}=1072$ $B_{8-15}=1094$	$A_{9-15}=1072$ $B_{9-15}=1094$	$A_{10-15}=446.1$ $B_{10-15}=549.0$	$A_{11-15}=411.8$ $B_{11-15}=-308.8$	NOT AVAILABLE	NOT AVAILABLE	$A_{14-15}=-77.21$ $B_{14-15}=156.1$	0	-	-	-	-	-
Group 16	$A_{1-16}=3557$ $B_{1-16}=11195$	$A_{2-16}=4324$ $B_{2-16}=12126$	$A_{3-16}=971.4$ $B_{3-16}=567.6$	NOT AVAILABLE	$A_{5-16}=2265$ $B_{5-16}=4722$	$A_{6-16}=2333$ $B_{6-16}=5147$	$A_{7-16}=2268$ $B_{7-16}=6218$	$A_{8-16}=543.5$ $B_{8-16}=411.8$	$A_{9-16}=1340$ $B_{9-16}=-65.88$	$A_{10-16}=4211$ $B_{10-16}=13031$	$A_{11-16}=244.0$ $B_{11-16}=-60.39$	$A_{12-16}=559.3$ $B_{12-16}=277.9$	$A_{13-16}=2574$ $B_{13-16}=5490$	$A_{14-16}=603.9$ $B_{14-16}=599.1$	$A_{15-16}=30.88$ $B_{15-16}=-113.6$	0	-	-	-	-
Group 17	$A_{1-17}=7.206$ $B_{1-17}=39.12$	$A_{2-17}=59.71$ $B_{2-17}=78.58$	$A_{3-17}=176.7$ $B_{3-17}=118.0$	$A_{4-17}=319.5$ $B_{4-17}=-248.1$	$A_{5-17}=14.69$ $B_{5-17}=30.20$	$A_{6-17}=7.549$ $B_{6-17}=19.22$	$A_{7-17}=20.25$ $B_{7-17}=94.02$	$A_{8-17}=65.20$ $B_{8-17}=125.2$	$A_{9-17}=199.0$ $B_{9-17}=3820$	$A_{10-17}=35.34$ $B_{10-17}=52.50$	$A_{11-17}=-38.43$ $B_{11-17}=-688.4$	$A_{12-17}=73.09$ $B_{12-17}=115.3$	$A_{13-17}=88.53$ $B_{13-17}=109.1$	NOT AVAILABLE	NOT AVAILABLE	$A_{16-17}=1632$ $B_{16-17}=1612$	0	-	-	-
Group 18	$A_{1-18}=54.22$ $B_{1-18}=142.4$	$A_{2-18}=11.67$ $B_{2-18}=29.51$	$A_{3-18}=118.4$ $B_{3-18}=158.9$	$A_{4-18}=50.79$ $B_{4-18}=-284.5$	$A_{5-18}=52.84$ $B_{5-18}=110.5$	$A_{6-18}=26.42$ $B_{6-18}=50.44$	$A_{7-18}=8.579$ $B_{7-18}=-7.549$	$A_{8-18}=-70.69$ $B_{8-18}=36.72$	NOT AVAILABLE	$A_{10-18}=27.11$ $B_{10-18}=454.7$	NOT AVAILABLE	$A_{12-18}=59.71$ $B_{12-18}=210.3$	$A_{13-18}=125.2$ $B_{13-18}=285.5$	NOT AVAILABLE	NOT AVAILABLE	$A_{16-18}=2243$ $B_{16-18}=5199$	$A_{17-18}=17.16$ $B_{17-18}=36.72$	0	-	-
Group 19	$A_{1-19}=115.6$ $B_{1-19}=118.4$	$A_{2-19}=60.39$ $B_{2-19}=272.8$	$A_{3-19}=103.6$ $B_{3-19}=430.6$	NOT AVAILABLE	NOT AVAILABLE	NOT AVAILABLE	$A_{7-19}=1.029$ $B_{7-19}=-16.81$	$A_{8-19}=217.5$ $B_{8-19}=-170.2$	NOT AVAILABLE	$A_{10-19}=-3.088$ $B_{10-19}=-168.5$	NOT AVAILABLE	$A_{12-19}=23.68$ $B_{12-19}=-186.0$	$A_{13-19}=455.7$ $B_{13-19}=30.54$	NOT AVAILABLE	NOT AVAILABLE	NOT AVAILABLE	$A_{17-19}=0.137$ $B_{17-19}=31.23$	$A_{18-19}=21.62$ $B_{18-19}=134.9$	0	-
Group 20	$A_{1-20}=117.1$ $B_{1-20}=358.9$	$A_{2-20}=0.000$ $B_{2-20}=-7.206$	$A_{3-20}=39.12$ $B_{3-20}=1038$	NOT AVAILABLE	NOT AVAILABLE	NOT AVAILABLE	$A_{7-20}=6.177$ $B_{7-20}=5.490$	$A_{8-20}=3.431$ $B_{8-20}=-37.75$	NOT AVAILABLE	$A_{10-20}=51.47$ $B_{10-20}=254.6$	$A_{11-20}=37.75$ $B_{11-20}=4173$	$A_{12-20}=87.85$ $B_{12-20}=94.37$	NOT AVAILABLE	NOT AVAILABLE	NOT AVAILABLE	NOT AVAILABLE	$A_{17-20}=-43.24$ $B_{17-20}=1798$	$A_{18-20}=119.8$ $B_{18-20}=-72.06$	$A_{19-20}=-392.9$ $B_{19-20}=-197.3$	0
Group 21	$A_{1-21}=202.8$ $B_{1-21}=317.4$	$A_{2-21}=132.5$ $B_{2-21}=147.2$	$A_{3-21}=415.2$ $B_{3-21}=726.4$	$A_{4-21}=226.5$ $B_{4-21}=1812$	$A_{5-21}=156.1$ $B_{5-21}=92.99$	$A_{6-21}=137.6$ $B_{6-21}=150.0$	$A_{7-21}=284.8$ $B_{7-21}=175.0$	$A_{8-21}=377.5$ $B_{8-21}=1201$	$A_{9-21}=549.0$ $B_{9-21}=1476$	$A_{10-21}=232.0$ $B_{10-21}=167.5$	$A_{11-21}=-314.0$ $B_{11-21}=-225.8$	$A_{12-21}=265.9$ $B_{12-21}=268.3$	$A_{13-21}=65.20$ $B_{13-21}=70.10$	$A_{14-21}=145.8$ $B_{14-21}=823.5$	NOT AVAILABLE	$A_{16-21}=830.8$ $B_{16-21}=-137.9$	$A_{17-21}=151.3$ $B_{17-21}=165.1$	$A_{18-21}=163.0$ $B_{18-21}=322.2$	$A_{19-21}=630.0$ $B_{19-21}=573.1$	$A_{20-21}=483.1$ $B_{20-21}=2417$

Table 4. Group interaction parameters of the PPR78 model: $(A_{kl} = A_{lk})/MPa$ and $(B_{kl} = B_{lk})/MPa$.

$$\text{with:}\begin{cases} F_{obj,bubble} = 100\sum_{i=1}^{n_{bubble}} 0.5\left(\frac{|\Delta x|}{x_{1,\exp}} + \frac{|\Delta x|}{x_{2,\exp}}\right)_i & with\ |\Delta x| = |x_{1,\exp} - x_{1,cal}| = |x_{2,\exp} - x_{2,cal}| \\ F_{obj,dew} = 100\sum_{i=1}^{n_{dew}} 0.5\left(\frac{|\Delta y|}{y_{1,\exp}} + \frac{|\Delta y|}{y_{2,\exp}}\right)_i & with\ |\Delta y| = |y_{1,\exp} - y_{1,cal}| = |y_{2,\exp} - y_{2,cal}| \\ F_{obj,crit.\ comp} = 100\sum_{i=1}^{n_{crit}} 0.5\left(\frac{|\Delta x_c|}{x_{c1,\exp}} + \frac{|\Delta x_c|}{x_{c2,\exp}}\right)_i & with\ |\Delta x_c| = |x_{c1,\exp} - x_{c1,cal}| = |x_{c2,\exp} - x_{c2,cal}| \\ F_{obj,crit.\ pressure} = 100\sum_{i=1}^{n_{crit}} \left(\frac{|P_{cm,\exp} - P_{cm,cal}|}{P_{cm,\exp}}\right)_i & \end{cases}$$

n_{bubble}, n_{dew} and n_{crit} are the number of bubble points, dew points and mixture critical points respectively. x_1 is the mole fraction in the liquid phase of the most volatile component and x_2 the mole fraction of the heaviest component (it is obvious that $x_2 = 1 - x_1$). Similarly, y_1 is the mole fraction in the gas phase of the most volatile component and y_2 the mole fraction of the heaviest component (it is obvious that $y_2 = 1 - y_1$). x_{c1} is the critical mole fraction of the most volatile component and x_{c2} the critical mole fraction of the heaviest component. P_{cm} is the binary critical pressure. For all the data points included in our database, the objective function defined by Eq. (36) is only:

$$F_{obj} = 7.6\ \% \tag{37}$$

The average overall deviation on the liquid phase composition is:

$$\overline{\Delta x}\% = \frac{\overline{\Delta x_1}\% + \overline{\Delta x_2}\%}{2} = \frac{F_{obj,bubble}}{n_{bubble}} = \underline{\underline{7.4\ \%}} \tag{38}$$

The average overall deviation on the gas phase composition is:

$$\overline{\Delta y}\% = \frac{\overline{\Delta y_1}\% + \overline{\Delta y_2}\%}{2} = \frac{F_{obj,dew}}{n_{dew}} = \underline{\underline{8.0\ \%}} \tag{39}$$

The average overall deviation on the critical composition is:

$$\overline{\Delta x_c}\% = \frac{\overline{\Delta x_{c1}}\% + \overline{\Delta x_{c2}}\%}{2} = \frac{F_{obj,crit.\ comp}}{n_{crit}} = \underline{\underline{7.1\ \%}} \tag{40}$$

The average overall deviation on the binary critical pressure is:

$$\overline{\Delta P_c}\% = \frac{F_{obj,crit.\ pressure}}{n_{crit}} = \underline{\underline{4.9\ \%}} \tag{41}$$

Taking into account the scatter of the experimental data which inevitably makes increase the objective function value, we can assert that the PPR78 model is a very accurate thermodynamic model. In conclusion, thanks to this predictive model it is today possible to estimate the k_{ij} for any mixture containing alkanes, aromatics, naphthenes, CO_2, N_2, H_2S, H_2, mercaptans, water and alkenes for any temperature. We thus can say that the PPR78 model is able to cover all the compounds that an engineer of a petroleum company is likely to encounter. We are proud to announce that the PPR78 model is now integrated in two famous process simulators: PRO/II commercialized by Invensys and ProSimPlus developed by the French company PROSIM. Figure 2 graphically illustrates the accuracy of the PPR78 model.

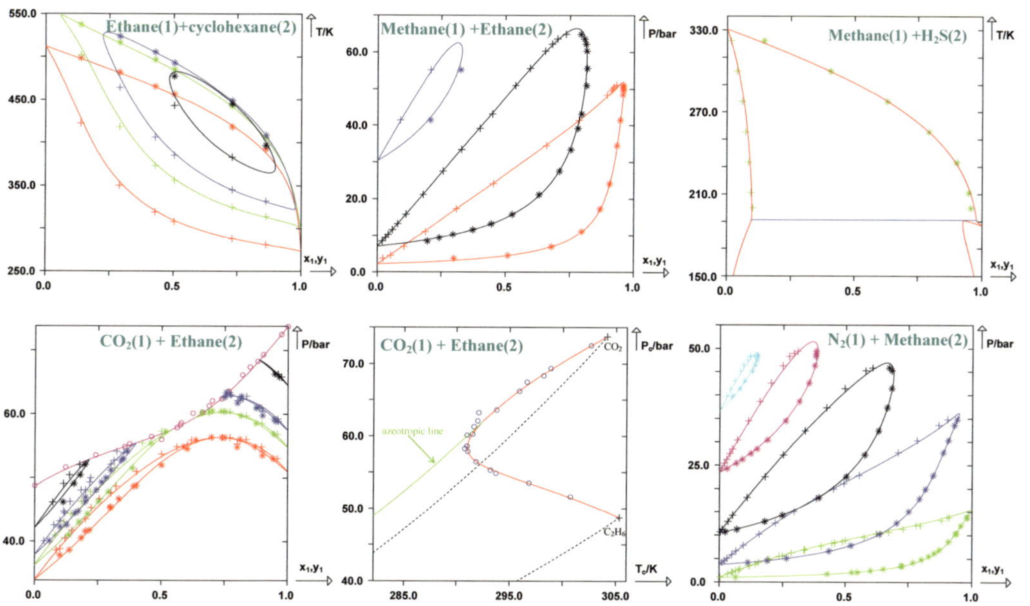

Fig. 2. Illustration of the accuracy of the PPR78 model. The symbols are the experimental data points. The full lines are the predictions with the PPR78 model.

5.2.2.2 On the temperature dependence of the k_{ij} parameter

In mixtures encountered in the petroleum and gas processing industries it is today accepted that the binary interaction parameter k_{ij} depends on temperature. This temperature dependence has been described by a few authors. The very good paper by (Coutinho et al., 1994) gives an interesting review of the different publications dealing with this subject. The same authors give a theoretical explanation for the temperature dependence of the k_{ij} and conclude that this parameter varies quadratically with the inverse temperature ($1/T$). Using Eq. (35), it is simple to plot k_{ij} versus temperature for a given binary system. As an illustration, Figure 3 presents plots of k_{ij} with respect to the reduced temperature of the heavy n-alkane for three binary systems: methane/propane, methane/n-hexane and methane/n-decane. The shapes of the curves are similar to the ones published by Coutinho et al. At low temperature, k_{ij} is a decreasing function of temperature. With increasing the temperature, the k_{ij} reaches a minimum and then increases again. The minimum is located at a reduced temperature close to 0.55, independent of the binary system. This minimum moves to $T_r = 0.6$ for the system methane/n-C_{30} (results not shown in Figure 3). From Figure 3, we can unambiguously conclude that it is necessary to work with temperature-dependent k_{ij}.

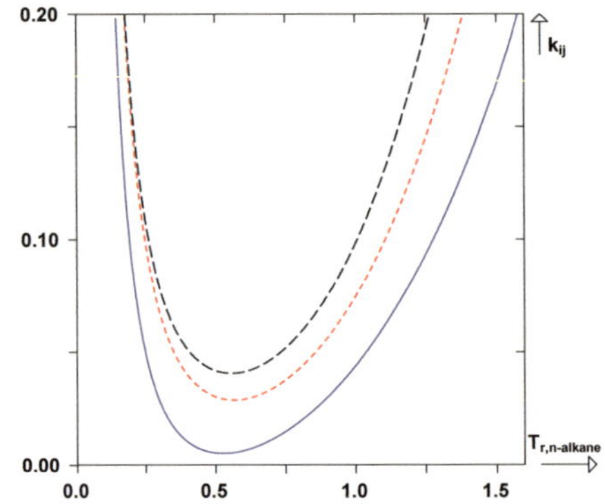

Fig. 3. Temperature dependence of predicted k_{ij} by means of Eq. (35). Solid line: system methane/propane. Short dashed line: system methane/n-hexane. Long dashed line: system methane/n-decane. In abscissa, T_r is the reduced temperature of the heavy n-alkane (respectively propane, n-hexane, n-decane).

5.2.3 Soave's group-contribution method (GCM)

In a recent paper, (Soave et al., 2010) also developed a GCM aimed at predicting temperature-dependent k_{ij} for the well known SRK EoS. Six groups are defined: group1 = CH_4, group2 = CO_2, group3 = N_2, group4 = H_2S, group5 = *alkyl group* and group6 = *aromatic group*. The last two groups are introduced to model alkanes and aromatic compounds. This small number of groups is an advantage since it reduces the number of group parameters to be estimated and it allows a faster estimation of the k_{ij}. Indeed, the calculation of a double sum as the one contained in Eq. (35) is time consuming and strongly affected by the number of groups. Very accurate results are obtained on binary systems. As shown by (Jaubert & Privat, 2010), the accuracy of Soave's GCM - although not tested on many experimental data - is similar to what is observed with the PPR78 model. Whether a naphthenic group was added, this GCM could be applied to predict the phase behaviour of reservoir fluids.

5.3 $k_{ij}(T)$ values: how to switch from a cubic EoS to another one?

As explained in the previous sections, the key point when using cubic EoS to describe complex mixtures like petroleum fluids is to give appropriate values to the binary interaction parameters (k_{ij}). We however know by experience that the k_{ij}, suitable for a given EoS (e.g. the PR EoS) cannot be directly used for another one (e.g. the SRK EoS). Moreover, numerical values of k_{ij} are not only specific to the considered EoS but they also depend on the alpha-function (Soave, Twu, Mathias-Copeman, etc.) involved in the mathematical expression of the a_i parameter. This assessment makes it impossible for petroleum engineers to use various equations of state and to test different alpha functions. Indeed, they usually have tables containing the numerical values of the k_{ij} only for the most widely used EoS and alpha function in their company. To overcome this limitation, (Jaubert & Privat, 2010) had the idea to establish a relationship between the k_{ij} of a first EoS (k_{ij}^{EoS1}) and those of a second one (k_{ij}^{EoS2}). As a consequence, knowing the numerical values of the k_{ij} for the first EoS makes it possible to deduce the corresponding values for any other cubic EoS. The obtained relationship is given just below (Eq. (43)). To understand the notations, let us consider two cubic equations of state (EoS1 and EoS2) deriving from Van der Waals' equation i.e. having the general form given in Eq. (12). At this step, we define the following quantities:

$$\begin{cases} C_{EoS} = \dfrac{1}{r_1 - r_2} \cdot \ln\left(\dfrac{1-r_2}{1-r_1}\right) \text{ if } r_1 \neq r_2 \text{ and } C_{EoS} = \dfrac{1}{1-r_1} \text{ if } r_1 = r_2 \\ \xi_{1\to 2} = \dfrac{C_{EoS1} \cdot \Omega_b^{EoS1}}{C_{EoS2} \cdot \Omega_b^{EoS2}} \\ \delta_i^{EoS} = \dfrac{\sqrt{a_i^{EoS}}}{b_i^{EoS}} \end{cases} \tag{42}$$

After some derivation, the obtained relationship is:

$$k_{ij}^{EoS2} = \frac{2\xi_{1\to 2} k_{ij}^{EoS1} \delta_i^{EoS1} \delta_j^{EoS1} + \xi_{1\to 2}\left(\delta_i^{EoS1} - \delta_j^{EoS1}\right)^2 - \left(\delta_i^{EoS2} - \delta_j^{EoS2}\right)^2}{2\delta_i^{EoS2}\delta_j^{EoS2}} \tag{43}$$

Eq. (43) can also be used if we work with the same EoS (let us say the PR EoS) but if we decide to only change the $a_i(T)$ function (e.g. we initially work with a Soave-type function for which the k_{ij} are known and we decide to work with a Mathias and Copeman function for which the k_{ij} are unknown). In this latter case, $\xi_{1\to 2} = 1$, but Eq. (43) in which the δ parameters depend on the $a(T)$ function will lead to a relationship between the k_{ij} to be used with the first $a(T)$ function and those to be used with the second one.

As previously explained, the PPR78 model is a GCM designed to predict the k_{ij} of the PR EoS. The coupling of this GCM with Eq. (43) makes it possible to predict the temperature-dependent k_{ij} of any desired EoS using the GC concept. Let us indeed consider a cubic EoS (see Eq. (12)) noted EoS1 hereafter, and let us define:

$$\xi_{PR\to EoS1} = \frac{C_{PR} \cdot \Omega_b^{PR}}{C_{EoS1} \cdot \Omega_b^{EoS1}} \tag{44}$$

By combining Eqs. (35) and (43), we can write:

$$k_{ij}^{EoS1}(T) = \frac{-\dfrac{1}{2}\left[\displaystyle\sum_{k=1}^{N_g}\sum_{l=1}^{N_g}(\alpha_{ik} - \alpha_{jk})(\alpha_{il} - \alpha_{jl})\xi_{PR\to EoS1} A_{kl}^{PR} \cdot \left(\frac{298.15}{T/K}\right)^{\left(\frac{B_{kl}^{PR}}{A_{kl}^{PR}} - 1\right)}\right] - \left(\dfrac{\sqrt{a_i^{EoS1}(T)}}{b_i^{EoS1}} - \dfrac{\sqrt{a_j^{EoS1}(T)}}{b_j^{EoS1}}\right)^2}{2\dfrac{\sqrt{a_i^{EoS1}(T) \cdot a_j^{EoS1}(T)}}{b_i^{EoS1} \cdot b_j^{EoS1}}} \tag{45}$$

Using Eq. (45), it is thus possible to calculate by GC, the temperature-dependent k_{ij} for any desired cubic EoS (EoS1), with any desired $a_i(T)$ function, using the group contribution parameters (A_{kl}^{PR} and B_{kl}^{PR}) we determined for the PPR78 model. Eq. (45) can also be used if we work with the PR EoS but with a different $a_i(T)$ function than the one defined by Eq. (10). In this latter case, $\xi_{PR\to EoS1} = 1$, but Eq. (45) will lead to k_{ij} values different of those obtained from the PPR78 model. Eq. (45) was extensively used by (Jaubert & Privat, 2010) in the particular case where EoS1 is the SRK EoS. They concluded that the accuracy obtained with the SRK EoS was similar to the one obtained with the PPR78 model. The resulting model, based on the SRK EoS and for which the k_{ij} are estimated from the GCM developed for the PPR78 model was called PR2SRK. In conclusion, we can claim that the PPR78 model is a *universal* GCM since it can predict the k_{ij} for any desired EoS with any desired $a_i(T)$ function at any temperature for any mixture containing hydrocarbons, permanent gases, and water.

5.4 Other mixing rules

Cubic EoS with Van der Waals one-fluid mixing rules lead to very accurate results at low and high pressures for *simple* mixtures (few polar, hydrocarbons, gases). They also allow the prediction of many more properties than phase equilibria (e.g. excess properties, heat capacities, etc.). Such mixing rules can however not be applied with success to polar mixtures. In return, g^E models (activity-coefficient models) are applicable to low pressures and are able to correlate polar mixtures. It thus seems a good idea to combine the strengths of both approaches, i.e. the cubic EoS and the activity coefficient models and thus to have a single model suitable for phase equilibria of polar and non-polar mixtures and at both low and high pressures. This combination of EoS and g^E models is possible via the so-called EoS/g^E models which are essentially mixing rules for the energy parameter of cubic EoS. As explained in the recent book by (Kontogeorgis & Folas, 2010), the starting point for deriving EoS/g^E models is the equality of the excess Gibbs energies from an EoS and from an explicit activity coefficient model at a suitable reference pressure. The activity coefficient model may be chosen among the classical forms of molar excess Gibbs energy functions (Redlich-Kister, Margules, Wilson, Van Laar, NRTL, UNIQUAC, UNIFAC...). Such models are pressure-independent (they only depend on temperature and composition) but the same quantity from an EoS depends on pressure, temperature and composition explaining why a reference pressure needs to be selected before equating the two quantities. In order to avoid confusion, we will write with a special font ($\mathcal{G}^E$) the selected activity coefficient model and with a classical font (g^E) the excess Gibbs energy calculated from an EoS by:

$$\frac{g^E}{RT} = \sum_{i=1}^{p} z_i \cdot \ln\left(\frac{\hat{\varphi}_i}{\varphi_{pure\,i}}\right) \tag{46}$$

where $\hat{\varphi}_i$ is the fugacity coefficient of component i in the mixture and $\varphi_{pure\,i}$ the fugacity coefficient of the pure compound. The starting equation to derive EoS/g^E models is thus:

$$\left[g^E/(RT)\right]_P = \mathcal{G}^E/(RT) \tag{47}$$

where subscript P indicates that a reference pressure has to be chosen.

5.4.1 The infinite pressure reference

The basic assumption of the method is the use of the infinite pressure as the reference pressure.

5.4.1.1 The Huron-Vidal mixing rules

The first systematic successful effort in developing an EoS/g^E model is that of (Huron & Vidal, 1979). Starting from Eq. (47), they obtained:

$$\begin{cases} \dfrac{a(T,\mathbf{x})}{b(\mathbf{x})} = \sum_{i=1}^{p} x_i \dfrac{a_i(T)}{b_i} - \dfrac{g^E}{C_{EoS}} \\ b(\mathbf{x}) = \sum_{i=1}^{p} x_i b_i \end{cases} \tag{48}$$

where C_{EoS} is defined by Eq. (42). A positive feature of the Huron-Vidal mixing rule includes an excellent correlation of binary systems. A limitation is that it does not permit use of the large collections of interaction parameters of g^E models which are based on low-pressure VLE data (e.g. the UNIFAC tables). Indeed, the excess Gibbs energy at high pressures is, in general, different from the value at low pressures at which the parameters of the g^E models are typically estimated.

5.4.1.2 The Van der Waals one-fluid (VdW1f) mixing rules

The classical VdW1f mixing rules used in the PPR78 and PR2SRK model write:

$$a(T,\mathbf{x}) = \sum_{j=1}^{p} x_i x_j \sqrt{a_i a_j}\left[1 - k_{ij}(T)\right] \quad \text{and} \quad b(\mathbf{x}) = \sum_{i=1}^{p} x_i b_i \tag{49}$$

We here want to give proof that such mixing rules are in fact strictly equivalent to the Huron-Vidal mixing rules if a Van-Laar type g^E model is selected in Eq. (48). Indeed, by sending Eq. (49) in Eq. (48), one has:

$$\begin{aligned} \frac{g^E}{C_{EoS}} &= \sum_{i=1}^{p} x_i \frac{a_i(T)}{b_i} - \frac{a(T,\mathbf{x})}{b(\mathbf{x})} = \sum_{i=1}^{p} x_i \cdot \frac{a_i(T)}{b_i} - \frac{\sum_{i=1}^{p}\sum_{j=1}^{p} x_i \cdot x_j \sqrt{a_i \cdot a_j} \cdot \left[1 - k_{ij}(T)\right]}{\sum_{j=1}^{p} x_j \cdot b_j} \\ &= \frac{\sum_{i=1}^{p} x_i \cdot \frac{a_i(T)}{b_i} \times \sum_{j=1}^{p} x_j \cdot b_j - \sum_{i=1}^{p}\sum_{j=1}^{p} x_i \cdot x_j \sqrt{a_i \cdot a_j} \cdot \left[1 - k_{ij}(T)\right]}{\sum_{j=1}^{p} x_j \cdot b_j} \end{aligned} \tag{50}$$

We can write:

$$\sum_{i=1}^{p} x_i \cdot \frac{a_i(T)}{b_i} \times \sum_{j=1}^{p} x_j \cdot b_j = \sum_{i=1}^{p}\sum_{j=1}^{p} x_i \cdot x_j \frac{a_i \cdot b_j}{b_i} = \frac{1}{2}\sum_{i=1}^{p}\sum_{j=1}^{p} x_i \cdot x_j \frac{a_i \cdot b_j}{b_i} + \frac{1}{2}\sum_{i=1}^{p}\sum_{j=1}^{p} x_i \cdot x_j \frac{a_j \cdot b_i}{b_j} \tag{51}$$

Thus,

$$\frac{g^E}{C_{EoS}} = \frac{\sum_{i=1}^{p}\sum_{j=1}^{p} x_i \cdot x_j \left[\frac{1}{2}\frac{a_i \cdot b_j}{b_i} + \frac{1}{2}\frac{a_j \cdot b_i}{b_j} - \sqrt{a_i \cdot a_j} \cdot \left[1 - k_{ij}(T)\right]\right]}{\sum_{j=1}^{p} x_j \cdot b_j}$$
$$= \frac{\sum_{i=1}^{p}\sum_{j=1}^{p} x_i \cdot x_j \cdot b_i \cdot b_j \left[\frac{1}{2}\frac{a_i}{b_i^2} + \frac{1}{2}\frac{a_j}{b_j^2} - \frac{\sqrt{a_i}}{b_i} \cdot \frac{\sqrt{a_j}}{b_j} \cdot \left[1 - k_{ij}(T)\right]\right]}{\sum_{j=1}^{p} x_j \cdot b_j} \tag{52}$$

At this step, we introduce for clarity: $\delta_i = \sqrt{a_i}/b_i$ which has the Scatchard-Hildebrand solubility parameter feature, and we define the parameter E_{ij} by:

$$E_{ij} = \delta_i^2 + \delta_j^2 - 2\delta_i\delta_j\left[1 - k_{ij}(T)\right] \tag{53}$$

Eq. (52) thus writes:

$$\frac{g^E}{C_{EoS}} = \frac{1}{2} \cdot \frac{\sum_{i=1}^{p}\sum_{j=1}^{p} x_i \cdot x_j \cdot b_i \cdot b_j \cdot E_{ij}(T)}{\sum_{j=1}^{p} x_j \cdot b_j} \tag{54}$$

Eq. (54) is the mathematical expression of a Van Laar-type g^E model. We thus demonstrated that it is rigorously equivalent to use a Van Laar-type g^E model in the Huron-Vidal mixing rules or to use classical mixing rules with temperature-dependent k_{ij}. From Eq. (53) we have:

$$k_{ij}(T) = \frac{E_{ij}(T) - (\delta_i - \delta_j)^2}{2\delta_i\delta_j} \tag{55}$$

Eq. (55) thus establishes a connection between E_{ij} of the Van Laar-type g^E model and k_{ij} of the classical mixing rules.

5.4.1.3 The Wong-Sandler mixing rules

It can be demonstrated that the Huron Vidal mixing rules violate the imposed by statistical thermodynamics quadratic composition dependency of the second virial coefficient:

$$B = \sum_{i=1}^{p}\sum_{j=1}^{p} x_i x_j B_{ij} \tag{56}$$

To satisfy Eq. (56), (Wong and Sandler, 1992) decided to revisit the Huron-Vidal mixing rules. Since they made use of the infinite pressure as the reference pressure, like Huron and Vidal, they obtained:

$$a(T,\mathbf{x}) = b(\mathbf{x})\sum_{i=1}^{p} x_i \frac{a_i(T)}{b_i} - \frac{g^E}{C_{EoS}} \tag{57}$$

However, knowing that the second virial coefficient from a cubic EoS is given as:

$$B = b - \frac{a}{RT} \tag{58}$$

eq. (56) writes:

$$b(\mathbf{x}) = \frac{a(T,\mathbf{x})}{RT} + \sum_{i=1}^{p}\sum_{j=1}^{p} x_i x_j B_{ij} \tag{59}$$

Substituting Eq. (57) in Eq. (59), we get:

$$b(\mathbf{x}) = \frac{\sum_{i=1}^{p}\sum_{j=1}^{p} x_i x_j B_{ij}}{1 + \frac{g^E}{C_{EoS}} - \sum_{i=1}^{p} x_i \frac{a_i(T)}{b_i}} \tag{60}$$

The following choice for the cross virial coefficient is often used:

$$B_{ij} = \frac{1}{2}(B_i + B_j)(1 - k_{ij}) \tag{61}$$

Eq. (61) makes unfortunately appear an extra binary interaction parameter (k_{ij}), the value of which can be estimated through various approaches.

Substituting Eq. (58) in Eq.(61), one has:

$$B_{ij} = \frac{\left(b_i - \frac{a_i}{RT}\right) + \left(b_j - \frac{a_j}{RT}\right)}{2}(1 - k_{ij}) \tag{62}$$

The Wang-Sandler mixing rules thus write:

$$\begin{cases} b(T,\mathbf{x}) = \dfrac{\sum_{i=1}^{p}\sum_{j=1}^{p} x_i x_j \dfrac{\left(b_i - \frac{a_i}{RT}\right) + \left(b_j - \frac{a_j}{RT}\right)}{2}(1 - k_{ij})}{1 + \dfrac{g^E}{C_{EoS}} - \sum_{i=1}^{p} x_i \dfrac{a_i(T)}{b_i}} \\ a(T,\mathbf{x}) = b(\mathbf{x})\sum_{i=1}^{p} x_i \dfrac{a_i(T)}{b_i} - \dfrac{g^E}{C_{EoS}} \end{cases} \tag{63}$$

We can thus conclude that the Wong-Sandler mixing rules differ from the Huron-Vidal mixing rules in the way of estimating the covolume. In Eq. (63), b has become temperature-dependent. Many papers illustrate the key success of the Wong-Sandler mixing rules to predict VLE using existing low pressure parameters from activity coefficients models. However parameters for gas-containing systems are not available in activity coefficient models like UNIFAC which limits the applicability of these mixing rules to such systems.

5.4.2 The zero-pressure reference

The zero-pressure reference permits a direct use of g^E interaction parameter tables. Starting from Eq. (47) and by setting $P = 0$, one obtains:

$$\begin{cases} \dfrac{g^E}{RT} = Q(\alpha) - \sum_{i=1}^{p} x_i \cdot Q(\alpha_i) + \sum_{i=1}^{p} x_i \cdot \ln\left(\dfrac{b_i}{b}\right) \text{ with } \alpha = \dfrac{a}{bRT} \\ Q(\alpha) = -\ln\left(\dfrac{1-\eta_0(\alpha)}{\eta_0(\alpha)}\right) - \dfrac{\alpha}{r_1 - r_2}\ln\left(\dfrac{1-\eta_0(\alpha)\cdot r_2}{1-\eta_0(\alpha)\cdot r_1}\right) \\ \eta_0(\alpha) = \dfrac{r_1 + r_2 + \alpha + \sqrt{(r_1 + r_2 + \alpha)^2 - 4(r_1 r_2 + \alpha)}}{2(r_1 r_2 + \alpha)} \text{ submitted to: } \alpha \geq \alpha_{\lim} \\ \alpha_{\lim} = -r_1 - r_2 + 2 + 2\sqrt{1 - r_1 - r_2 + r_1 r_2} \end{cases} \quad (64)$$

After selecting a mixing rule for the covolume, Eq. (64) becomes an implicit mixing rule for the energy parameter, which means that an iterative procedure is needed for calculating the energy parameter.

5.4.2.1 The MHV-1 mixing rule

In order to obtain an explicit mixing rule and to address the limitation introduced by the presence of $\alpha_{\lim}$ (Michelsen, 1990) proposed to define a linear approximation of the Q function by:

$$Q(\alpha) = q_0 + q_1 \cdot \alpha \quad (65)$$

Doing so, Eq. (64) writes:

$$\alpha = \sum_{i=1}^{p} x_i \cdot \alpha_i + \frac{1}{q_1}\left[\frac{g^E}{RT} - \sum_{i=1}^{p} x_i \cdot \ln\left(\frac{b_i}{b}\right)\right] \quad (66)$$

Eq. (66) is the so called MHV-1 (modified Huron-Vidal first order) mixing rule usually used with a linear mixing rule for the covolume parameter. Michelsen advices to use $q_1 = -0.593$ for the SRK EoS, $q_1 = -0.53$ for the PR EoS and $q_1 = -0.85$ for the VdW EoS.

5.4.2.2 The PSRK model

(Holderbaum & Gmehling, 1991) proposed the PSRK (predictive SRK) model based on the MHV-1 mixing rule. These authors however use a slightly different q_1 value than the one proposed by Michelsen. They select: $q_1 = -0.64663$. In order to make their model predictive, Holderbaum and Gmehling combine the SRK EoS with a predictive g^E model (the original or the modified Dortmund UNIFAC). Moreover they developed extensive parameter table, including parameters for gas-containing mixtures. The PSRK model may thus be used to model petroleum fluids. No comparison was performed between PSRK and PPR78 but we can expect similar results.

5.4.2.3 The UMR-PR and VTPR models

The UMR-PR (universal mixing rule-Peng Robinson) and the VTPR (volume translated Peng Robinson) models, both use the MVH-1 mixing rule. They were respectively developed by (Ahlers & Gmehling, 2001, 2002) and by (Voutsas et al., 2004). In both cases, the same translated form of the PR EoS is used. The Twu *a(T)* function is however used in the VTPR model whereas the Mathias-Copeman expression is used in the UMR-PR model. Both models incorporate the UNIFAC g^E model in Eq. (66). However, in order to be able to properly correlate asymmetric systems, only the residual part of UNIFAC is used in the VTPR model. These authors indeed assume that the combinatorial part of UNIFAC and $\sum x_i \cdot \ln(b_i/b)$ in Eq. (66) cancel each other. In the UMR-PR model the residual part of UNIFAC but also the Staverman-Guggenheim contribution of the combinatorial term is used. These authors indeed assume that the Flory-Huggins combinatorial part of UNIFAC and $\sum x_i \cdot \ln(b_i/b)$ in Eq. (66) cancel each other. A novel aspect of these models is the mixing rule used for the covolume parameter. Both models give better results than PSRK. The equations to be used are:

$$\underline{\underline{VTPR:}}$$
$$\alpha = \sum_{i=1}^{p} x_i \cdot \alpha_i + \frac{1}{q_1}\left[\frac{g_{residual}^{E,UNIFAC}}{RT}\right] \qquad (67)$$
$$b = \sum_{i=1}^{p}\sum_{j=1}^{p} x_i x_j b_{ij} \text{ with: } b_{ij} = \left(\frac{b_i^{1/s} + b_j^{1/s}}{2}\right) \text{ and } s = \frac{4}{3}$$

$$\underline{\underline{UMP-PR:}}$$
$$\alpha = \sum_{i=1}^{p} x_i \cdot \alpha_i + \frac{1}{q_1}\left[\frac{g_{residual}^{E,UNIFAC}}{RT} + \frac{g_{\text{Staverman-Guggenheim combinatorial contribution}}^{E,UNIFAC}}{RT}\right] \qquad (68)$$
$$b = \sum_{i=1}^{p}\sum_{j=1}^{p} x_i x_j b_{ij} \text{ with: } b_{ij} = \left(\frac{b_i^{1/s} + b_j^{1/s}}{2}\right) \text{ and } s = 2$$

The mixing rule described by Eq. (67), was also applied to the SRK EoS (Chen et al., 2002) in order to define a new version of the PSRK model. In this latter case, the SRK EoS was combined with a Mathias-Copeman alpha function.

5.4.2.4 The LCVM model

The LCVM model (Boukouvalas et al., 1994) is based on a mixing rule which is a **L**inear **C**ombination of the **V**idal and **M**ichelsen (MHV-1) mixing rules:

$$\alpha_{LCVM} = \lambda \cdot \alpha_{Huron-Vidal} + (1-\lambda) \cdot \alpha_{Michelsen} \tag{69}$$

From Eq. (48), one has:

$$\alpha_{Huron-Vidal} = \sum_{i=1}^{p} x_i \cdot \alpha_i - \frac{g^E}{R \cdot T \cdot C_{EoS}} \tag{70}$$

whereas $\alpha_{Michelsen}$ is given by Eq. (66). The LCVM as used today is based on the original UNIFAC g^E model and the value $\lambda = 0.36$ should be used in all applications. Their authors use a translated form of the PR EoS and obtain accurate results especially for asymmetric systems.

5.4.2.5 The MHV-2 mixing rule

In order to increase the accuracy of the MHV-1 mixing rule, (Dahl & Michelsen, 1990) proposed a quadratic approximation of the Q function by:

$$Q(\alpha) = q_0 + q_1 \cdot \alpha + q_2 \cdot \alpha^2 \tag{71}$$

thus defining the MHV-2 model. It is advised to use $q_1 = -0.4783$ and $q_2 = -0.0047$ for the SRK EoS; $q_1 = -0.4347$ and $q_2 = -0.003654$ for the PR EoS. Doing so, Eq. (64) writes:

$$\frac{g^E}{RT} = q_1\left(\alpha - \sum_{i=1}^{p} x_i \cdot \alpha_i\right) + q_2\left(\alpha^2 - \sum_{i=1}^{p} x_i \cdot \alpha_i^2\right) + \sum_{i=1}^{p} x_i \cdot \ln\left(\frac{b_i}{b}\right) \tag{72}$$

Eq. (72) does not yield anymore to an explicit mixing rule but instead has to be solved in order to determine α. For such a mixing rule, parameter tables are available for many gases. As a general rule, MHV-2 provides a better reproduction of the low-pressure VLE data than MHV-1.

6. Energetic aspects: estimation of enthalpies from cubic EoS

Engineers use principles drawn from thermodynamics to analyze and design industrial processes. The application of the first principle (also named *energy rate balance*) to an open multi-component system at steady state writes:

$$\dot{W} + \dot{Q} = \sum \dot{n}_{\text{out}} h_{\text{out}} - \sum \dot{n}_{\text{in}} h_{\text{in}} \tag{73}$$

where $\dot{W}$ and $\dot{Q}$ are the net rates of energy transfer resp. by work and by heat; $\dot{n}$ is the molar flowrate and h denotes the molar enthalpy of a stream. Subscripts *in* and *out* resp. mean *inlet* and *outlet* streams. Note that kinetic-energy and potential-energy terms are supposed to be zero in Eq. (73). According to classical thermodynamics, the molar enthalpy of a p-component homogeneous system at a given temperature T, pressure P and composition **z** (mole fraction vector) is:

$$\underbrace{h(T,P,\mathbf{z})}_{\substack{\text{molar enthalpy}\\ \text{of an inlet or}\\ \text{an outlet stream}}} = \sum_{i=1}^{p} z_i\left[h_{pure\,i}(T,P) - h_{pure\,i}^{ref}\right] \quad + \quad h^M(T,P,\mathbf{z}) \tag{74}$$

wherein $h_{pure\,i}(T,P)$ is the molar enthalpy of pure component i at the same temperature and pressure as the mixture, $h_{pure\,i}^{ref}$ is the molar enthalpy of pure component i in its reference state, i.e. at a reference temperature T_{ref}, pressure P_{ref} and aggregation state. Note that this term is specific to each component i and does not depend on the temperature and pressure of the stream. $h^M(T,P,\mathbf{z})$ is the molar enthalpy change on isothermal and isobaric mixing. This section is dedicated to explain how to calculate these terms when a cubic equation of state (as defined in Eq. (12)) is used with Soave's alpha function (see Eq. (15)) and classical mixing rules involving a temperature-dependent k_{ij} (see Eq. (49)).

6.1 Calculation of pure-component enthalpies

At this step, the concept of *residual molar enthalpy* h^{res} needs to be introduced: h^{res} is a difference measure for how a substance deviates from the behaviour of a perfect gas **having the same temperature T** as the real substance. The molar enthalpy of a pure fluid can thus be written as the summation of the molar enthalpy of a perfect gas having the same temperature as the real fluid plus a residual term:

$$h_{pure\,i}(T,P) = h_{pure\,i}^{pg}(T) + h_{pure\,i}^{res}\left(T, v_{pure\,i}(T,P)\right) \tag{75}$$

where the superscript *pg* stands for *perfect gas*; $v_{pure\,i}(T,P)$ is the molar volume of the pure fluid i at temperature T and pressure P; it can be calculated by solving the cubic EoS (see Eq. (13)) at given T and P. Since cubic EoS are explicit in pressure (i.e. they give the pressure as an explicit function of variables T and v), the expression of the residual molar enthalpy can be naturally written in variables T and v:

$$h_{pure\,i}^{res}(T,v) = \frac{RTb_i}{v-b_i} - \frac{a_i(T)\,v}{(v-r_1b_i)(v-r_2b_i)} + \frac{1}{b(r_1-r_2)}\left(a_i - T\frac{da_i}{dT}\right)\cdot\ln\left(\frac{v-r_1b_i}{v-r_2b_i}\right) \tag{76}$$

If employing Soave's alpha function, then:

$$\frac{da_i}{dT} = -a_{c,i}m_i\left[1+m_i\left(1-\sqrt{T/T_{c,i}}\right)\right]/\sqrt{T\cdot T_{c,i}} \tag{77}$$

Finally, according to Eq. (75), the difference of pure-fluid enthalpy terms in Eq. (74) writes:

$$\begin{aligned} h_{pure\,i}(T,P) - h_{pure\,i}^{ref} &= \left[h_{pure\,i}^{pg}(T) - h_{pure\,i}^{pg}\left(T_{ref}\right)\right] \\ &+ \left[h_{pure\,i}^{res}\left(T, v_{pure\,i}(T,P)\right) - h_{pure\,i}^{res}\left(T, v_{pure\,i}(T_{ref},P_{ref})\right)\right] \end{aligned} \tag{78}$$

wherein the h_i^{res} function is given by Eq. (76). Let us recall that:

$$\left[h_{pure\,i}^{pg}(T) - h_{pure\,i}^{pg}\left(T_{ref}\right)\right] = \int_{T_{ref}}^{T} c_{P,pure\,i}^{pg}(T)\,dT \tag{79}$$

where $c_{P,i}^{pg}(T)$ denotes the molar heat capacity at constant pressure of the pure perfect gas i.

6.2 Calculation of the enthalpy change on mixing

By definition, the molar enthalpy change on mixing h^M is the difference between the molar enthalpy of a solution and the sum of the molar enthalpies of the components which make it up, all at the same temperature and pressure as the solution, *in their actual state* (see Eq. (74)) weighted by their mole fractions z_i. Consequently to this definition, h^M can be expressed in terms of residual molar enthalpies:

$$h^M(T,P,\mathbf{z}) = h^{res}(T,v,\mathbf{z}) - \sum_{i=1}^{p} z_i \, h_{pure\ i}^{res}(T, v_{pure\ i}(T,P)) \tag{80}$$

where v is the molar volume of the mixture at T, P and $\mathbf{z}$. To calculate this molar volume, Eq. (13) has to be solved. The residual molar enthalpy of pure component i is given by Eq. (76) and the residual molar enthalpy of the mixture is given by Eq. (81):

$$h^{res}(T,v,\mathbf{z}) = \frac{RT\, b(\mathbf{z})}{v - b(\mathbf{z})} - \frac{a(T,\mathbf{z})\, v}{[v - r_1 b(\mathbf{z})][v - r_2 b(\mathbf{z})]} + \frac{1}{b(\mathbf{z})(r_1 - r_2)}\left[a(T,\mathbf{z}) - T\left(\frac{\partial a}{\partial T}\right)_{\mathbf{z}}\right] \cdot \ln\left[\frac{v - r_1 b(\mathbf{z})}{v - r_2 b(\mathbf{z})}\right] \tag{81}$$

If classical mixing rules with temperature-dependent k_{ij} are considered, then:

$$\left(\frac{\partial a}{\partial T}\right)_{\mathbf{z}} = \sum_{i=1}^{p}\sum_{j=1}^{p} z_i z_j \left[\left[1 - k_{ij}(T)\right] \frac{a_j(T) \cdot \frac{da_i}{dT} + a_i(T) \cdot \frac{da_j}{dT}}{2\sqrt{a_i(T) a_j(T)}} - \frac{dk_{ij}}{dT}\sqrt{a_i(T) a_j(T)}\right] \tag{82}$$

From Eqs. (80), (81) and (82), it appears that the use of a temperature-dependent k_{ij} allows a better flexibility of the h^M function (see the term dk_{ij}/dT in Eq. (82)). As a consequence, a better accuracy on the estimation of the molar enthalpies is expected when temperature-dependent k_{ij} rather than constant k_{ij} are used.

6.3 Practical use of enthalpies of mixing and illustration with the PPR78 model

As previously explained, the molar enthalpy of a multi-component phase is obtained by adding a pure-component term and a molar enthalpy change on mixing term (see Eq. (74)). When molecules are few polar and few associated (and this is often the case within petroleum blends), pure-component terms provide an excellent estimation of the molar enthalpy of the mixture. Therefore, the enthalpy-of-mixing term can be seen as a correction, just aimed at improving the first estimation given by pure-component ground terms. In other words, with few-polar and few-associated molecules, h^M terms are generally nearly negligible with respect to pure-component terms in the energy rate balance. Typically, h^M terms are very small in alkane mixtures and are not negligible in petroleum mixtures containing CO_2, H_2O or alcohols. When parameters involved in k_{ij} correlations are not directly fitted on enthalpy-of-mixing data (and this is, for instance, the case with the PPR78 model), the relative deviations between calculated and experimental h^M data can be very important and reach values sometimes greater than 100 %. However, as explained in the introduction part of this section, since h^M quantities are only used to evaluate the molar enthalpies, h_{in} and h_{out}, involved in the energy rate balance, only absolute deviations and their effect on the accuracy of the energy balance are of interest. When experimental and calculated h^M values are very low (typically < 100 *J/mol*), the energy rate balance is not significantly affected by high relative deviations. On the contrary, if h^M values are very important (e.g. > 3000 *J/mol*), important absolute deviations on h^M can have a detrimental impact on the energy rate balance even if the corresponding relative deviation remains low. Note that in such a case, h^M terms can become dominant with respect to pure-component enthalpy terms. As an illustration, the PPR78 was used to predict isothermal and isobaric curves h^M vs. z_1 of two binary mixtures: n-hexane + n-decane and N_2 + CO_2 (see Figure 4).

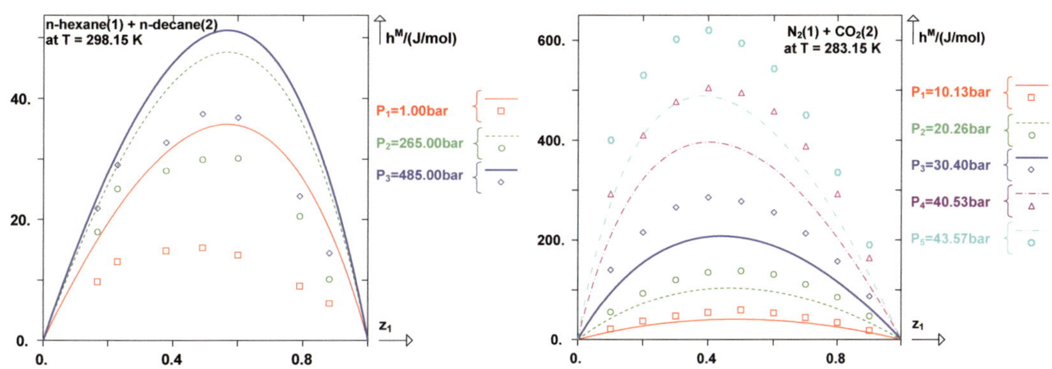

Fig. 4. Representation of molar enthalpy change on isothermal isobaric mixing vs. mole fraction z_1. *Symbols:* experimental data. *Full lines:* predicted curves with the PPR78 model.

One observes that enthalpies of mixing of an alkane mixture (not too much dissymmetric in size) are very low. The PPR78 model predicts h^M with an acceptable order of magnitude (and as a consequence, only pure-component enthalpy terms will govern the energy rate balance). Regarding the binary mixture N_2 + CO_2, it clearly appears that the orders of magnitude of h^M are around ten times bigger than those with the n-hexane + n-heptane system. Deviations between predicted values and experimental data are around 100 - 150 J/mol at the most, which remains acceptable and should very few affect the energy rate balance. When at the considered temperature and pressure, a liquid-vapour phase equilibrium occurs, the corresponding h^M vs. z_1 curve is made up of three different parts: a homogeneous liquid part, a liquid-vapour part and a homogeneous gas part. The liquid-vapour part is a straight line, framed by the two other parts. Figure 5 gives an illustration of the kind of curves observed in such a case. For system exhibiting vapour-liquid equilibrium (VLE) at given T and P, it is possible to show that the essential part of the enthalpy-of-mixing value is due to the vaporization enthalpies of the pure compounds. As a consequence, a good agreement between experimental and predicted h^M vs. z_1 curves of binary systems exhibiting VLE, mainly attests of the capacity of the EoS to model vaporization enthalpies of pure components rather than its capacity to estimate h^M.

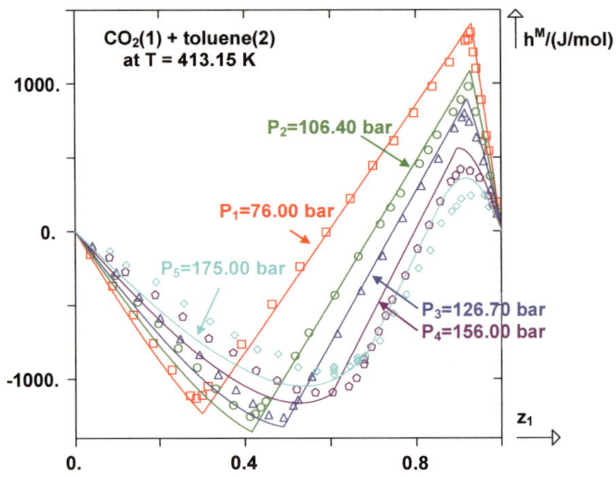

Fig. 5. System $CO_2(1)$ + toluene(2): representation of molar enthalpy change on isothermal isobaric mixing vs. mole fraction z_1. *Symbols:* experimental data. *Full lines:* predicted curves with the PPR78 model.

7. Prediction of the thermodynamic behaviour of petroleum fluids

Dealing with petroleum fluids, many difficulties appear. Indeed, such mixtures contain a huge number of various compounds, such as paraffins, naphthenes, aromatics, gases (CO_2, H_2S, N_2, …), mercaptans and so on. A proper representation involves to accurately quantifying the interactions between each pair of molecules, which is obviously becoming increasingly difficult if not impossible as the number of molecules is growing. To avoid such a fastidious work, an alternative solution lies in using a predictive model, able to estimate the interactions from mere knowledge of the structure of molecules within the petroleum blend. For this reason, it is advised to use predictive cubic EoS (PPR78, PR_2SRK, PSRK) to model petroleum fluid phase behaviour. As an illustration of the capabilities of such models, some phase envelops of petroleum fluids predicted with the PPR78 model are shown and commented hereafter.

7.1 Prediction of natural gases

As an example, (Jarne et al., 2004) measured 110 upper and lower dew-point pressures for two natural gases containing nitrogen, carbon dioxide and alkanes up to n-C_6. The composition of the fluids and the accuracy of the PPR78 model can be seen in Figure 6. The average deviation on these 110 pressures is only 2.0 bar which is close to the experimental uncertainty.

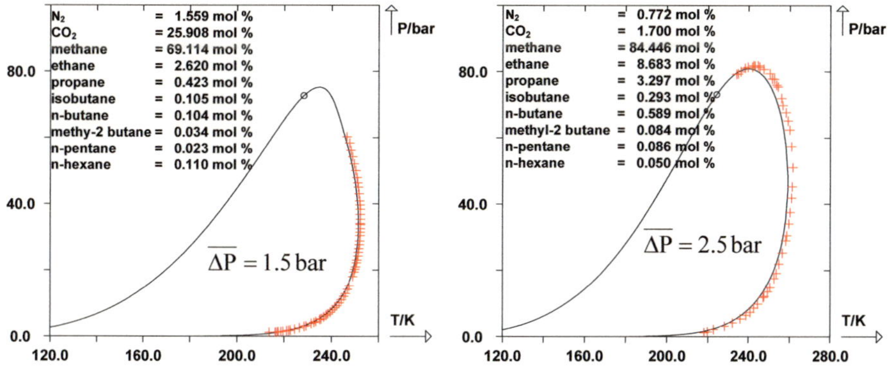

Fig. 6. Solid line: (P,T) phase envelopes of Jarne et al.'s natural gases predicted with the PPR78 model. +: experimental upper and lower dew-point pressures. ○: predicted critical point.

7.2 Prediction of gas condensates

(Gozalpour et al., 2003) measured 6 dew-point pressures for a gas condensate containing 5 normal alkanes ranging from methane to n-hexadecane. Figure 7 puts in evidence that with an average deviation of 3.0 %, the PPR78 model is able to accurately predict these data.

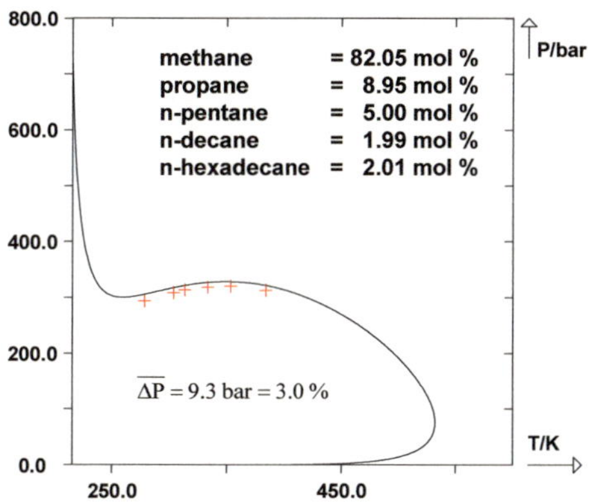

Fig. 7. Solid line: (P,T) phase envelope of Gozalpour et al.'s gas condensates predicted with the PPR78 model. +: experimental dew-point pressures. ○: predicted critical point.

7.3 Prediction of gas injection experiments

(Turek et al., 1984) performed swelling tests at two temperatures on a crude oil containing 10 n-alkanes ranging from methane to n-tetradecane. The injected gas is pure CO_2. 22 mixture saturation pressures were measured. The composition of the crude oil along with the accuracy of the PPR78 model to predict these data are shown in Figure 8. With an average deviation of 2.8 bar (i.e. 2.3 %), we can conclude that the PPR78 model is able to predict these data with high accuracy. It is here important to recall that no parameter is fitted on the experimental data.

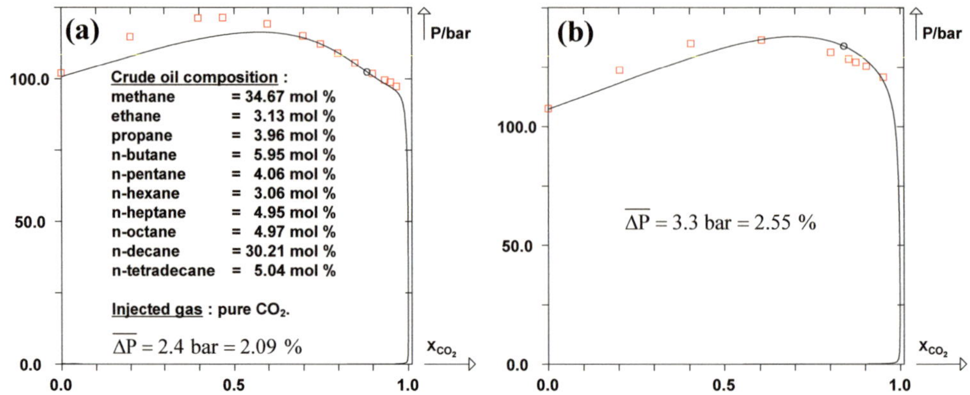

Fig. 8. Solid line: variation of mixture saturation pressure with added CO_2 to a synthetic crude oil (swelling test) predicted with the PPR78 model. a) T/K = 322 and b) T/K = 338.7. b): experimental bubble-point and dew-point pressures. ○: predicted critical point.

8. Conclusion

A complete account of all cubic equations of state is not easy to provide; however, according to (Valderrama, 2003), it is not adventurous to estimate that there must be about 150 RK-type equations and a total of 400 cubic EoS proposed to date in the literature. There are in general small differences in the VLE correlation among all these cubic EoS provided the same way of obtaining the pure parameters and the same mixing/combining rules are used. This is why, in the petroleum industry, only two cubic EoS are generally used: the SRK and the PR EoS. Such cubic EoS have many advantages but also shortcomings. The main advantages are:

- they are simple and capable of fast calculations
- they apply in both liquid and vapour phases
- they are applicable over wide ranges of pressures and temperatures
- they allow a good correlation for non-polar systems encountered in the petroleum industry
- they estimate accurate densities if a volume translation is used
- accurate correlations and GCM are available to estimate the k_{ij}

The outstanding book by (Kontogeorgis & Folas, 2010), cites the following sentence by (Tsonopoulos & Heidman, 1986) which summarizes well the advantages of such models: *cubic EoS are simple, reliable, and allow the direct incorporation of critical conditions. We, in the petroleum industry, continue to find that simple cubic EoS such as RKS and PR are very reliable high-pressure VLE models, and we have not yet found in our work any strong incentive for using non-cubic EoS*. Among the shortcomings, we can cite:

- often a temperature-dependent interaction parameter is needed
- poor correlation of polar/associating systems. The use of two interaction parameters (k_{ij} and l_{ij}) highly improves the results but such parameters can not be easily estimated knowing only the structure of the molecules
- Unsatisfactory correlation of LLE especially with highly immiscible systems (e.g. water or glycols with alkanes)

To conclude, predictive cubic EoS (PPR78, PR2SRK, PSRK, VTPR, UMR-PR) make a perfect job to simulate the phase behaviour of crude oils, gas condensate and natural gases. For processes in which water and/or glycol are present (e.g. transportation processes), it is advised to use more complex EoS like the CPA (Cubic-Plus-Association) by (Derawi et al., 2003) or equation deriving from the SAFT (Statistical Associating Fluid Theory) which are however non predictive (many parameters have to be fitted on experimental data).

9. Acknowledgment

We thank Dr. P. Thibault and Dr. J. Hadrien warmly for the helpful discussions we had while doing this research.

10. References

Ahlers, J. & Gmehling, J. (2001). Development of an universal group contribution equation of state. I. Prediction of liquid densities for pure compounds with a volume translated Peng-Robinson equation of state. *Fluid Phase Equilib.*, Vol.191, pp. 177 - 188

Ahlers, J. & Gmehling, J. (2002). Development of an universal group contribution equation of state. 2. Prediction of vapor-liquid equilibria for asymmetric systems, *Ind. Eng. Chem. Res.*, Vol.41, pp. 3489 - 3498

Abdoul, W.; Rauzy, E. & Péneloux, A. (1991). Group-contribution equation of state for correlating and predicting thermodynamic properties of weakly polar and non-associating mixtures. Binary and multicomponent systems. *Fluid Phase Equilib.*, Vol.68, pp. 47-102

Avlonitis, G.; Mourikas, G.; Stamataki, S. & Tassios, D. (1994). A generalized correlation for the interaction coefficients of nitrogen hydrocarbon binary mixtures. *Fluid Phase Equilib.*, Vol.101, pp. 53-68

Bartle, K.D.; Clifford, A.A. & Shilstone, G.F. (1992). Estimation of solubilities in supercritical carbon dioxide: A correlation for the Peng-Robinson interaction parameters. *J. Supercrit. Fluids.*, Vol.5, pp. 220-225

Boukouvalas, C.; Spiliotis, N.; Coutsikos, P.; Tzouvaras, N. & Tassios, D. (1994). Prediction of vapor-liquid equilibrium with the LCVM model: a linear combination of the Vidal and Michelsen mixing rules coupled with the original UNIFAC and the t-mPR equation of state. *Fluid Phase Equilib.*, Vol.92, pp. 75 - 106

Carrier, B.; Rogalski, M. & Péneloux, A. (1998). Correlation and prediction of physical properties of hydrocarbons with the modified Peng-Robinson equation of state. 1. Low and medium vapor pressures. *Ind. Eng. Chem. Res.*, Vol.27, No.9, pp. 1714-1721

Chen, J.; Fischer, K. & Gmehling, J. (2002). Modification of PSRK mixing rules and results for vapor–liquid equilibria, enthalpy of mixing and activity coefficients at infinite dilution. *Fluid Phase Equilib.*, Vol.200, pp. 411–429

Chueh, P.L. & Prausnitz, J.M. (1967). Vapor-liquid equilibria at high pressures: calculation of partial molar volumes in non polar liquid mixtures. *AIChE J.*, Vol.13, pp. 1099-1107

Clausius, R. (1880). Ueber das Verhalten der Kohlensäure in Bezug auf Druck, Volumen und Temperatur (German title). *Annalen der Physic und Chemie.*, Vol.IX, pp. 337-358

Coutinho, J.A.P.; Kontogeorgis, G.M. & Stenby, E.H. (1994). Binary interaction parameters for nonpolar systems with cubic equations of state: a theoretical approach 1. CO2/hydrocarbons using SRK EoS. *Fluid Phase Equilib.*, Vol.102, No.1, pp. 31-60

Dahl, S. & Michelsen, M.L. (1990). High pressure vapour-liquid equilibrium with a UNIFAC-based equation of state. *AIChE J.*, Vol.36, No.12, pp. 1829 - 1836

Derawi, S.O.; Kontogeorgis, G.M.; Michelsen, M.L. & Stenby, E.H. (2003). Extension of the Cubic-Plus-Association Equation of State to Glycol-Water Cross-Associating Systems. *Ind. Eng. Chem. Res.* Vol.42, pp. 1470-1477

Gao, G.; Daridon, J.L.; Saint-Guirons, H.; Xans, P. & Montel, F. (1992). A simple correlation to evaluate binary interaction parameters of the Peng-Robinson equation of state: binary light hydrocarbon systems. *Fluid Phase Equilib.*, Vol.74, pp. 85-93

Gozalpour, F.; Danesh, A.; Todd, A.; Tehrani, D.H. & Tohidi, B. (2003). Vapour-liquid equilibrium volume and density measurements of a five-component gas condensate at 278.15 – 383.15 K. *Fluid Phase Equilib.*, Vol. 26, pp. 95–104

Graboski, M.S. & Daubert, T.E. (1978). A modified Soave equation of state for phase equilibrium calculations. 2. Systems containing CO_2, H_2S, N_2, and CO. *Ind. Eng. Chem. Process Des. Dev.*, Vol.17, pp. 448-454

Holderbaum, T. & Gmehling, J. (1991). PSRK : A group contribution equation of state based on UNIFAC. *Fluid Phase Equilib.*, Vol.70, pp. 251 - 265

Huron, M.J. & Vidal, J. (1979). New mixing rules in simple equations of state for strongly non-ideal mixtures. *Fluid Phase Equilib.*, Vol.3, pp. 255 - 271

Jarne, C.; Avila, S.; Bianco, S.; Rauzy, E.; Otín, S. & Velasco, I. (2004). Thermodynamic properties of synthetic natural gases. 5. dew point curves of synthetic natural gases and their mixtures with water and with water and methanol : measurement and correlation. *Ind. Eng. Chem. Res.*, Vol. 43, pp. 209–217

Jaubert, J.N. & Mutelet, F. (2004). VLE predictions with the Peng-Robinson equation of state and temperature dependent k_{ij} calculated through a group contribution method. *Fluid Phase Equilib.*, Vol.224, No.2, pp. 285-304

Jaubert, J.N. & Privat, R. (2010). Relationship between the binary interaction parameters (k_{ij}) of the Peng-Robinson and those of the Soave-Redlich-Kwong equations of state. Application to the definition of the PR2SRK model. *Fluid Phase Equilib.*, Vol.295, pp. 26-37

Jaubert, J.N. & Privat, R. (2011). Péneloux's mixing rules: 25 years ago and now. *Fluid Phase Equilib.*, Vol.308, pp. 164-167

Jaubert, J.N.; Privat, R. & Mutelet, F. (2010). Predicting the phase equilibria of synthetic petroleum fluids with the PPR78 approach. *AIChE J.*, Vol.56, No.12, pp. 3225-3235

Jaubert, J.N.; Vitu, S.; Mutelet, F. & Corriou, J.P. (2005).Extension of the PPR78 model to systems containing aromatic compounds. *Fluid Phase Equilib.*, Vol.237,No.1-2, pp 193-211

Kato, K.; Nagahama, K. & Hirata, M. (1981). Generalized interaction parameters for the Peng-Robinson equation of state: Carbon dioxide n-paraffin binary systems. *Fluid Phase Equilib.*, Vol.7, pp. 219-231

Kontogeorgis, G. & Folas, G. (2010). *Thermodynamic models for industrial applications*, Wiley, ISBN 978-0-470-69726-9, United Kingdom

Kordas, A.; Magoulas, K.; Stamataki, S. & Tassios, D. (1995). Methane hydrocarbon interaction parameters correlation for the Peng-Robinson and the t-mPR equation of state. *Fluid Phase Equilib.*, Vol.112, pp. 33-44

Kordas, A.; Tsoutsouras, K.; Stamataki, S. & Tassios, D. (1994). A generalized correlation for the interaction coefficients of CO_2-hydrocarbon binary mixtures. *Fluid Phase Equilib.*, Vol.93, pp. 141-166

Mathias, P.M. & Copeman, T.W. (1983). Extension of the Peng-Robinson EoS to Complex Mixtures: Evaluation of the Various Forms of the Local Composition Concept. *Fluid Phase Equilib.*, Vol.13, pp. 91-108

Michelsen, M.L. (1990). A modified Huron-Vidal mixing rule for cubic equations of state, *Fluid Phase Equilib.*, Vol.60, pp. 213-219

Moysan, J.M.; Paradowski, H. & Vidal, J. (1986). Prediction of phase behaviour of gas-containing systems with cubic equations of state. *Chem. Eng. Sci.*, Vol.41, pp. 2069-2074

Nishiumi, H.; Arai, T & Takeuchi, K. (1988). Generalization of the binary interaction parameter of the Peng-Robinson equation of state by component family. *Fluid Phase Equilib.*, Vol.42, pp. 43-62

Peneloux, A.; Rauzy, E. & Freze, R. (1982). A Consistent Correction for Redlich-Kwong-Soave Volumes. *Fluid Phase Equilib.*, Vol.8, pp. 7-23

Peng, D.Y. & Robinson, D.B. (1976). A New Two-Constant Equation of State. *Ind. Eng. Chem. Fundam.*, Vol.15, pp. 59-64

Privat, R.; Jaubert, J.N. & Mutelet, F. (2008a). Addition of the Nitrogen group to the PPR78 model (Predictive 1978, Peng Robinson EoS with temperature dependent k_{ij} calculated through a group contribution method). *Ind. Eng. Chem. Res.*, Vol.47, No.6, pp. 2033-2048

Privat, R.; Jaubert, J.N. & Mutelet, F. (2008b). Addition of the sulfhydryl group (-SH) to the PPR78 model (Predictive 1978, Peng-Robinson EoS with temperature dependent k_{ij} calculated through a group contribution method). *J. Chem. Thermodyn.*, Vol. 40, No.9, pp. 1331-1341

Privat, R.; Jaubert, J.N. & Mutelet, F. (2008c). Use of the PPR78 model to predict new equilibrium data of binary systems involving hydrocarbons and nitrogen. Comparison with other GCEOS. *Ind. Eng. Chem. Res.*, Vol.47, No.19, pp. 7483-7489

Privat, R.; Mutelet, F. & Jaubert, J.N. (2008d). Addition of the Hydrogen Sulfide group to the PPR78 model (Predictive 1978, Peng Robinson EoS with temperature dependent k_{ij} calculated through a group contribution method). *Ind. Eng. Chem. Res.*, Vol.47, No.24, pp. 10041-10052

Rauzy, E. (1982). Les méthodes simples de calcul des équilibres liquide-vapeur sous pression. Ph.D. Dissertation. The French University of Aix-Marseille II.

Redlich, O. & Kwong, J.N.S. (1949). On the Thermodynamics of Solutions. V. An Equation of State. Fugacities of Gaseous Solutions. *Chem. Rev.*, Vol.44, pp. 233-244

Robinson, D.B. & Peng, D.Y. (1978). *The characterization of the heptanes and heavier fractions for the GPA Peng–Robinson programs*, Gas processors association, Research report RR-28

Soave, G. (1972). Equilibrium Constants from a Modified Redlich-Kwong Equation of State. *Chem. Eng. Sci.*, Vol.27, pp. 1197-1203

Soave, G.; Gamba, S. & Pellegrini, L.A. (2010). SRK equation of state: Predicting binary interaction parameters of hydrocarbons and related compounds. *Fluid Phase Equilib.*, Vol.299, pp. 285-293

Stryjek, R. & Vera, J.H. (1986). PRSV: An Improved Peng-Robinson Equation of State for Pure Compounds and Mixtures. *Can. J. Chem. Eng.*, Vol.64, pp. 323-340

Stryjek, R. (1990). Correlation and prediction of VLE data for n-alkane mixtures. *Fluid Phase Equilib.*, Vol.56, pp. 141-152

Tsonopoulos C., & Heidman, J.L. (1986). High-pressure vapor-liquid equilibria with cubic equations of state. *Fluid Phase Equilib.*, Vol.29, pp. 391-414

Turek, E.; Metcalfe, R.; Yarborough, L. & Robinson, R. (1984). Phase equilibria in CO_2-multicomponent hydrocarbon systems: experimental data and an improved prediction technique. *Society of petroleum engineers journal.*, Vol.24, pp. 308–324

Twu, C.H.; Bluck, D. & Cunningham, J.R. (1991). A cubic equation of state with a new alpha function and a new mixing rule. *Fluid Phase Equilib.*, Vol.69, pp. 33-50

Twu, C.H.; Coon, J.E. & Cunningham, J.R. (1995a). A new generalized alpha function for a cubic equation of state Part 1. Peng-Robinson equation, *Fluid Phase Equilib.*, Vol.105, pp. 49-59

Twu, C.H.; Coon, J.E. & Cunningham, J.R. (1995b). A new generalized alpha function for a cubic equation of state Part 2. Redlich-Kwong equation, *Fluid Phase Equilib.*, Vol.105, pp. 61-69

Valderrama, J.O.; Arce, P.F. & Ibrahim, A.A. (1999). Vapour-liquid equilibrium of H_2S-hydrocarbon mixtures using a generalized cubic equation of state. *Can. J. Chem. Eng.*, Vol.77, pp. 1239-1243

Valderrama, J.O.; Ibrahim, A.A. & Cisternas, L.A. (1990). Temperature-dependent interaction parameters in cubic equations of state for nitrogen-containing mixtures. *Fluid Phase Equilib.*, Vol.59, pp. 195-205

Valderrama, J.O. (2003). The State of the Cubic Equations of State. *Ind. Eng. Chem. Res.*, Vol.42, pp. 1603 - 1618

Van der Waals, J.D. (1873). *On the Continuity of the Gaseous and Liquid States.* Ph.D. Dissertation, Universiteit Leiden, Leiden, The Netherlands

Vitu, S.; Jaubert, J.N. & Mutelet, F. (2006). Extension of the PPR78 model (Predictive 1978, Peng Robinson EoS with temperature dependent k_{ij} calculated through a group contribution method) to systems containing naphtenic compounds. *Fluid Phase Equilib.*, Vol.243, pp. 9-28

Vitu, S.; Privat, R.; Jaubert, J.N. & Mutelet, F. (2008). Predicting the phase equilibria of CO_2 + hydrocarbon systems with the PPR78 model (PR EoS and k_{ij} calculated through a group contribution method). *J. Supercrit. Fluids.*, Vol.45, No.1, pp. 1-26

Voutsas, E.; Magoulas, K. & Tassios, D. (2004). Universal mixing rule for cubic equations of state applicable to symmetric and asymmetric systems: results with the Peng-Robinson equation of state. *Ind. Eng. Chem. Res.*, Vol.43, pp. 6238 - 6246

Wong, D.S.H. & Sandler, S.I.A. (1992). A theoretically correct mixing rule for cubic equations of state. *AIChE J.*, Vol.38, pp. 671 - 680

6

Analysis of Polar Components in Crude Oil by Ambient Mass Spectrometry

Chu-Nian Cheng[1], Jia-Hong Lai[1], Min-Zong Huang[1], Jung-Nan Oung[2] and Jentaie Shiea[1]
[1]*Department of Chemistry, National SunYat-Sen University*
[2] *Exploration and Production Business Division, Chinese Petroleum Co.*
Taiwan

1. Introduction

Crude oil is a naturally generated material comprising a very complex mixture of coexisting hydrocarbons and polar organic compounds. It is found in geologic formations below the Earth's surface and recovered mostly through oil drilling. It is refined and separated by distillation according to the various boiling points of the components resulting in a number of products, such as petrol, kerosene, and numerous chemical reagents that can be used to produce plastics, pharmaceuticals, and a wide variety of other materials. Due to the extreme complexity of the components of crude oil samples, the characterization of these constituents or their products has been a challenging research topic for analytical chemists.

Gas chromatography/mass spectrometry (GC/MS) is routinely used for identifying volatile and non-polar components in crude oil, and the characterization of trace polar components is usually achieved by liquid chromatography/mass spectrometry (LC/MS). However, recent advances in mass spectrometry have enabled the development of novel ionization techniques that are potentially useful in addressing some issues associated with conventional mass spectrometric technologies.

As shown in Table 1, conventional mass spectrometric ionization techniques, such as electron impact ionization (EI), field ionization (FI) (Beckey *et al.*, 1969; Hsu *et al.*, 2001), and field desorption (FD) (Stanford *et al.*, 2007), are suitable for the analysis of volatile or semivolatile compounds in crude oil. For compounds with higher boiling points (>500 °C), pyrolysis (Py) combined with GC/MS to produce characteristic fragments for identification is necessary (Shute *et al.*, 1984). For the purpose of characterizing polar components that cannot be achieved by Py/GC/MS, approaches that couple a HPLC system with atmospheric ionization sources, such as electrospray ionization (ESI) (Fenn *et al.*, 1989 & 1990), atmospheric pressure chemical ionization (APCI) (Carroll *et al.*, 1975), and atmospheric pressure photoionization (APPI) (Robb *et al.*, 2000), have been developed. Although some of the polar fragments can be separated and detected by the LC/MS approach, strong interaction between certain polar components with the stationary phase in the chromatographic system is still a problem. Directly introducing samples into the ionization source without passage through a chromatographic system may be a solution.

The direct analysis of trace polar components containing N, O, and S in crude oil has been achieved by ESI/MS and the interference with the ionization processes resulting from the presence of a large amount of non-polar components in the sample is not observed (Zhan *et al.*, 2000).

For ambient mass spectrometric approaches, techniques such as electrospray-assisted pyrolysis ionization (ESA-Py) (Hsu *et al.*, 2005), desorption electrospray ionization (DESI) (Takáts *et al.*, 2004), easy ambient sonic-spray ionization (EASI) (Haddad *et al.*, 2008), and atmospheric pressure laser-induced acoustic desorption chemical ionization (AP/LIAD-CI) (Nyadong *et al.*, 2011) have been used for the direct analysis of crude oil with minimal sample pretreatment. Such approaches prevent unexpected effects on the composition of crude oil samples during preparation. Another attractive feature of performing analyses under ambient conditions is the capacity for rapid sampling, thereby enabling opportunities for high-throughput analysis.

Within this context, ambient mass spectrometry is regarded as a potential analytical tool for "petroleomics" applications owing to its specific features that differ from conventional mass spectrometry. Such features support the selective characterization of trace polar components by constructing variable sampling methods and ionization models. In the present chapter, we focus on a recently developed ambient mass spectrometric approach for the rapid characterization of polar components in crude oil.

Technique name	Acronym	Environment	Analyte polarity	Reference
Pyrolysis/mass spectrometry	Py/MS[I]	Vacuum	Non-polar/Polar	Shute *et al.*
Gas chromatography x gas chromatography/mass spectrometry	GC×GC/MS[I]	Vacuum	Non-polar/Polar	Blomberg *et al.*
Pyrolysis/gas chromatography/mass spectrometry	Py/GC/MS[I]	Vacuum	Non-polar/Polar	Snyder *et al.*
Matrix-assisted laser desorption/ionization mass spectrometry	MALDI	Vacuum	Polar	Robins *et al.*
Electrospray ionization	ESI	Ambient	Polar	Zhan *et al.*, Klein *et al.* & Hsu *et al.*
Atmospheric pressure photoionization	APPI	Ambient	Polar/Non-polar	Purcell *et al.*
Atmospheric pressure chemical ionization	APCI	Ambient	Polar/Non-polar	Hsu *et al.*
Atmospheric pressure laser-induced acoustic desorption chemical ionization	AP/LIAD-CI	Ambient	Polar/Non-polar	Nyadong *et al.*
Easy ambient sonic-spray ionization	EASI	Ambient	Polar	Corilo *et al.*
Desorption electrospray ionization	DESI	Ambient	Polar/Non-polar [II]	Wu *et al.*
Electrospray-assisted pyrolysis ionization	ESA-Py	Ambient	Polar	Hsu *et al.*

I: The ionization sources: electron impact (EI), chemical ionization (CI), field desorption (FD), and field ionization (FI).
II: Analyses performed through discharge-induced oxidation reactions.

Table 1. Summary of ionization methods used in the analyses of crude oil.

2. Conventional analytical approaches for characterizing polar components in crude oil

Although gas chromatography/mass spectrometry (GC/MS) has been widely used in characterizing volatile and semivolatile components in different samples, many natural macromolecules are still not amenable to be characterized by GC/MS. However, these materials often yield volatile, gas chromatographable products upon controlled thermal degradation induced by pyrolysis (Anhalt *et al.*, 1975; Gutteridge *et al.*, 1987; Yang *et al.*, 2003). These volatile components characterized by Py/MS or Py/GC/MS may then serve as fingerprints for classifying or studying the composition of different species of macromolecules (Snyder *et al.*, 1990; Goodacre *et al.*, 1991; DeLuca *et al.*, 1993; Smith *et al.*, 1993). In general, the pyrolysis of a complicated mixture like crude oil produces a wide variety of chemical compounds ranging in polarity from non-polar to highly polar. The non-polar components, for example, saturated acyclic or cyclic terpenoids and aromatic hydrocarbons, are often used as biomarkers in organic geochemistry. The polar components, which are difficult to characterize by conventional Py/MS or Py/GC/MS, are seldom used for sample diagnosis or classification. Moreover, these complex polar compounds are nearly impossible to resolve by GC and are often referred to as unresolved complex mixtures (UCMs) (Panda *et al.*, 2007). With the aim of resolving UCMs, two-dimensional GC/MS (GCxGC/MS) was developed to provide higher resolution for the separation of these compounds. This approach has been applied to the characterization of many compounds, including the different hydrocarbon isomers in crude oil (Blomberg *et al.*, 1997; Hua *et al.*, 2004; von Mühlen *et al.*, 2006).

For certain materials (e.g., synthetic polar polymers), the polar pyrolysates may contain useful diagnostic structural information (Williamson *et al.*, 1980; Marshall *et al.*, 1983; Shiea *et al.*, 1996; Galipo *et al.*, 1998). Unfortunately, during pyrolysis processes, many polar macromolecules are broken down to fragments that cannot be detected by GC/MS because they are retained in the pyrolysis zone, the injection system, or the capillary column owing to their high polarity or high molecular weight (Moss *et al.*, 1980; Holzer *et al.*, 1989). Even when the polar pyrolysates do enter the separation column, they often display peak tailing, poor reproducibility, and long elution times. In many cases, no chromatographic peaks are even observed in this fraction (Derbyshire *et al.*, 1989; Manion *et al.*, 1996).

Directly introducing the gaseous pyrolysates into the ionization source of a mass spectrometer may be a way to circumvent the chromatographic problems that affect the analysis of polar pyrolysates, but the ionization of trace polar pyrolysates in samples that contain a large amount of non-polar compounds still remains a challenge. As is generally known, when electron impact ionization (EI) is used to ionize samples, non-polar ion signals always overwhelm those of polar components—the so-called "ion suppression effect" (Tang *et al.*, 1991).

Electrospray ionization (ESI), developed by Fenn (for which he received the 2002 Nobel Prize in Chemistry), is an atmospheric ionization method used to characterize polar compounds through continuous infusion of a solution to which a high DC voltage (ca. 3–4 kV) is applied. By electric field forces, a Taylor cone is induced leading to the generation of a large number of charged droplets forming the protonated analyte ions from an ESI emitter. In this way, compounds consisting of functional groups with high proton affinity can be effectively ionized, in contrast to less or non-polar compounds. The characterization of polar hydrocarbons containing N, S, and O in crude oil samples by low or high resolution electrospray ionization mass spectrometry has been reported (Klein *et al.*, 2003; Hsu *et al.*, 2011).

In addition to ESI, matrix-assisted laser desorption/ionization (MALDI) developed by Tanaka in 1987 (for which he received the 2002 Nobel Prize in Chemistry) is an alternative method for ionizing large biomolecules, such as proteins and peptides. In MALDI, the sample solution mixes with the UV-absorbing organic matrix solution (usually in equal volumes). After drying, the crystals containing a large amount of matrix and sample molecules are irradiated with a pulsed laser. The matrix absorbs the energy supplied by the pulsed laser to assist the ionization of analytes. Although the MALDI analysis is performed under vacuum, a characterization of the chemical components in crude oil has been reported (Robins *et al.*, 2003). Unlike the traditional polar and acidic matrices, such as sinapinic acid, α-cyano-3-hydroxycinnamic acid, and 2, 5-dihydroxybenzoic acid, non-polar matrices, including anthracite and 9-cyanoanthracene, are more effective for characterizing crude oil fractions as they provide higher quality MALDI mass spectra. In addition, no interfering matrix ions are observable in the resulting spectra.

Recently, Marshall *et al.* presented results for the characterization of polar compounds in crude oil samples by laser desorption/ionization-ion mobility/mass spectrometry (LDI/IM/MS). In this approach, a Fourier transform ion cyclotron resonance (FT-ICR) was used to obtain high resolution mass spectra (Fernandez-Lima *et al.*, 2009). The distribution of polar compounds in the crude oil samples was studied. Atmospheric pressure chemical ionization (APCI) combined with HPLC has been used to characterize naphthenic acids, which are known to be corrosive to the containers and pipelines used in the petroleum industry (Hsu *et al.*, 2000). Atmospheric pressure photoionization (APPI) combined with a 9.4 T FT-ICR and HPLC has demonstrated utility for characterizing both polar and non-polar components that cannot be ionized by ESI (Purcell *et al.*, 2006). The results indicate that APPI is capable of ionizing both polar and non-polar components in crude oil. In the mean time, the problem of the high complexity of the crude oil components can be solved by using a high resolution FT-ICR to detect the ions generated by APPI. Although APCI and APPI are useful tools for ionizing chemical compounds with different polarities, complicated sample pretreatments are needed to remove insoluble particles in the sample prior to HPLC/MS analysis.

3. Ambient mass spectrometric approaches

Ambient ionization mass spectrometry is another recently developed ionization method and possesses the following features: (1) rapid sample switching, (2) minimal or no sample pretreatment, and (3) ability to analyze solid, liquid, or gaseous samples (Huang *et al.*, 2010). To date, certain ambient ionization mass spectrometric techniques have been used to characterize either non-polar or polar components in petroleum and crude oil without tedious pretreatments.

3.1 Desorption electrospray ionization mass spectrometry (DESI-MS) for crude oil analysis

Desorption electrospray ionization mass spectrometry (DESI-MS), an alternative to ESI-based ambient mass spectrometry, has been used to analyze petroleum samples. Under typical DESI conditions, the technique is only suitable for the detection of polar components. However, in a separate study, Cooks *et al.* reported that discharge-induced oxidation reactions occurred for non-polar components when betaine aldehyde was added to the DESI spraying solution (Wu *et al.*, 2010). The oxidation reactions transformed hydrocarbons to oxidized components of low-polarity, such as alcohols or ketones. In the mean time, the so-called reactive-DESI approach was developed to enhance the detection sensitivity for these oxides. This was achieved by replacing the spray solution with one that contains functional groups with higher proton-affinity (Cotte-Rodríguez *et al.*, 2005). The study indicated that saturated hydrocarbons can still be ionized by DESI under ambient conditions. Although accompanying dehydrogenation may make the method unsuitable for characterizing the extent of unsaturation, a rapid and accurate determination of the carbon distribution in the saturated hydrocarbons of petroleum distillates is demonstrated.

3.2 Easy ambient sonic-spray ionization mass spectrometry (EASI-MS) for crude oil analysis

Easy ambient sonic-spray ionization mass spectrometry (EASI-MS), developed by Eberlin *et al.*, has also been applied to the analysis of crude oil samples. The samples were directly exposed to a flow of nitrogen gas along with the spray reagent. The moist surfaces of the crude oil samples were then desorbed and subsequently ionized to generate analyte ions that were detected by a mass spectrometer (Corilo *et al.*, 2010). The process of ion formation in EASI-MS is mainly supported by the production of bipolar solvent droplets using a nebulizer to generate a supersonic spray, and the high electric voltage applied in conventional ESI is not used to form the charged droplets. The performance of EASI-FT-ICR MS for characterizing crude oil samples has been found to be almost as fast as ESI/FT-ICR-MS and provide similar compositional information on the polar components and comparable spectral quality to that of a commercial ESI device.

3.3 Atmospheric pressure laser-induced acoustic desorption chemical ionization/mass spectrometry (AP-LIAD/CI/MS) for crude oil analysis

A laser-induced acoustic desorption (LIAD) device combined with a chemical ionization source was employed for the analysis of crude oil distillates under atmospheric pressure. In general, LIAD, a matrix-free and laser-based approach, is usually performed under vacuum conditions. The desorption process in LIAD is induced by the action of a shockwave that is generated as a pulsed laser irradiated on the backside of a metal foil. As the energy is transferred from the metal foil to the sample, which is deposited on another side of the foil, it induces the desorption of analytes. By the interaction of the analyte with an ion cloud generated by a chemical ionization (CI) process, analytes with a wide range of polarity are successfully ionized. Marshall *et al.* have combined an atmospheric pressure AP-LIAD/CI with a 9.4 T FT-ICR/MS to perform high resolution chemical analyses under ambient conditions. It was demonstrated that not only polar but also non-polar compounds in the crude oil distillates could be successfully characterized by this AP-LIAD/CI/FT-ICR/MS approach.

3.4 Electrospray-assisted pyrolysis ionization/mass spectrometry (ESA-Py/MS) for crude oil analysis

To characterize polymers and trace polar components in crude oil samples, we previously developed an interface to combine electrospray ionization mass spectrometry (ESI/MS) with a pyrolytic probe. This technique successfully detects the polar pyrolysates that are released from synthetic polymers, which are constructed from polar units and crude oil. We refer to this technique as "electrospray-assisted pyrolysis ionization/mass spectrometry (ESA-Py/MS)" (Hsu *et al.*, 2005 & 2007).

The pyrolyte products generated by a commercial Curie-point pyroprobe are conducted to the tip of a capillary where charged droplets are continuously produced by electrospraying an acidic methanol solution. The ionization of the polar pyrolysates is suggested to occur through (1) ion–molecule reactions (IMRs) between the gaseous pyrolysate molecules (M) and protons (H^+) or protonated methanol species [e.g., $(H_3O)^+$, $(MeOH)H^+$, $(MeOH)_2H^+$] and/or (2) polar pyrolysate molecules dissolving (or fusing) in the charged methanol droplets, followed by electrospray ionization from the droplets to generate protonated analyte molecules (MH^+).

The ESA-Py mass spectra are then used to rapidly distinguish synthetic polymer standards that differ in the nature of their building units, degree of polymerization, and copolymerization coefficients. In addition, a petroleomic application of ESA-Py/MS was also demonstrated. Trace polar compounds that coexist with large amounts of non-polar hydrocarbons in crude oil, amber, humic substances, and rubber samples were selectively ionized without any chromatographic separation or complicated pretreatment processes.

According to the ionization features of ESI, the ionization of polar pyrolysates in an ESA-Py source may go through IMRs and/or an ESI process. Only those compounds containing polar functional groups can be ionized. Non-polar compounds, such as saturated, unsaturated, cyclic, acyclic, or aromatic hydrocarbons possess no functional groups that receive a proton in the source and cannot be detected during the ESA-Py/MS analysis. Thus, this technique may be useful for a rapid characterization of polar components within fossil fuels and other materials that generate traces of polar pyrolysates together with large amounts of non-polar hydrocarbons. Herein, we show the results that were obtained using ESA-Py/MS to selectively detect trace polar components in the pyrolysates of different crude oil samples. The samples of different origins are then rapidly distinguished by their ESA-Py mass spectra. In addition, analyses of samples using ESI/MS were also performed, and the results were compared with those obtained using the ESA-Py/MS approach.

Other ambient mass spectrometric approaches, such as low temperature plasma mass spectrometry (LTP/MS) and direct analysis in real time mass spectrometry (DART/MS), have been used to characterize less polar components in olive oil samples. The use of organic solvents and/or additional reagents in the source is unnecessary, except for the use of helium gas for discharging (García-Reyes *et al.*, 2009; Vaclavik *et al.*, 2009). The results indicate that both techniques are potentially useful for characterizing polar components in petroleomics applications.

3.5 Electrospray-assisted pyrolysis ionization/mass spectrometry (ESA-Py/MS) combined with statistical methods for crude oil analysis

Recently, we modified the existing ESA-Py source to simplify the operation. During this ESA-Py analysis, a drop of untreated crude oil sample (10 μL) was deposited on the Teflon block. The analytes in the sample were desorbed by inserting an electric soldering iron probe heated at 350 °C. The desorbed gaseous analytes then moved into an ESI plume located 5 mm above the top of the sample. Trace polar compounds (M) in the gaseous analytes were then ionized. The schematic of the modified ESA-Py/MS system is displayed in Fig. 1. An acidic methanol solution (50% aqueous methanol with 1% acetic acid) was continuously electrosprayed from an ESI emitter at high voltage (4 kV).

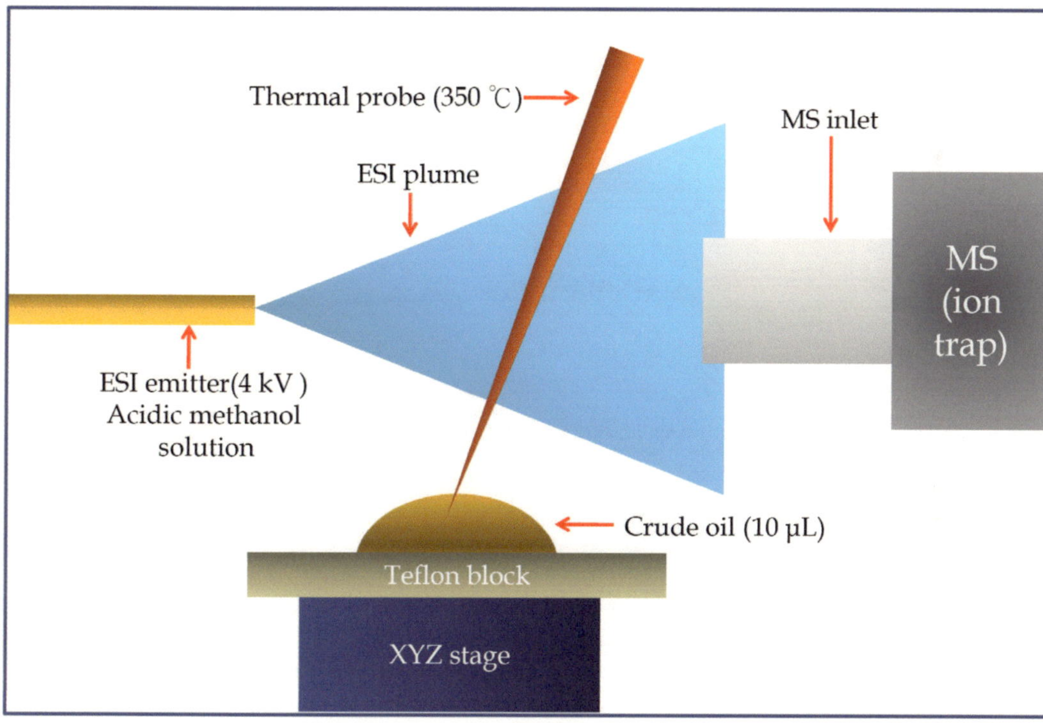

Fig. 1. Schematic illustration of the ESA-Py/MS system for crude oil analysis.

As described previously, the major components of the desorbed analytes are non-polar hydrocarbons, and their lack of functional groups that can accept a proton in the ESA-Py source results in the absence of hydrocarbon ion signals. Instead, the ions produced in the source are from trace polar components in the crude oil samples. Fig. 2 displays the ESA-Py/MS spectra of six crude oil samples from different origins. Samples A1 and A2 are from different depths of a well drilled in Africa, samples B1 and B2 are from different fields in a sedimentary basin of Asia, and samples C1 and C2 are from different basins in Taiwan. As shown in Fig. 2, all samples show significant responses for lower molecular weight components in the mass spectra. These low molecular weight compounds are presumed to be polar compounds owing to their high volatility. For sample correlation, it is demonstrated that oil samples of different origins are rapidly distinguished by their positive ion ESA-Py mass spectra. It is noted that samples A1 and A2, which are from the same well, show high similarity in their ESA-Py/MS mass spectra.

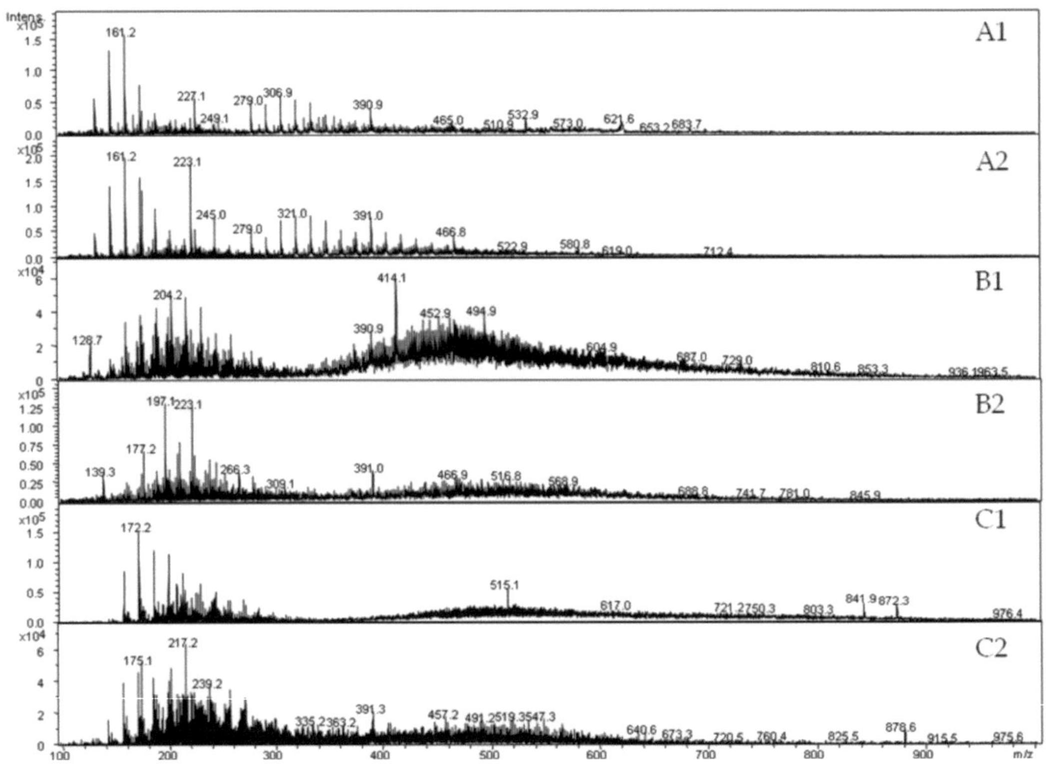

Fig. 2. The ESA-Py/MS spectra recorded from six crude oil samples. Samples A1 and A2: different depth of a well drilled in Africa; samples B1 and B2: different fields in a sedimentary basin of Asia; samples C1 and C2: different basins in Taiwan.

ESA-Py/MS analyses of each crude oil sample were performed in triplicate, and the results were further processed using a multivariate statistical method, principle component analysis (PCA). The results of the statistical analysis are shown in Fig. 3. In the sample groupings, A1 and A2 overlap, indicating that these two samples are closely related. This result is in agreement with the sample origins in which samples A1 and A2 are from the same drilling well but acquired at different depths. Although sample B2 was collected from a different geological region than samples A1 and A2, it contains components that are more similar to those in samples A1 and A2 than to those in sample B1. The remaining samples (B1, C1, and C2) form individual groups independent from each other, reflecting their different origins.

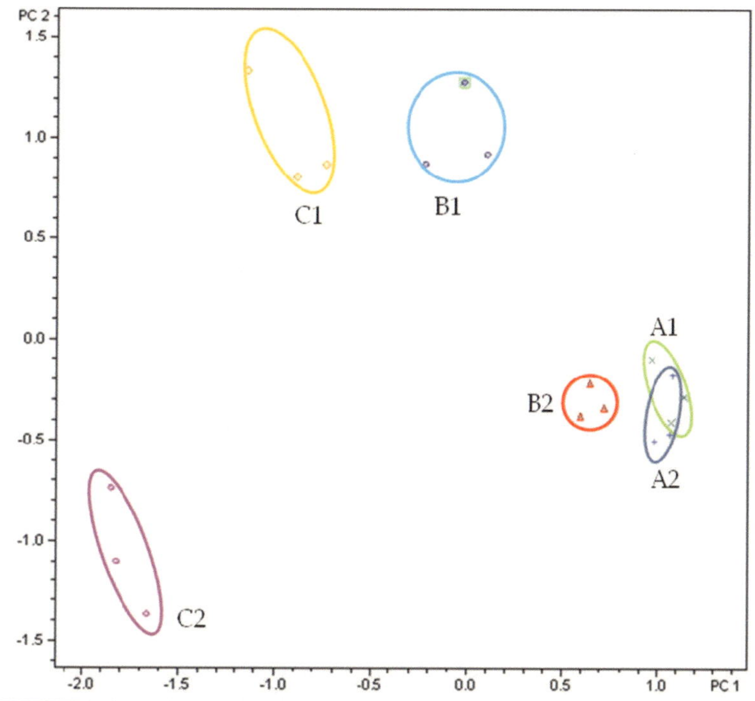

Fig. 3. The PCA diagram of ESA-Py/MS results from six crude oil samples.

In this study, a low resolution ion trap mass spectrometer (Bruker Esquire 3000+ ion trap mass spectrometer) equipped with a conventional ESI source was also used to analyze the methanol-soluble fraction of the same crude oil samples. The crude oil samples were ultrasonically mixed with methanol to extract the polar components. The methanol extract was concentrated and analyzed using ESI/MS. The ESI/MS spectra of the crude oil samples are shown in Fig. 4. As can be seen in the figure, several ion signals were observed in the ESI mass spectra, probably resulting from the dissolution of more polar components in the methanol solvent by the action of ultrasonication. Additionally, for the same reasons, the signals for polar components of higher molecular weight are found to be more intense than those for lower molecular weight components.

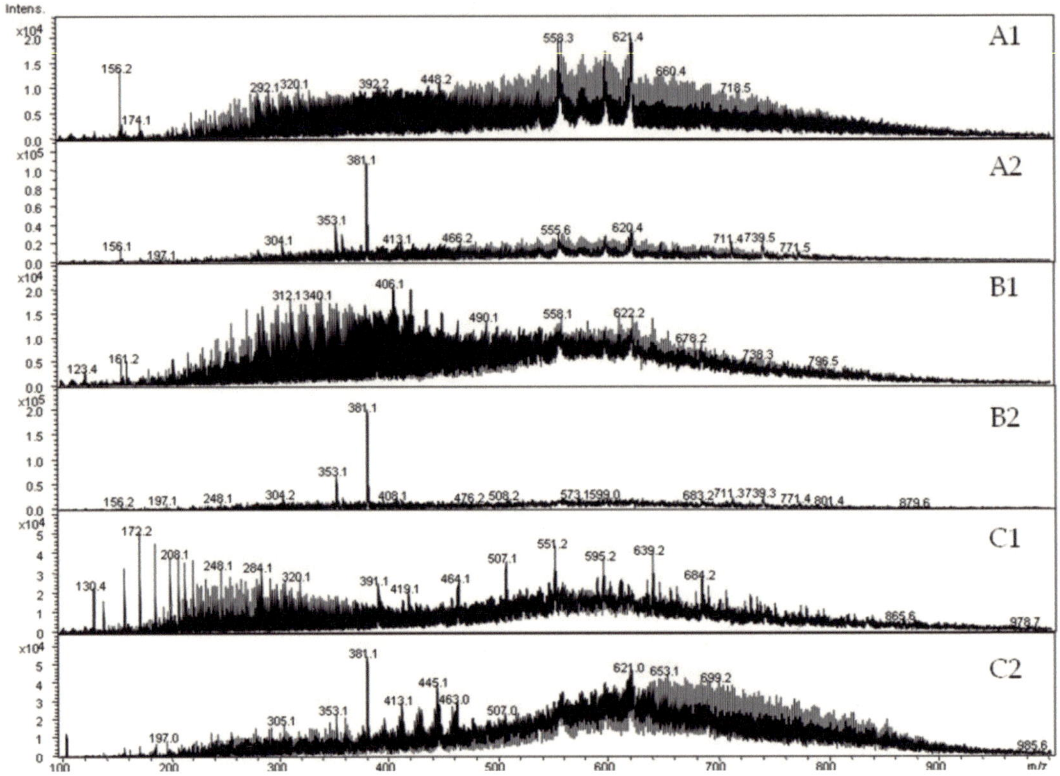

Fig. 4. The ESI mass spectra recorded from six crude oil samples.

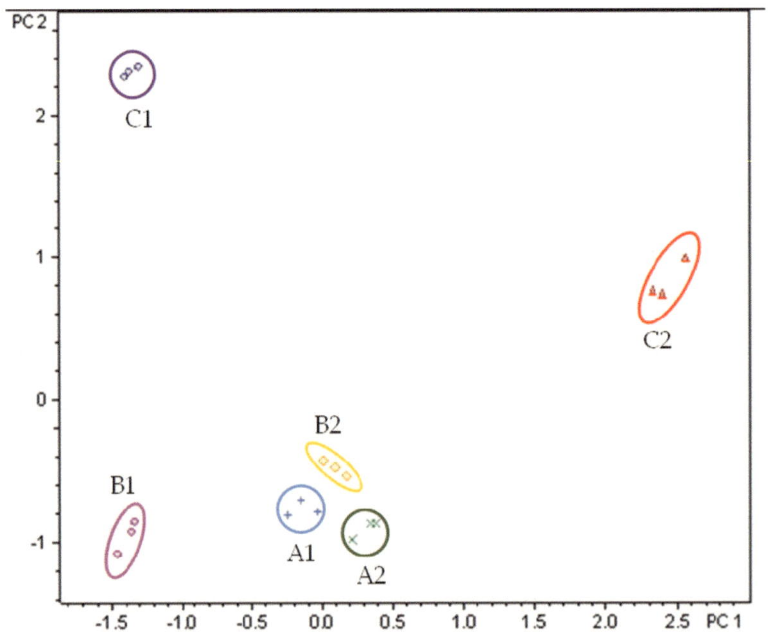

Fig. 5. The PCA diagram of ESI mass spectra results from six crude oil samples.

The PCA diagram based on the ESI mass spectra results from six crude oil samples is shown in Figure 5. In general, the distribution of samples on the PCA diagram is similar to that shown in Figure 3, excluding samples A-1 and A-2, which overlap in the ESA-Py/MS and PCA analyses. Although ESI/MS can be used to ionize more trace polar components, the required sample pretreatment and potential for blocking the ESI emitter (a capillary column) should be considered as possible difficulties associated with this analytical technique.

4. Conclusions

One of the most promising advantages of crude oil characterization by ambient mass spectrometry combined with a statistical processing method is the potential to rapidly analyze and distinguish the components and origins of the crude oil samples with minimal or no sample pretreatment. The technique also helps to reduce unexpected modifications to the composition of oil samples that may occur during sample pretreatment. The results obtained indicate that ESA-Py/MS, an ambient mass spectrometry method, is a useful technique for rapidly characterizing trace polar components in crude oil. With the assistance of modern statistical analysis (PCA), the crude oil samples from different origins are well classified.

5. References

Anhalt, J.P. & Fenselau, C. (1975). Identification of bacteria using mass spectrometry. *Analytical Chemistry*, Vol. 47, No. 2, pp. 219-225, ISSN 1520-6882

Blomberg, J.; Schoenmakers, P.J.; Beens, J. & Tijssen, R. (1997). Compehensive two-dimensional gas chromatography (GCxGC) and its applicability to the characterization of complex (petrochemical) mixtures. *Journal of High Resolution Chromatography*, Vol.20, No.10, pp.539–544, ISSN 1615-9314

Carroll, D.I.; Dzidic, I.; Stillwell, R.N.; Haegele, K.D. & Horning. E.C. (1975). Atmospheric pressure ionization mass spectrometry. Corona discharge ion source for use in a liquid chromatograph-mass spectrometer-computer analytical system. *Analytical Chemistry*, Vol.47, No.14, pp. 2369–2373, ISSN 1520-6882.

Cooks, R.G.; Ouyang, Z.; Takáts, Z. & Wiseman, J.M. (2006). Ambient mass spectrometry. *Science*, Vol.311, No.5767, pp.1566-1570, ISSN 0036-8075.

Cotte-Rodríguez, I.; Takáts, Z.; Talaty, N.; Chen, H. & Cooks, R.G. (2005). Desorption electrospray ionization of explosives on surfaces: sensitivity and selectivity enhancement by reactive desorption electrospray ionization. *Analytical Chemistry*, Vol.77, No.21, pp. 6755-6764, ISSN 1520-6882

Derbyshire, F.; Davis, A. & Lin, R. (1989). Considerations of physicochemical phenomena in coal processing. *Energy & Fuels*, Vol.3, No.4, pp. 431 - 437, ISSN 1520-5029.

Fenn, J.B.;Mann, M.; Meng, C.K. & Wong, S.F. (1990). Electrospray ionization - principles and practice. *Mass Spectrometry Reviews*, Vol.9, No.1, pp. 37-70. ISSN 1098-2787.

Fenn, J.B.; Mann, M.; Meng, C.K.; Wong, S.F. & Whithouse, C.M. (1989). Electrospray ionization for mass spectrometry of large biomolecules. *Science*, Vol.246, No.4926, pp. 64-71, ISSN 0036-8075.

Fernandez-Lima,F.A.; Becker, C.; McKenna, A.M.; Rodgers, R.P.; Marshall, A.G. & Russell, D.H. (2009). Petroleum crude oil characterization by IMS-MS and FTICR MS. *Analytical Chemistry*, Vol. 81, No. 24, pp. 9941-8847, ISSN 1520-6882.

Goodacre, R.; Berkeley, R.C. & Beringer, J.E. (1991). The use of pyrolysis—mass spectrometry to detect the fimbrial adhesive antigen F41 from Escherichia coli HB101 (pSLM204). *Journal of Analytical and Applied Pyrolysis*, Vol. 22, No. 1-2, pp. 19-28, ISSN 0165-2370.

Galipo, R.C.; Egan, W.J.; Aust, J.F.; Myrick, M.L. & Morgan, S.L. (1998). Pyrolysis gas chromatography/mass spectrometry investigation of a thermally cured polymer. *Journal of Analytical and Applied Pyrolysis*, Vol.45, No.1, pp.23-40, ISSN 0165-2370.

García-Reyes, J.F.; Mazzoti, F.; Harper, J.D.; Charipar, N.A.; Oradu, S.; Ouyang, Z.; Sindona, G. & Cooks, R.G. (2009). Direct olive oil analysis by low-temperature plasma (LTP) ambient ionization mass spectrometry. *Rapid Communications in Mass Spectrometry*, Vol.23, No.19, pp. 3057–3062, ISSN 1097-0231

Haddad, R.; Sparrapan, R.; Kotiaho, T. & Eberlin, M.N. (2008). Easy ambient sonic-spray ionization-membrane interface mass spectrometry for direct analysis of solution constituents. *Analytical Chemistry*, Vol.80, No.3, pp. 898-903, ISSN 1520-6882.

Holzer, G.; Bourne, T.F. & Bertsch, W. (1989). Analysis of in situ methylated microbial fatty acid constituents by Curie-point pyrolysis—gas chromatography—mass spectrometry. *Journal of Chromatography A*, Vol.468, No.181, pp.181-190, ISSN 0021-9673.

Hsu, C.S.; Dechert, G.J.; Robbins, W.K. & Fukuda, E.K. (2000). Naphthenic Acids in Crude Oils Characterized by Mass Spectrometry. *Energy & Fuels*, Vol.14, No.1, pp. 217-223, ISSN 1520-5029.

Hsu, C.S. & Green, M. (2001). Fragment-free accurate mass measurement of complex mixture components by gas chromatography/field ionization-orthogonal acceleration time-of-flight mass spectrometry: an unprecedented capability for mixture analysis. *Rapid Communications in Mass Spectrometry*, Vol.15, No.3, pp. 236–239, ISSN 1097-0231

Hsu, C.S.; Hendrickson, C.L.; Rodgers, R.P.; McKenna, A.M. & Marshalla, A.G. (2011). Petroleomics: advancedmolecular probe for petroleum heavy ends. *Journal of Mass Spectrometry*, Vol.46, No.4, pp.337-343, ISSN 1096-9888

Hsu, H.J.; Kuo, T.L.; Wu, S.H.; Oung, J.N. & Shiea, J. (2005). Characterization of Synthetic Polymers by Electrospray-Assisted Pyrolysis Ionization-Mass Spectrometry. *Analytical Chemistry*, Vol.77, No.23, pp. 7744-7749, ISSN 1520-6882.

Hsu, H.J.; Oung, J.N.; Kuo, T.L.; Wu, S.H. & Shiea, J. (2007). Using electrospray-assisted pyrolysis ionization/mass spectrometry for the rapid characterization of trace polar components in crude oil, amber, humic substances, and rubber samples. *Rapid Communications in Mass Spectrometry*, Vol.21, No.3, pp. 375-384, ISSN 1520-6882.

Hua, R.X.; Wang, J.H.; Kong, H.W.; Liu, J.; Lu, X. & Xu, G.W. (2004). Analysis of sulfur-containing compounds in crude oils by comprehensive two-dimensional gas chromatography with sulfur chemiluminescence detection. *Journal of Separation Science*, Vol.27, No.9, pp.691–698, ISSN 1615-9314

Huang, M.Z.; Yuan, C.H.; Cheng, S.C.; Cho, Y.T. & Shiea, J. (2010). Ambient ionization mass spectrometry, *Annual Review of Analytical Chemistry*, Vol.3, pp. 43-65, ISSN 1936-1335.

Klein, G.C.; Rodgers, R.P.; Teixeirac, M.A.G.; Teixeirad, A.M.R.F. & Marshallb, A.G. (2003). Petroleomics: Electrospray ionization FT-ICR mass analysis of NSO compounds for correlation between total acid number, corrosivity, and elemental composition. *Fuel Chemistry Division Preprints*, Vol.48, No.1, pp. 14–15, ISSN 0569-3772.

Marshall, G.L. (1983). Pyrolysis-mass spectrometry of polymers—II. Polyurethanes. *European Polymer Journal*, Vol.19, No.5, pp. 439-444, ISSN 0014-3057.

Moss, C.W.; Dees, S.B. & Guerrant, G.O. (1980). Gas-liquid chromatography of bacterial fatty acids with a fused-silica capillary column. *Journal of Clinical Microbiology*. Vol.12, No.1, pp.127-130, ISSN 1098-660X.

Manion, J.A.; McMillen, D.F. & Malhorta, R. (1996). Decarboxylation and coupling reactions of aromatic acids under coal-liquefaction conditions. *Energy & Fuels*, Vol.10, No.3, pp.776-788, ISSN 1520-5029.

Nyadong, L.; McKenna, A.M.; Hendrickson,C.L.; Rodgers, R.P. & Marshall, A.G. (2011). Atmospheric pressure laser-induced acoustic sesorption chemical ionization fourier transform ion cyclotron resonance mass spectrometry for the analysis of complex mixtures. *Analytical Chemistry*, Vol.83, No.5, pp. 1616-1623, ISSN 1520-6882.

Panda, S.K.; Andersson, J.Y. & Schrader, W. (2007). Mass-spectrometric analysis of complex volatile and nonvolatile crude oil components: a challenge. *Analytical and Bioanalytical Chemistry*, Vol.389, No.2, pp.1329–1339, ISSN 1618-2650

Purcell, J.M.; Hendrickson,C.L.; Rodgers, R.P. & Marshall, A.G. (2006). Atmospheric pressure photoionization fourier transform ion cyclotron resonance mass spectrometry for complex mixture analysis. *Analytical Chemistry*, Vol.78, No.16, pp. 5906-5912, ISSN 1520-6882.

Robb, D.B.; Covey, T.R. & Bruins. A.P. (2000). Atmospheric pressure photoionization: an ionization method for liquid chromatography-mass Spectrometry. *Analytical Chemistry*, Vol.72, No.15, pp. 3653–3659, ISSN 1520-6882.

Robins, C. & Limbach, P.A. (2003). The use of nonpolar matrices for matrix-assisted laser desorption/ionization mass spectrometric analysis of high boiling crude oil fractions. *Rapid Communications in Mass Spectrometry*, Vol.17, No.24, pp. 2839–2845, ISSN 1097-0231

Snyder, A.P.; McClennen, W.H.; Dworzanski, J.P. & Meuzelaar, H.L.C. (1990). Characterization of underivatized lipid biomarkers from microorganisms with pyrolysis short-column gas chromatography/ion trap mass spectrometry. *Analytical Chemistry*, Vol.62, No.23, pp.2565-2573, ISSN 1520-6882.

Smith, P.B. & Snyder, A.P. (1993). Characterization of bacteria by oxidative and non-oxidative pyrolysis—gas chromatography/ion trap mass spectrometry. *Journal of Analytical and Applied Pyrolysis*. Vol. 24, No.3, pp.199-210, ISSN 0165-2370.

Shute, L.A.; Gutteridge, C.S.; Norris, J.R. & Berkeley, R.C.W. (1984) Curie-point pyrolysis mass spectrometry applied to characterization and identification of selected bacillus species. *Journal of General Microbiology*. Vol.130, pp.343-355, ISSN 0022-1287.

Shiea, J.; Wang. W.S.; Chen, C.H. & Chou, C.H. (1996). Analysis of a reactive dimethylenedihydrothiophene in methylene chloride by low-temperature atmospheric pressure ionization mass spectrometry. *Analytical Chemistry*. Vol.68, No.6, pp.1062-1066, ISSN 1520-6882.

Stanford, L.A.; Kim, S.H.; Klein, G.C.; Smith, D.F.; Rodgers, R.P. & Marshall. A.G. (2007). Identification of water-soluble heavy crude oil organic-acids, bases, and neutrals by electrospray ionization and field desorption ionization fourier transform ion cyclotron resonance mass spectrometry. *Environmental Science & Technology*, Vol.41, No.8, pp. 2696–2702, ISSN 1520-5851

Tang, L. & Kebarle, P. (1991). Effect of the conductivity of the electrosprayed solution on the electrospray current. Factors determining analyte sensitivity in electrospray mass spectrometry. *Analytical Chemistry*. Vol.63, No.23, pp. 2709-2715, ISSN 1520-6882.

Vaclavik, L.; Cajka, T.; Hrbek, V. & Jana Hajslova, J. (2009). Ambient mass spectrometry employing direct analysis in real time (DART) ion source for olive oil quality and authenticity assessment. *Analytica Chimica Acta*, Vol.645, No.1-2, pp.56–63, ISSN 0003-2670.

Von Mühlen, C.; Zini, C.A.; Caramao, E.B. & Marriott, P.J. (2006). Applications of comprehensive two-dimensional gas chromatography to the characterization of petrochemical and related samples. *Journal of Chromatography A*, Vol.1105, No.1-2, pp.39–50 ISSN 0021-9673

Williamson, J.E.; Cocksedge, M.J. & Evans, N. (1980). Analysis of polyurethane and epoxy based materials by pyrolysis—mass spectrometry. *Journal of Analytical and Applied Pyrolysis*. Vol.2, No.3, pp. 195-205, ISSN 0165-2370.

Wu, C.; Qian, K.; Nefliu, M. & Cooks, R.G. (2010). Ambient analysis of saturated hydrocarbons using discharge-induced oxidation in desorption electrospray ionization. *Journal of the American Society for Mass Spectrometry*, Vol. 21. No.2, pp. 261-267, ISSN 1044-0305.

Yang, R.; Liu, Y.; Wang, K. & Yu, J. (2003). Characterization of surface interaction of inorganic fillers with silane coupling agents. *Journal of Analytical and Applied Pyrolysis*. Vol.70, No.2, pp.413-425, ISSN 0165-2370.

Yuri, E. Corilo, Y.E.; Vaz, B.G.; Simas, R.C.; Nascimento, H.D.L.; Klitzke, C.F. & Pereira, R.C.L.; Bastos, W.L.; Neto, E.V.S.; Ryan P. Rodgers, R.P. & Eberlin, M.N. (2010). Petroleomics by EASI(±) FT-ICR MS. *Analytical Chemistry*. Vol.82, No.10, pp.3990–3996, ISSN 1520-6882.

Zhan, D. & Fenn, J.B. (2000). Electrospray mass spectrometry of fossil fuels. *International Journal of Mass Spectrometry*, Vol.194, No.2-3, pp. 197-208, ISSN 1387-3806.

7

Determination of Metal Ions in Crude Oils

M.Y. Khuhawar[1], M. Aslam Mirza[2] and T.M. Jahangir[1]
[1]*Institute of Advanced Research Studies in Chemical Sciences, University of Sindh, Jamshoro,*
[2]*Mirpur University of Science & Technology (MUST), Mirpur, AJ&K, Pakistan*

1. Introduction

Crude oil is complex mixture of hydrocarbons that occur in the earth in liquid form. It constitutes an important part of primary fossil fuels. Crude oil was used as a medicine by the ancient Egyptians, presumably as wound dressing, liniment and laxative. Several centuries later, Spanish explorers discovered crude oil in Cuba, Mexico, Boliva and Peru. The industrial revolution brought increasing demand for cheaper and convenient source of energy. Crude oil (liquid petroleum) was easily transportable source of energy, concentrated and flexible from of fuel. At the beginning of the 20th century the industrial revolution had progressed to the extent that the oil industry became the major supplier of the energy, largely because of the advent of automobile. The oil achieved a primary importance as an energy source on which the world economy depends. The growth in the energy production during the 20th century was unprecedented and is the major contributor to the growth. On the time scale within the human history, the utilization of oil as a major source of energy will be affair of a few centres, but it will have profound effect on world industrialization.

The crude oils are mostly based on two elements carbon and hydrogen and almost all crude oil ranges from 82-87% carbon and 12-15% hydrogen. Crude oil contains three basic chemical series: paraffins, naphthenes, and aromatics. The crude oils from different sources may not be completely identical (Evans et al, 1971).

The paraffins are also called methane series, and comprises most common hydrocarbons in crude oil. The paraffins that are liquid at normal temperature boil between 40-200 ^{o}C. The naphthenes are saturated closed ring series and are important part of all liquid refinery products. The aromatics are unsaturated closed ring series. Benzene is most common of the series and is present in most of the crude oils, but aromatics constitute a small fraction of all crudes.

The crude oil also contains sulphur, nitrogen and oxygen in small quantities. Sulphur is the third most abundant constituent of crude oil. The total sulphur in crude oil varies from below 0.05% up to 5% or more. Generally greater the specific gravity of the crude oil, higher is its sulphur content. The oxygen contents of the crude oil are usually less than 2%. Nitrogen is present in most of the crude oils, usually in quantities of less than 0.1% (Britannica online).

Preliminary fractionation of crude oil according to chemical class is carried out before identification of individual components. Several such fractionation and isolation schemes are available (Rudzinski and Aminabhavi, 2000) depending on the type of crude oil under investigation. One of the separation scheme is based on SARA method, which has name from the fractions produced, namely saturates (S), aromatics (A) resins (R) and asphaltenes (A). The sample is adsorbed on the silica (Isitas-Flores et al., 2005), (Andersen et al.,1997); (Goreli et al., 2008) or alumina, followed by the selective elution of the components with increasingly polar solvent (Seidl et al., 2004), (Sharma et al., 1998), (Seidl et al., 2004). Mansfield et al (1999) have reviewed the crude oil separation and identification including SARA method. HPLC and infra red spectroscopy have also been used for SARA characterization (Fan and Buckley, 2002), (Aske et al ., 2001).

Asphaltenes consist of polar fraction of the crude oil comprising polyaromatics, heteroaromatics and various metals (Kaminski et al., 2000).

2. Metal ions in crude oil

The metals present in the crude oils are mostly Ni(II) and VO(II) porphyrins and non-porphyrins. Other metal ions reported form crude oils, include copper, lead, iron, magnesium, sodium, molybdenum, zinc, cadmium, titanium, manganese, chromium, cobalt, antimony, uranium, aluminum, tin, barium, gallium, silver and arsenic. Metalloporphyrins are among the first compounds identified to belong to biological origin. Treibs et.al (1936) proposed that plant chlorophylls transformed into the geoporphyrins. Metalloporphyrins in crude oils are of fundamental interest from geochemical context for better understanding geochemical origin of petroleum source. The information could be useful for catagenetic oil formation, maturation of organic matter, correlation, depositional and environmental studies. Vanadium and nickel metalloporphyrins are present in large quantity in heavy crude oils. Their presence cause many problems because such metals have a deleterious effect on the hydrogenation catalysts used in upgrading processes (Pena et al., 1996).

Among the porphyrins encountered in the crude oils, etioporphyrins (etio) and dexophylloerithroetioporphyrin (DPEP), and their homologues are more frequently observed (Baker and Louda, 1988), (Barwise and Roberts, 1984). The complexity of porphyrin mixtures have made the isolation of these pigments difficult, but the improved chromatographic and spectroscopic techniques have made possible the separation and identification of a number of metalloporphyrins (Les Ebdon et al.,1994). The identification of Ni and V porphyrin was quite earlier (Treibs et al., 1936) but the organic forms of other metals in crude oils was achieved only later, with the advent of hyphenated techniques, e.g. HPLC or GC coupled to AAS or ICP-MS for elemental detection. The porphyrins of Co, Cr, Ti and Zn were identified in oil shales by HPLC-ICP-MS (Les Ebdon et al.,1994).

It was earlier observed that V/Ni or V/(V+Ni) ratio was constant in crude oils of common rock source and was dependent on geological age of the rocks (Ball et al., 1960) and this ratio was used for tracing source effects (Shahristani and Al-Alyia,1972), (Gayer et al., 2002). The studies on the thermal evolution of the major VO complex of DPEP to etio indicated a maturity dependence (Didyk et al., 1975), (Barwise et al., 1987), thus suggesting these compounds as biomarkers (Peters et al., 2004), (Duyck et al., 2007).

The organic forms of metalloporphyrins are described as tetrapyrrolic complexes with structure similar to chlorophyll and heme (Treibs et al.,1936), but the chemical nature of nonporphyrins is not well established. These are polar compounds and largely exist as cations of organic acids. Nitrogen, oxygen and sulphur can all act as donor atoms in various combinations in nonporphyrins (Amorin et al., 2007). These are predominantly associated with the asphaltenes. Some of the elements in crude oils may be present in associated mineral matter or entrained formation waters. A number of metal complexes may be associated with humic substances that have large capacities for metal complexations (Choudhry et al., 1983) which may be precursors to kerogen.

3. Determination of metals in crude oils

There is a need to determine the trace metals in the crude oils quantitatively because of their importance in the geochemical characterization of its source and origin. Trace metals have been used as a tool to understand the depositional environments and source rock (Alberdi-Genolet and Tocco, 1999). The metal ions and their ratios have been observed as a valuable tool in oil-oil correction and oil-source rock correlation studies (Barwise et al., 1990), (Akinlua et al., 2007). The trace metals are also indicated as biomarkers of the source rocks (Odermatt and Cruriale, 1991). The determination of metal ions in crude oils has environmental and industrial importance. The metal ions like vanadium, nickel, copper and iron, behave as catalyst poisons during catalytic cracking process in refining of crude oil. The metal ions are released in the environment during exploration, production and refining of crude oil. The determination of mercury content in crude oil is also important for petroleum industry, because the metal can deposit in the equipment, which could affect the maintenance and operation (Wllhelm et al., 2006). It is therefore considered necessary to know the concentration of metals in the oils for meaningful impact assessment.

Metals and metalloids may be naturally found in the crude oils and these could be added during production, transportation and storage. In general these elements are present in the crude oils as inorganic salts (mainly as chloride and sulphate of K, Mg, Na and Ca), associated with water phase of crude oil emulsions, or as organometallic compounds of Ca, Cu, Cr, Mg, Fe, Ni, Ti, V and Zn adsorbed in water-soil interface acting as emulsion stabilizers (Speight et al., 2001).

Molecular absorption spectrophotometry (Milner et al., 1952), atomic absorption spectrometry (AAS), (Langmyhr and Aadalen,1980), inductively coupled plasma-optical emission spectrometry (ICP-OES) (Fabbe and Ruschak, 1985), inductively coupled plasma mass spectrometry (ICP-MS) (Lord et al., 1991), high performance liquid chromatography (HPLC) (Khuhawar and Lanjwani, 1996), gas chromatography (GC) (Delli and Patsalide, 1981), capillary electrophoreses, (Mirza et al., 2009) and X-ray fluorescence spectroscopy (XFS) (Vilhunen et al., 1997) methods have been reported for the determination of metals in crude oils.

3.1 Sample preparation

Crude oil is a complex matrix of varying viscosities and mixed phases (organic, water and particulate matter) and therefore not an ideal matrix for analysis. The determination of the metals in crude oil requires pretreatment to the sample before presentation to the instrument. This is the stage where most of the errors occur and is time consuming. The selection of a particular procedure depends upon (1) analytical technique to be employed, (2) nature and the number of the samples to be analyzed, (3) desired degree of precision and accuracy required, (4) availability of the equipment, materials and reagents and (5) the cost of analysis (Oliveira et al., 2003). It is generally desired that analysis is completed within shortest time with minimum contamination, using smallest quantities of the reagents and the samples and little residues and waste generation (Amorin et al., 2007).

Metal determination in crude oil is carried out by using dry ashing, dilution in organic solvents (Annual Book of ASTM, 2000, 2002) or using micro-emulsions (Souza et al., 2006), (Santella et al., 2008). The use of micro-wave radiation as a potential sample preparation technique has been applied due to high efficiency of heat transfer and sample digestion efficiency (Mello et al., 2009), (Perira et al., 2010). Literature data concerning metal and metalloid determinations in the crude oil by direct sample introduction is also available (Anselmi et al., 2002), (de Oliveira et al., 2006).

3.1.1 Sample decomposition by ashing

Drug ashing is used for complete elimination of organic matter, before analytical determination and is based on the ignition of the organic matrix in air or in the stream of oxygen, followed by the dissolution of the residue in an acid medium. This is one of the cheapest sample preparation procedure. Larger quantities of the sample could be used and the analyte could be concentrated into small volume of dilute mineral acid (HCl or HNO_3). This also make possible the use of aqueous standards for the calibration of equipment. The main disadvantages of the dry ashing procedure for crude oils are the risks of contamination or loss of the analyte due to the formation of volatile compounds. (Ekanem et al., 1998). The addition of sulphur containing compounds has long been used for avoiding losses of Ni and V by volatization during ashing (Udoh et al., 1992). The dry ashing takes longer time for sample preparation with low sample preparation frequency.

3.1.2 Sample decomposition by wet digestion

The sample decomposition of organic constituents by wet digestion is achieved by the use of oxidizing agents prior to analyte determination. Normally concentrated acids are applied under heating, and the important aspects for consideration are the strength of the acids, their oxidizing and complexing power, their boiling points, the solubility of the resulting salts, safely in manipulation and purity (Amorim et al., 2007). The acids and oxidizing agents used for oil samples include mixtures of nitric, hydrochloric and sulphuric acids and hydrogen peroxide. This is also one way to overcome the difference in response caused by the presence of different analyte compounds in the fuel by their conversion in water soluble salts. The procedure is generally performed with larger volume of oxidizing acids (e.g. 10 ml of acid per 0.5 g of sample) and time for complete decomposition is long (up to several days). The use of sulphuric acid in wet digestion procedure for trace metal determination in oils, suggested in literature has been reapproved as standard methods for the determination of Ni, V and Fe in crude oils and residual fuels (Standard test methods, 2005).

3.1.3 Wet digestion assisted by microwave radiation

Wet digestions assisted by microwave radiation have been observed as safer and efficient. The procedures minimize contamination and amount of reagents for sample preparations (Ozcan and Akman, 2005), (Trindade et al., 2006). Microwave heating enhances the efficiency of acid digestion. US-EPA method 3051 (1994) reported microwave assisted acid digestion procedure for 0.5 g of oil and other samples with 10 ml of nitric acid with 10 min. of heating. Alvarado et al (1990) optimized a procedure for the microwave digestion of crude oil samples using different proportions of HNO_3 and H_2SO_4 for the analysis of Cr, Cu, Fe, Mn, Ni, Na and Zn. Bettinelli et al (1995) examined microwave digestion of fuel oils in a high pressure closed vessel and observed a decrease in sample decomposition time due to higher temperature achieved. Using the conditions, difficult samples were also decomposed completely. Munoz et al (2007) evaluated the use of different microwave ovens for the decomposition of crude oils and diesel fuel to determine the contents of Cu, Pb, Hg and Zn in the diagestates. A focused-microwave (FM) oven using $H_2SO_4/HNO_3/H_2O_2$ operated at atmospheric pressure, and a closed vessel microwave (CVM) oven using HNO_3/H_2O_2, operated under pressure in a vessel, were evaluated. Better detection limits were reported for FM digested solutions with 0.8 - 1.0 g of sample, in contrast low quantities (0.10 - 0.28 g) were used when pressurized vessels were used. However the loss of Hg was verified when samples were decomposed in the FM oven. Sant' Ana et al (2007) reported focused microwave assisted procedure for the wet acid dissolution of diesel oil for the determination of metals in samples. The dissolution process was monitored by measuring residual carbon content (RCC) after application of digestion program. The dissolution program comprised three steps: (1) carbonization with H_2SO_4, (2) oxidation with HNO_3 and (3) final oxidation with H_2O_2. It was reported that the first step was important on the dissolution process. At optimized conditions it was possible to digest 2.5 g diesel oil with a 40 min. heating program. At these conditions, residual carbon content was lower than 5%. Optimized methodology was used in the determination of Al, Cu, Fe, Ni and Zn in three diesel oil samples. Pereira et al (2010) described a method for light and heavy crude oil digestion by microwave induced combustion in the closed vessel for the determination of Ag, As, Ba, Bi, Ca, Cd, Cr, Fe, K, Mg, Li, Mn, Mo, Ni, Pb, Rb, Se, Sr, Tl, V and Zn. Conventional microwave assisted digestion in pressurized vessels were also used for results comparison. Accuracy of microwave-induced combustion method was evaluated for As, Ba, Ni, Se, V and Zn using certified reference material with similar matrix. Recovery tests were better than 97% using 2 mol/L nitric acid as absorbing solution. Both sample preparation techniques were suitable for crude oil digestion, but microwave induced combustion was preferable in the view of possibility of using diluted nitric acid as absorbing solution. Mello et al (2009) applied microwave induced combustion for the determination of Ni, V, S from crude oil distillation residues. The results obtained agreed with certified values for Ni, V and S within 99 to 101% using 2 mol/L HNO_3 as absorbing solution.

3.1.4 Dilution with organic solvents

The dilution of crude oils and derivatives with organic solvents is an attractive sample preparation method, because it is simple and rapid and could be used for the determination of the metals by spectro-analytical techniques. The solvents commonly used are xylene, kerosene, methyl isobutyl ketone (MIBK), n-hexane, dimethylbenzene, 1-propanol and mixture of these solvents. The procedure is widely used in industry (Batho et al., 1993), (Botto and Zhu, 1996). Direct dilution of crude oils and residual fuels with an organic solvent for the determination of Ni, V, Fe and Na is proposed in ASTM standard test method (Standard test methods ASTM D 5863-00 a (2005). Ni and V in crudes and heavy crude fractions were determined after dilution in xylene (Fabec and Ruschak, 1985). The solubilisation of crudes and its burning residues was carried out with MIBK for the determination of V as well as Cd and Ni (Guidr and Sneddon, 2002), (Hammond et al., 1998). Bethinelli and Tillarelli (1994) validated a procedure for the determination of Ni and V in fuel oils based on the 1 + 9 dilution with xylene and calibration was with constant organometallic standards, using base oil for matching the viscosity on fuel oil with known metal contents. The procedure was compared with a series of independent methods in the analysis of six samples with different metal contents. The results very close to the consensus values were reported.

Dilution with organic solvents, in spite of simplicity has number of limitations (1) Analyte concentration may change due to the evaporation of the solvent or due to the adsorption on the walls of container (Campos et al., 2002), (2) The problems of plasma destabilization or extinction in case of ICP technique and the contamination of the instrument with carbon residues persist, as the organic load is not reduced with dilution, (3) The toxicity of many organic solvents requires care to avoid and health hazard for laboratory personnel, (4) different metal - organic compounds often exhibit different sensitivity and require use of expensive organic standards for calibration (Vale et al., 2004). The standards may also show a sensitivity difference from that of the metal organic compounds present in the fuel (Teserovsky and Arpadjan, 1991).

3.1.5 Preparation of emulsion

A fuel sample may be modified by formation of emulsions or micro-emulsion (Three component system). When two immiscible liquids are stirred, a macro emulsion is obtained either oil in water (o/w, droplets of oil in water) or water in oil (w/o, droplets of water in oil), depending on the dispersed phase. In emulsion and micro-emulsion the fuel is dispersed in the aqueous phase as micro drops stabilized by micelles or vesicles generated by the addition of a detergent. The emulsion that is formed is mainly related to the formulation and to the lesser degree to the o/w ratio. In the case of micro-emulsion without detergent a co-solvent allows the formation of a homogenous and long term stable three component solution containing the aqueous and organic phase (das Gracas Andrade Koen et al., 2007), (Pelizzeti and Pramauro, 1985). The procedure enables to use aqueous standards for calibration without the need of sample mineralization. A surfactant with a suitable hydrophilic-lyophilic balance is used in the preparation of emulsion, which permits relatively high solubility between the immiscible phases. In the case of detergentless microemulsions, an alcohol of low molecular weight is added as co-solvent (Cardarelli et al., 1986). Emulsions and micro-emulsions have been successfully applied for the preparation of oil samples, due to homogenous dispersion and stabilization of the oil micro-droplets in aqueous phase, which reduces oil viscosity and the organic load of the system.

Kumat and Gangadharen (1999) applied Triton X-100 emulsions to the determination of V, Co, Ni, As, Hg and Pb in naphtha. Murillo and Chirinos (1994) examined non-ionic emulsifier, polyoxyethylene nonylphenylether for heavy crude oils because of its slightly higher hydrophile-lipophile balance, which enabled higher solubility in water through hydrogen bonding. Souza and da Silveira (2006) reported detergentless emulsions for the determination of elements in crude oils by using acidified water for element stabilization and propan-1-ol as a co-solvent. Meeravali and Kumar (2001) determined Ni and V in naphtha and fuel oils after emulsion formation. The oil samples were diluted in toluene, and this solution was emulsified by stirring with 3% Triton X-100 in water. Calibrations were prepared with organometallic standards following the same procedure. The emulsions were stable from 20 to 50 min. Good agreement between found and certified results was reported. Vale et al (2004) optimized the emulsification of petroleum for the determination of the nickel. They stabilized samples and analytic solutions as an o/w emulsion consisting of xylene, Triton X-100 and water. Ultrasonic bath was used in the emulsification process and the mixture was further homogenized just before the measurement by manually flushing them with a micropipette. Aucetio et al (2004) determined V in the asphaltene petroleum fraction. The analytic solution was stabilized by mixing with propan-1-ol and 6 mol/L HNO_3 forming a detergentless micro-emulsion. The micro-emulsion was immediately formed and was reported to be stable upto 80 h. Calibration was performed by spiking in organic V in the same micro-emulsion medium. A comparison with established methods (acid digestion or dilution in organic solvent) shows that emulsion or micro-emulsion methodology presents advantages in terms of simplicity of sample preparation, total analysis time, long term sample stability and the use of inorganic standards for calibration instead of expensive metal organic standards (Anorim et al., 2007).

3.1.6 Direct analysis of crude oil

Direct analysis using little or no sample preparation has the advantage of time saving and minimum risk of analytic loss. This technique has been applied for highly viscous liquids and has been examined for the determination of Ni (Brandao et al., 2006), Ni and V (Silva et al., 2007) and Cu, Fe and V (Brandao et al., 2007) in oil samples. However there are some general problems such as volatility, flammability and immiscibility with water. In addition to the problems related to the complexity of the matrix, organic standards, which are indispensable in case of direct sample introduction, are unstable and there are no certified reference materials available for these samples. It is therefore necessary to compare the accuracy of the developed method with results obtained with independent technique, particularly with respect to the sample preparation.

3.1.7 Analyte extraction

Extraction for the analyte from the fuel can be used for sample preparation, which combines the advantages of separating the analyte from the matrix, transferring it to an aqueous phase and may also result in preconcentration. Liquid-liquid extraction procedures present as main advantage for their simplicity. Akinlua and Smith (2010) reported the extraction of trace metals from petroleum source rock by superheated water and the conditions for maximum yield were determined. The optimum temperature for superheated water extraction of the metals from petroleum source rocks was 250°C. The extraction time was 30 min. The leaching of Cd, Cr, Mn and Ni had better yield with superheated water, while V had better yield with acid digestion.

Solid phase extraction is a useful separation and preconcentration procedure for the determination of trace metals in fuels. It is based on the partition between a liquid (sample) and a solid phase (sorbent), which can be unloaded, load on chemically modified with organofunctional groups (Koen et al., 2006). After pre-concentration the analyte is recovered by elution with an appropriate solvent or directly determined in the solid phase.

3.2 Atomic absorption spectrometry

3.2.1 Flame Atomic Absorption Spectrometry (FAAS)

FAAS indicates inherently low sensitivity for metal determinations and a few reports are available in the literature involving the direct analysis of crude oils by FAAS. A sufficiently large sample mass may compensate for lower FAAS sensitivity with longer analysis time. Platteau and Carriillo (1995) determined Fe, Na and Ni in crude oils by FAAS. Ni, V and Fe in crude oils and residual fuels have been determined by FAAS after ashing with H_2SO_4 (standard test methods ASTM 2005). Fabec and Ruschak (1985) determined Ni and V in crudes and heavy crude fraction by FAAS after dilution in xylene. Guidr and Sneddon (2002) determined V by nitrous oxide : acetylene flame AAS and (Hammond et al., 1998) analyzed Cd, Pb and Ni by FAAS in crudes and its burning residues after solublization in MIBK. Osibanjo et al (1984) determined Ni, Cu, Zn, Na, Pb, Cd and Fe by FAAS in petroleum crudes after dilution with toluene - acetic acid mixture. Calibration was performed with inorganic salts and by analyte addition. Sebor et al (1982) discussed FAAS analysis of crude oils using dilution methods with different solvents or solvent mixtures. Different organic compounds of the same element present different responses in the flame, no matter if an air or a nitrous oxide acetylene flame was used. This led to calibration difficulties as well different responses depending on the organometallic composition of the sample. De la Guardia and Lizondo (1993) determined Ni in fuel oil by FAAS using 4% v/v oil-in-water emulsion.

3.2.2 Electrothermal Atomic Absorption Spectrometry (ET-AAS)

ET-AAS with a graphite furnace is a useful analytical technique for metal analysis from crude oils, because of its high sensitivity and capability to deal with organic loads. In addition ET-AAS requires only small amount of sample. ET-AAS make possible direct analysis of crude oils, because it allows complete elimination of organic matrix, if an appropriate heating program and suitable chemical modifiers are used (das Graces Andrade Koren, 2007).

Turunen et al (1995) determined As, Cd, Cr, Cu, Mn, Ni, Pd and V in heavy oils by ET-AAS after acid digestion with HNO_3-H_2SO_4 mixture. Alvarado et al (1990) analyzed Cr, Cu, Fe, Mn, Ni and V by ET-ASS after micro-wave digestion of various crude oil samples. Bruhn and Cabalin (1983) proposed determination of Ni in gas oil after dilution with xylene by ET-AAS, analyte addition was used for quantitation. Gonzalez et al (1987) determined V, Ni, Fe and Pb in crude and fuel oils following dilution with xylene and MIBK. The quantitation was by analyte addition calibration curves. Bermejo-Barrera (1991) determined V in petroleum samples by ET-AAS after dissolution in xylene. They proposed that the crude oil is diluted to the extent that matrix interference is eliminated. Thomainidis and Piperaki (1996) examined the behavior of a series of chemical modifiers for the determination of V in a water and oil matrix. In the determination of V in a multi-element standard diluted with MIBK, the addition of Pt as modifier enhanced the pyrolysis temperature from 1000 to 1400 ^oC with improvement in the sensitivity. Stigter et al (2000) determined Cd, Cr and Cu in crude oils by ET-AAS using Zeeman-effect background correction as well as oxygen ashing during the pyrolysis step. Finally they used mixture of toluene and acetic acid 4:1 v/v as a solvent for dissolution of the samples. Nakamoto et al (2004) determined V in heavy fuel oils using tungsten coated graphite furnace. The effect of sulphur interference on the determination was examined. The sulphur content less than 1% can be tolerated. An agreement between proposed and comparative procedures was reported. Meeravali and Kumar (2001) analyzed Ni and V in naphtha and fuel oils after emulsion formation by ET-AAS using W-Ir as permanent modifier. The oil samples diluted in toluene was emulsified with 3% Triton X-100 in water. Calibration was performed with organometallic standards prepared in same manner. Burguera et al (2003) analyzed Cr in heavy crude oil and in bitumen in water emulsion by ET-AAS after sample emulsification. Vale et al (2004) optimized the determination of Ni in petroleum using both line source and high resolution continuum source ET-AAS. They stabilized sample and analyte solutions as an emulsion consisting of xylene, Triton X-100 and water. The authors observed a significant Ni loss at pyrolysis temperature of 500 °C, most probably due to the presence of volatile Ni species. However better results were reported at pyrolysis temperature of 400 °C. Aucelio et al (2004) determined V in the asphaltene petroleum fraction by ET-AAS. The solution of the sample was stabilized by mixing with 1-propanol and HNO_3, forming detergentless micro-emulsion. Calibration was performed by spiking inorganic V in the same micro-emulsion medium. Damin et al (2005) determined Ni and V in oil samples by line source ET-AAS, Pd (20 µg) was used as chemical modifier. Good agreement was reported between found and expected values. Quadros et al (2010) determined Ni and V simultaneously as their total and nonvolatile fractions in crude oil samples using high resolution continuum source graphite atomic absorption spectrometry. Determination was carried out at 305.432 nm and 305.633 nm for Ni and V respectively using linear charge - coupled device array detection. Oil-in-water emulsions were used for crude oil sample preparation. Nitric acid was added to emulsion for the determination of total Ni and V concentration. In the absence of acid volatile fraction was lost in the pyrolysis and thermally stable fractions were determined. The concentration of the volatile fraction was obtained by difference. Vale et al (2006) used ET-AAS to differentiate between volatile and nonvolatile Ni and V compounds in crude oil. Two crude oil samples were separated in two steps: firstly the asphaltenes were precipitated with n-heptane. Another portion was loaded on a silica column and eluted with solvents of increasing polarity. Four fractions 1, 2, 3 and 4 were separated. Oil in water emulsions were prepared for determination of Ni and V by ET-AAS. The analysis was carried out without chemical modifier (stable compounds) and with 20 µg Pd (Total Ni and V) and the volatile fraction was calculated by difference (Fig.1). Brandeo et al (2007) proposed a procedure for direct determination of Cu, Fe and V in petroleum samples by ET-AAS using a solid sampling accessory without any sample pretreatment or dilution. A Pd + Triton X-100 solution was used as chemical modifier. The limits of detection at the optimized conditions were 10, 200 and 800 pg for Cu, Fe and V respectively for sample masses ranging 0.10-3.0 mg. Aqueous calibrations were used for quantitation. Luz and Oliveira (2011) determined Cr, Fe, Ni and V in crude oil using emulsion sampling by ET-AAS. The emulsion was prepared in a mixture of n-hexane + Triton X-100, $Mg(NO_3)_2$ was used as chemical modifier. The reliability of the proposed method was checked by fuel oil standard reference material analysis. Dittert et al (2010) simultaneously determined Co and V in crude oils by high resolution continuum source ET-AAS, V and Co were determined at 240.674 nm and 240.725 nm. The samples were analyzed directly without dilution; Pd and Triton X-100 were added as chemical modifiers. Aqueous solutions were used for calibration. Two certified oil reference materials were analyzed and results were in agreement with certified and reported values.

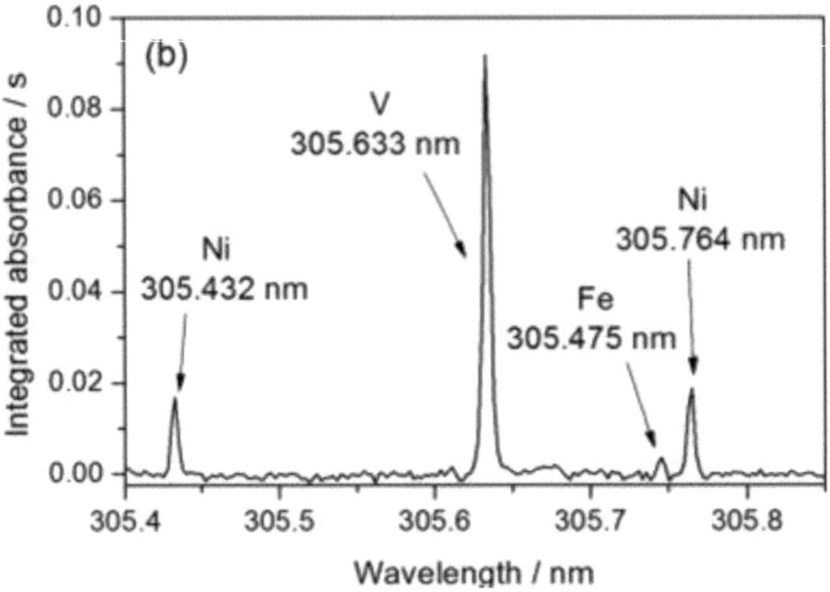

Fig. 1. Absorbance spectrum for the OB-2 crude oil sample emulsion in the vicinity of the vanadium secondary line at 305.633nm, with a pyrolysis temperature of 1000 °C, Wavelength resolved integrated absorbance spectrum (Quadros et al. 2010) with permission.

3.2.3 Chemical vapour generation AAS

The elements Hg, As and Sb are present in the fuels at low concentrations and chemical vapour generation (CVG) technique can be applied to provide required sensitivity. However CVG when applied to crude oils is susceptible to spectral interference from aromatic organic compounds in the gas stream. Cold vapour (CV) technique for mercury and hydride generation (HG) technique for As & Sb requires the analyte to be present in inorganic form in a defined oxidation state. Therefore the applications of CVG to the analysis of fuels generally requires complete mineralization of the metal-organic compounds and separation from organic phase (das Gracas Andrade Korn, 2007). Campbell and Karnel (1992) determined As, Sb and Se in oil waters by hydride generation AAS, after complete oxidation of the organic matrix using microwave-assisted digestion in the closed system. Puri and Irgolic (1989) determined As in crude oils after extraction of As in boiling aqueous nitric acid, followed by mineralization of the extract with concentrated HNO_3/H_2SO_4 and reduction of arsenate to arsine in hydride generator. Wilhelm and Bloom (2000) have reviewed mercury in petroleum and have described the use of cold vapour AAS for sensitive detection of mercury.

3.3 Inductively Coupled Plasma Optical Emission Spectroscopy (ICP-OES)

ICP-OES has the advantage of multi-elemental detection capability and offers a wide linear dynamic range. However, the introduction of organic solvents, such as fuels cause plasma destabilization or even plasma extinction and the use of ICP accessories may be necessary, such as direct injection nebulizer, ultrasonic nebulizer with micro-porous membrane dessolvator, a thermostated condenser between the spray chamber and the plasma torch or a chilled spray chamber. Alternatively the sample is digested to obtain aqueous solutions (Das Gragas Andrade Korn (2007).

Botto (1993) and Botto and Zhu (1996) determined metal by ICP-OES in petroleum oils after dilution with organic solvents. Souza et al (2006) determined Mo, Zn, Cd, Ti, Ni, V, Fe, Mn, Cr and Co in crude oil by ICP-OES using detergentless emulsions comprising of acidified water for element solublization and propan-1-ol as a cosolvent with the addition of oxygen to the nebulizer gas flow. Fabec and Ruschak (1985) determined Ni and V in crudes and heavy crude fractions by ICP-OES after dilution in xylene. Mello et al (2009) determined Ni, V and S in crude oils distillation residues by ICP-OES after digestion of sample by microwave induced combustion. De Souza et al (2006) determined Mo, Cr, V and Ti in diesel and in used fuel oil by ICP-OES. Detergent and detergentless emulsion sample preparation procedures were evaluated and better results were obtained for detergent emulsions with recoveries ranging from 90.1-106.5%. Borszeki et al (1992) determined Al, Cr, Cu, Fe, Mg, Ni and Pb in oil and petroleum products by ICP-OES using a minitorch. The samples were prepared as aqueous emulsions. Quantitation was with aqueous standards and the results were in good agreement with those obtained using oil standard solutions. Brenner et al (1996) determined lead in gasoline by ICP-OES using argon and argon-oxygen as plasma gas.

3.4 Inductively Coupled Plasma Mass Spectrometry (ICP-MS)

ICP-MS is highly sensitive multielement technique, but the introduction of organic solvents and the compounds in plasma requires special care similar to ICP-OES, because the organic load may destabilize or extinguish the plasma. ICP-MS encounters additional problems of formation of carbon deposit on the sampler and skimmer cone and in the ion lens of the mass spectrometer and spectral interferences owing to carbon based species (Lord et al., 1991), (Brenner et al., 1997). In ICP-MS spectral interferences may also occur as crude oil contains a number of elements at low concentrations (ng/g) levels (Tan and Horlick, 1986). The effect of polyatomic interferences may be minimized by the choice of alternate isotopes or the use of desolvation system to reduce the solvent load (Jakubowski et al., 1992). The problem of the deposition of carbon on the sampler and skimmer cone in ICP-MS was decreased by the addition of O_2 to the argon (Magyar (1986).

Duyck et al (2002) determined Ag, Al, Ba, Cd, Co, Cu, Fe, La, Mg, Mo and Mn in residual fuel oil and crude oils by ICP-MS after dilution of the samples in toluene, using ultrasonic nebulization. Good accuracy was reported for the determinations of the metals. Wondimu et al (2000) analysed residual fuel oil for Ag, Al, As, Ba, Bi, Ca, Cd, Co, Cr, Cu, Fe and Hg by ICP-MS after micro-wave acid decomposition. H_2O_2 was used after acid decomposition for better carbon removal. Lord (1991) determined Li, Al, Ti, V, Mn, Fe, Co, Ni, Cu, Zn, Sr, Mo, Ag, Cd, Sn, Sb, Ba and Pb in crude oils by ICP-MS with mciro-emulsion sample introduction. Kowalewska et al (2005) determined Cu in crude oils and crude oil distillation products by ICP-MS after ashing and micro-wave assisted decomposition of analyte and transferred to aqueous solution. Good recovery of Cu was reported. Kelly et al (2003) determined Hg in crude oils and refined products by cold Vapor ICP-MS after decomposition of the sample by closed system combustion. Botto (2002) analysed crude oil, petroleum naphthas and tars for Na, P, Ti, V, Mn, Fe, Co, Ni, Cu, Zn, As, Y, Mo, Cd, Sn, Sb,

Hg, Pb and Bi by ICP-MS using direct injection technique after the dilution of the samples in xylene. Al-Swaidan (1996) determined Pb, Ni and V in crude oil by coupling sequential injection with ICP-MS. A microemulsion crude oil sampling procedure was used. The quantitation was by standard addition technique using oil soluble organometallic elements. Pecira et al (2010) analyzed light and heavy crude oil for the determination of Ag, As, Ba, Bi, Ca, Cd, Cr, Fe, K, Mg, Li, Mn, Mo, Ni, Pb, Rb, Se, Sr, Ti, V and Zn by ICP-MS after micro-wave induced combustion in closed vessel. Akinlua et al (2008) determined rare earth elements of crude oil from the offshore-shallow water and onshore fields in the Niger delta by ICP-MS. The samples were prepared by acid digestion into colourless aqueous solutions. The concentrations of the detected elements La, Ce, Pr, Nb, Sm, Eu, Gd, Dy, Er and Yb ranged from 0.01-1.58 μg/L with an average of 0.98 μg/L (% RSD < 5).

3.5 Chromatographic techniques

Chromatography is a separation technique and has been used for the determination of metal ions in crude oils. The methods are based on determination of total metal ions, after complete mineralization of the crude oil or as metalloporphyrins using hyphenated techniques..

Quimby et al (1991) described selective detection of volatile Ni, V and Fe metalloporphyrins in crude oil samples. Gas chromatography connected with an atomic emission detector (AED) was used for the simultaneous detection of Ni 301.2 nm, V 292.4 nm and Fe 302.1 nm. Detection limit for these metals ranged from 0.05-5 pg/sec. The presence of volatile forms of these metals in several crude oil samples was confirmed.

Les Ebdon et al (1994) examined high temperature gas chromatography (HTGC) and high performance liquid chromatography (HPLC) coupled to ICP-MS for the determination of geoporphyrins from crude oil. HTGC-ICP-MS was used for rapid identification of Co, Cr, Fe, Ni, Ti, V and Zn metalloporphyrins. Quantitative data was obtained by HPLC-ICP-MS and HPLC with UV absorption detector. Levels of Ni geoporphyrins were in the range 15-20 ug/g. Khuhawar et al (1996) determined V in crude petroleum oils from lower Indus basin oil fields by GC after wet acid digestion. Tetradentate ligand bis(trifluoroacetylacetone)dl-stilbenediimine was used as chelating reagent for elution from GC column BP-1. Khuhawar and Arain (2006) reported liquid chromatographic method for the determination of V in the petroleum oils using 2-acetylpyridine-4-phenyl-3-thiosemicarbazone as derivatizing reagent from Kromasil 100 C-18 column. The vanadium contents were reported 0.32-2.3 ug/g with RSD 1.5-4.5%. Khuhawar et al reported normal phase HPLC procedure for the determination of V from crude petroleum oils using bis(salicylaldehyde) tetramethylethylenediimine as complexing reagent. The V contents in petroleum oils were reported 0.47-0.54 ug/g. Amoil et al (2006) analyzed V, Ni. Fe and Cu by HPLC using 10 cm reversed phase HPLC column. The metals were extracted with 8-hydroxyquinoline from acidic medium. The recoveries of metal ions were reported within 85-99% with HPLC separation time less than 4 min. Khuhawar and Lanjwani (1996) described simultaneous HPLC determination of Cu, Fe, Ni and V in crude petroleum oils based on pre-column complexation of analytes by bis(acetylpivalylmethane)ethylenediimine, followed by solvent extraction and HPLC separation on a reversed phase C-18 column with UV detection at 260 nm. The crude petroleum oils collected from the Indus south basin oil fields were analyzed for metal contents. Caumette et al (2010) coupled size exclusion chromatography (SEC) and normal phase (NP) HPLC with ICP-MS and investigated molecular distribution of Ni and V in crude oils. The metal species in SEC fractions were reported to be sufficiently stable to be collected and preconcentrated to allow the development of a bidimensional chromatography SEC-NP-HPLC -ICP-MS for the probing of the metal distribution in crude oils in terms of molecular weight and polarity. Ellis et al (2011) coupled GC with ICP-MS for the determination of Ni and V in crude oils and its fractions. The method required a transfer line and ICP injector heated at 350 °C for rapid transfer of separated species from GC to the ICP-MS in heated argon gas. Ni and V determination was carried out in different crude oils. Mirza et al (2009) reported the determination of Fe. Co, Ni and V from crude petroleum oil samples after chelation with bis(salicylaldehyde) tetramethylethylenediimine by miceller electrokinetic chromatography (MEKC). Uncoated fused silica capillary was used with an applied voltage of 30 kV with photodiode array detection at 228 nm. SDS was added as miceller medium at pH 8.2 with sodium tetraborate buffer (0.1M). The determinations of Fe, Cu, Co, Ni and V in crude petroleum oils were reported with RSD within 1.1-4.1% (n=3) (Fig.2). Zeng and Uden (1994) used high temperature GC coupled with micro-wave induced plasma atomic emission for the determination of V, Ni and Fe porphyrins in crude oils.

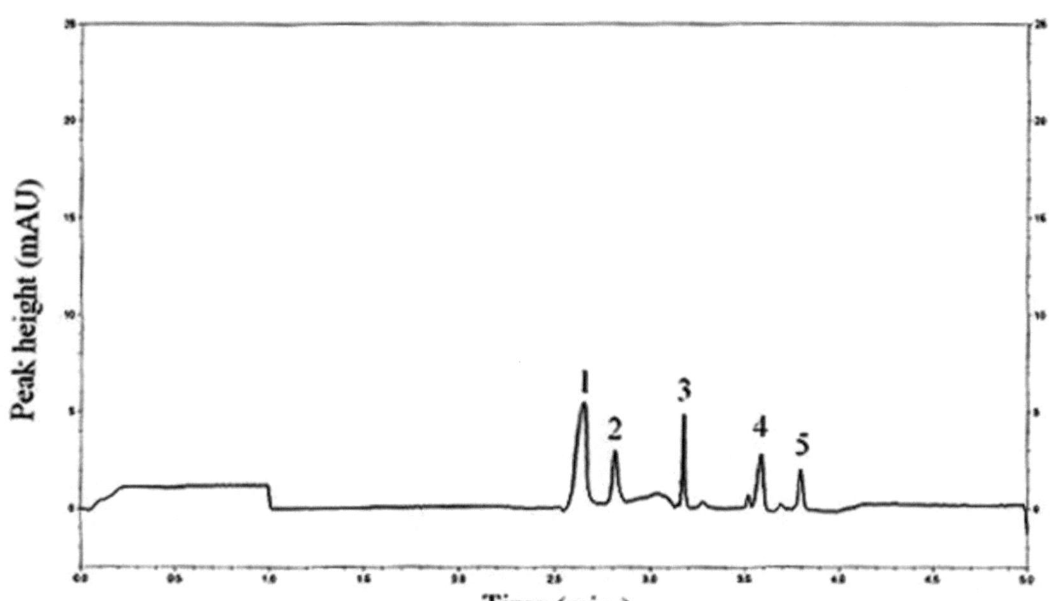

Fig. 2. Analysis of crude oil sample (1) H_2SA_2Ten, (2) Co(II), (3) V(IV), (4) Fe(II), and (5) Ni(II), as chelates of H_2SA_2Ten., on uncoated silica capillary with total length 50 cm and effective length 38.8 cm with 75 µm id at 25°C. Run buffer tetraborate (0.1 M), SDS (0.04 M) 3:1 v/v, voltage 30 kV, pH 8.2, and photodiode array detection at 228 nm (Mirza et al 2009) with permission.

3.6 Other techniques

Other techniques can be used, but they are not widely reported. Gondal et al (2006) examined laser induced breakdown spectroscopy (LIBS) for the analysis of Ca, Fe, Mg, Cu, Zn, Na, Ni, K and Mo from Arabian crude oil residue samples. The dependence of time delay and laser beam energy on the elemental spectra was investigated. Quantitation was carried out by calibration of LIBS system with standard samples containing these trace elements. Mumoz et al (2007) examined electroanalytical stripping techniques for determination of Cu, Pb, Hg and Zn from crude oil and petroleum based fuels. The samples were decomposed by different microwave ovens before determinations. Square-wave stripping voltametry (SWSW) and stripping chromopotentiometry (SCP) at gold film electrodes were applied for Cu, Pb and Hg. Potentiometric stripping analysis (PSA) at mercury film electrodes was applied for Cu, Pb and Zn. SCP presented higher sensitivity for Cu and Hg at gold electrodes. PSA at mercury electrodes was preferred for Pb and Zn determinations. Xu et al (2005) extracted Ni and V petroporphyrins from Chinese heavy crude oils and then purified by silica gel chromatography. The extraction and purification were monitored using ultraviolet visible spectroscopy and the total petroporphyrins were analyzed using laser desorption ionization coupled with time of flight mass spectrometry. Yang et al (2003) determined Fe, Cu, Zn and Pb of engine oil by mild acid digestion and energy dispersive X-ray fluorescence spectrometry. A small aliquot (0.5 ml) of the acid digested sample was spotted onto a C-18 solid phase and then analyzed directly by X-ray fluorescence spectrometry.

3.7 Speciation of metal ions

Speciation studies are also directed towards isolation and identification of elements in different forms in crude oils. More is reported on V and Ni species. In majority of these studies focus has been the organometallic forms. The information may be of help in development of metal removal methods and in understanding the molecular environment associated in crude oils. The hyphenated HPLC and GC with ICP-MS, ICP-OES, and AAS have been examined (Les Ebdon et al., 1994), (Kumar et al., 1994), (Fish and Komlenic, 1984), (Fish et al., 1984), (Tao et al.,1998) for speciation studies of V, Ni and Hg in heavy crude petroleum, asphaltenes and natural gas condensates. Margueza et al (1999) used FAAS in comparison of three analytical methods to isolate and characterize V and Ni porphyrin from heavy crude oils. Lepri et al (2006) have carried out speciation analysis of volatile and nonvolatile vanadium compounds using high resolution continuum source atomic absorption spectrometry with a graphite furnace, taking advantage of higher volatility of the vanedyl porphyrin complexes, compared to nonporphyrins. Ackley et al (2005) examined capillary electrophoresis for the separation of metalloporphyrins, however the separation was limited to ionizable metalloporphyrins (with COOH functional group).

4. Conclusion

A significant improvement in the analytical instrumentation has been observed during last decade which has enabled to improve the sensitivity and selectivity of the determinations. The choice of the methodology for the determination of elements should be based on the matrix, the elements to be determined and the equipment available. Sample decomposition by wet digestion, preferably with microwave assisted heating is more robust and accurate than direct sample introduction after dilution. Sample dilution is attractive because of its simplicity, but complete solublization may be difficult due to the complex nature of the crude oil. The concentration of the diluted solution may change rapidly due to the adsorption on the walls of the container. Alternatively dilution with the three component system (emulsions and micro emulsions) are reported to provide better precision and accuracy of results. The use of aqueous standards for calibration is an added advantage. Electrothermal AAS with graphite furnace atomizer and ICP-MS appear to be more sensitive techniques. In case of ET-AAS stabilization of volatile compounds during pyrolysis step must be considered. The use of appropriate modifier, even for the element known as a nonvolatile may be necessary. ICP-MS requires periodic cleaning of lens, sampler and skimmer due to carbon build up. Polyatomic interferences in quadruple ICP-MS must also be considered. Metal speciation using hyphenated techniques has important place in oil analysis for less explored crude oils and resins.

5. References

Ackley, K.L., Day, J.A., Caruso, J.A., (2005), Separation of metalloporphyrins by capillary electrophoresis with UV detection and inductively coupled plasma mass spectrometric detection, *J. Chromatogr. A.*, 888, 293-298.

Akinlu, A, Smith, M.S. (2010), Supercritical water extraction of trace metals from petroleum source rock, *Talanta*, 81, 1346-1349.

Akinlua, A., Ajayi, T.R., Adeleke, B.B., (2007), Organic and inorganic geochemistry of Northwestern Niger delta oils, *Geochem. J.* , 41, 271-281.

Akinlua, H., Torto, N., Ajayi, T.R., (2008), Determination of rare earth elements in Nigar delta crude oil by inductively coupled plasma-mass spectrometry, *Fuel*,87, 1469-1477.

Alberdi-Genolet, M., Tocco, R., (1999), Trace metals and organic geochemistry of the Machiques Member (Aptian–Albian) and La Luna Formation (Cenomanian–Campanian), Venezuela,*Chem. Geol.* 160, 19-38.

Al-Shahristani, H., Al-Atya, M.J., (1972), Vertical migration of oil in Iraqi oil fields: Evidence based on vanadium and nickel concentrations, *Geochim. Cosmochina Acta*, 36, 929-938.

Al-Swaidan, H.M. (1996), The determination of lead, nickel and vanadium in Saudi Arabian crude oil by sequential injection analysis inductively coupled plasma mass spectrometry, *Talanta*, 43, 1313-1319.

Alvarado, J., Picon, A.R., de Vecchi, C.M., (1990), Microwave wet acid digestion in the preparation of crude oil samples for AAS determination of their chromium, copper, iron manganese, sodium, nickel, vanadium and zinc contents, *Acta Chim. Ven.* 41, 306-319.

Amoli, H.S., Porgam, A., Sadr, Z.B., Mohanazadeh, F., (2006), Analysis of metal ions in crude oils by revered phase high performance liquid chromatography using short column, *J. Chromatogr, A.*, 1118, 82-84.

Amorin, F.A.C., Welz, B., Costa, A.C.S., Lepri, F.G., Vale, S.L.C., Ferreira, M.G.R.., (2007), Determination of vanadium in petroleum and petroleum products using atomic spectroscopic techniques, *Talanta*, 72, 349-359.

Andersen, S.I., Keul, A., Stenby .E., (1997), Variation in composition of sub fractions of petroleum asphaltenes, *Pet. Sci. Technol.* 15, 611-645.

Annual Book of ASTM Standard, ATSM-D5863 (2000), Standard test method for determination of nickel, vanadium, iron and sodium in crude oils and residual fuels by flame atomic absorption spectrometry, American Society for Testing and Materials, West Conshohockm, PA.

Annual Book of ATSM Standard, ASTM-D5708 (2002), Standard test method for determination of nickel, vanadium and iron in crude oils and residual fuels by inductively coupled plasma (ICP) atomic emission spectrometry, American Society for Testing and Materials, West Conshohockm, PA.

Annual Book of ATSM Standard, ASTM-D5863-00A (2005), Method for determination nickel, vanadium, iron and sodium in crude oils and residual fuels by flame atomic absorption spectrometry, American Society for Testing and Materials, West Conshohockm, PA.

Anselmi, A., Tittarelli, P., Katskov, D.A., (2002), Determination of trace elements in automotive fuels by filter furnace atomic absorption spectrometry *Spectrochim. Acta* Part B. 57, 403-411.

Aske, N., Hallevik, H., Sjoblom, J., (2001), Determination of Saturate, Aromatic, Resin, and Asphaltenic (SARA) components in crude oils by means of infrared and Near-Infrared Spectroscopy, *Energy Fuels*, 15, 1304-1312.

Aucelio, R.Q., Doyle, A., Pizzorano, B.S., Tristao, M.L.B., Campos, R.C., (2004), Electrothermal atomic absorption spectrometric method for the determination of vanadium in diesel and asphaltene prepared as detergentless micro-emulsions, *Microchem. J.* 78, 21-26.

Baker, E.W., Louda, J.W., (1986), Porphyrin geochemistry of Atlantic Jurassic-Cretaceous black shales , *Adv. Organic Geochem.* 10, 905-914.

Ball, J.S., Wenger, W.J., Hyden, H.J., Horr, C.A., Myers, A.T., (1960), Metal content of twenty-four petroleums, *J. Chem. Eng. Data* 5, 553-557.

Barwise, A.J.G., (1987), Mechanisms involved in altering Deoxophylloerythroetioporphyrin-Etioporphyrin ratios in sediments and oils, *J. Am. Chem. Soc. Spec. Pub.* 344, 100-109.

Barwise, A.J.G., (1990), Role of nickel and vanadium in petroleum classification , *Energy Fuel*, 4, 647-652.

Barwise, A.J.G., Roberts, I., (1984), Diagenetic and catagenetic pathways for porphyrins in sediments, *Organic Geochem.* 6, 167-176.

Bermejo-Barrera, P., Pita-Calvo, C., Bermejo-Marinez, F., (1991), Simple preparation procedures for vanadium determination in petroleum by atomic absorption spectrometry with electrothemal atomization, *Anal. Lett.*, 24, 447-458.

Bethinetti, M., Spezia, S., Baroni, U., Bizzarri, G. (1995), Determination of trace elements in fuel oils by inductively coupled plasma mass spectrometry after acid mineralization of sample in microwave oven, *J. Anal. At. Spectrom.*, 10, 555-560.

Bettinelli, M., Tittarelli, P., (1994), Evaluation and validation of Instrumental procedures for the determination of nickel and vanadium in fuel oils, *J. Anal. At. Spectrom.* 9, 805-812.

Borszeki, J., Knapp, G., Halmos, P., Bartha, L., (1992), Sample preparation procedure for the determination of sulphur and trace metals in oil products by the ICP with a minitorch using emulsions, *Microchim. Acta*, 108, 157-161.

Botto, R,I., (2002), Trace element analysis of petroleum naphthas and tars using direct injection ICP-MS, *Can. J. Anal. Sci. Spectrosc.* 47, 1-13.

Botto, R.I., (1993), Applications of ultrasonic neublization in the analysis of petroleum petrochemical by inductively coupled plasma atomic emission spectrometry, *J. Anal. At. Spectrom.* 8, 51-57.

Botto, R.I., Zhu, J., (1996), Universal calibration for analysis of organic solution by inductively coupled plasma atomic emission spectrometry, *J. Anal. At. Spectrom.* 11, 675-681.

Brandao, G.P., Camps, R.C., de Castro, E.V.R., de Jenus, H.C., (2006), Direct determination of nickel in petroleum by solid sampling-graphite furnace atomic absorption spectrometry, *Anal. Bioanal. Chem.* 386, 2249-2253.

Brandas, G.P., de Campos, R.C., de Castro; E.V.R., de Jesus H.C. (2007), Direct determination of copper, iron and vanadium in petroleum by solid sampling graphite furnace atomic absorption spectrometry, *Spectrochim. Acta*, Part B, 62, 962-969.

Brener, I.B., Zander, A., Kim, S. Shkolnik, J., (1996), Direct determination of lead in gasoline using emulsification and argon and argon–oxygen inductively coupled plasma atomic emission spectrometry, *J. Anal. Atom. Spectrom.* 11, 91-97.

Brenner, I.B., Zander, A., Plantz, M., Zhu, J., (1997), Characterization of an ultrasonic nebulizer - member separation interface with inductively coupled plasma mass spectrometry for the determination of trace elements by solvent extraction, *J. Anal. At. Spectrom.* 12, 273-279.

Bruhn, C.F., Cabalin, V.G., (1983), Direct determination of nickel in gas oil by atomic absorption spectrometry with electrothermal atomization, *Anal. Chim. Acta*, 143, 193-203.

Burguera, J.L., Avila-Gomez, R.M., Burgera, M., de Salaguer, R.A., Salaguer, J.L., Bracho, C.L., Burguera-Pascu, M., Burguera-Pascu, C., Brunetto, R., Gallignani, M., de Pena, Y.P., (2003), Optimum phase-behavior formulation of surfactant oil/water system for the determination of chromium in heavy crude oil and in bitumen in water emulsion, *Talanta*, 61, 353-361.

Cardarelli, E., Cifani, M., Mecozzi, M., Sechi, G., (1986), Analytical application of emulsions: Determination of lead in gasoline by atomic-absorption spectrophotometry, *Talanta*, 33, 279-280.

Caumette, G., Lienemann, C.P. Merdrignae, I., Bouyssiere, B., Lobinski, R., (2010), Fractionation and speciation of nickel and vanadium in crude oils by size exclusion chromatography ICP-MS and normal phase HPLC-ICP-MS, *J. Anal. At. Spectrom.* 25, 1123-1129.

Choudhry, G.G., (1983), Humic substances: Sorptive interactions with environmental chemicals ,*Toxicol. Envir. Chem.* 6, 127-171.

Damin, I.C.F., Vale, M.G.R., Silva, M.M., Welz, B., Lepri, F.G., Santos, W.N.I., Ferricira, M., (2005), Palladium as chemical modifier for the stabilization of volatile nickel and vanadium compounds in crude oil using graphite furnace atomic absorption spectrometry. *J. Anal. At. Spectrom.* 20, 1332-1336.

das Gracas Andrade Korn, M., Sodre dos Santos, D.S., Welz, B., Rodrigues Vale, M.G., Teixeira, A.P., de Castrolima, D., Ferreira, S.L.C., (2007), Atomic spectrometric methods for the determination of metals and metalloids in automotive fuels. A review, *Talanta*, 73, 1-11.

de Campos, R.C., dos Santos, H.R., Grinberg, P., (2002), Determination of copper, iron, lead and nickel in gasoline by electrothermal atomic absorption spectrometry using three-component solutions, *Spectrochim. Acta,* Part B 57, 15-28.

de la Guardia, M., Lizondo, M.J., (1993), Direct determination of nickel in fuel oil by atomic absorption spectrometry using emulsion, *At. Spectr.*, 6, 208-211.

de Oliveira, A.P., Gomes Neto, J.A., Ferreira, M.M.C., (2006), Use of exploratory data analysis in the evaluation of chemical modifier for direct determination and simultaneous metals in fuel ethanol by GFAAS, *Eclet. Quim.*31,7-12.

de Souza, R.M., da Silverira, C.L.P., (2006), Determination of Mo, Zn, Cd, Ti, Ni, V, Fe, Mn, Cr and Co in crude oil using inductively coupled plasma optical emission spectrometry and sample introduction as detergentless micro-emulsions, *Microchem. J.*, 82, 137-141.

de, B.M Souza, R.M., Mathias., Searminio, I.S, de Silveira, C.L.P Aucelio, R.Q., (2006), Comparison between two sample emulsification procedures for the determination of Mo, Cr, V and Ti in diesel and fuel oil by ICP OES along with factorial design, *Microchim. Acta,* 153, 219-225.

Dilli, S., Patsalides, E., (1981), Determination of vanadium in petroleum crudes and fuel oils by gas chromatography, *Anal. Chem. Acta,* 128, 109-119.

Dittert, I.M., Silva, J.S.A., Araujo, R.G.O., Curtius, A.J., Welz, B., Becker-Ross, H., (2010), Simultaneous determination of cobalt and vanadium in undiluted crude oil using high resolution continuum source graphite furnace atomic absorption spectrometry. *J. Anal. At. Spectrom.* 25, 590-595.

Duyck, C., Miekeley, N., Porto da Silveria, C.L., Aucelio, R.Q., Campos, R.C., Grinberg, P., Brandao, G. P., (2007), The determination of trace elements in crude oil and its heavy fractions by atomic spectrometry, *Spectrochim. Acta, Part B, At. Spectrom.* 62, 939-951.

Duyck, C.; Miekcley, N. Silveira, C.L.P. Szatmari, P., (2002), Trace element determination in crude oil and its fractions by inductively coupled plasma mass spectrometry using ultrasonic nebulization of toluene solutions, *Spectrochim. Acta,* 57B, 1979-1990.

Ebdon, E.L., Evans, H., Pretorius, W. G., Rowland, S.J., (1994), Analysis of geoporphyrins by high-temperature gas chromatography ion chromatography with inductively coupled plasma mass spectrometry., *J. Anal. At. Spectrom.* 9, 939-943.

Ekanem, E.J., Lori, J.A., Thomas, S.A., (1998), Ashing procedure for the determination of metals in petroleum fuels, *Bull. Chem. Soc. Ethiop.* 2, 9-16.

El-Gayar, M., Mostafa, M.S., Abdelfallah, A.E., Barakat, A.O., (2002), Application of geochemical parameters for classification of crude oils from Egypt into source-related types , *Fuel Process. Technol.* 79, 13-28.

Ellis, J., Rechsteiner, C., Moir, M., Wilbur, S. (2011), Determination of volatile nickel and vanadium species in crude oil and crude oil fractions by gas chromatography coupled to inductively coupled plasma mass spectrometry, *J. Anal. At. Spectrom.*, 26, 1674-1678.

Encyclopedia Britannica, Available at http://www.britannica.com., Accessed October 16,2011.

Evans, C.R., Rogers, M.A., Baily N.J.I., (1971), Evolution and alteration of petroleum in western Canada ,*Chem. Geol.* 8,147-170.

Fabec, J.L., Ruschak, M.L., (1985), Determination of nickel, vanadium and sulphur in crude and heavy crude fractions by inductively coupled argon plasma atomic emission spectrometry and flame atomic absorption spectrometry, *Anal. Chem.* 57, 1853-1863.

Fan, T.G., Buckley, J.S., (2002), Rapid and accurate SARA analysis of medium gravity crude oils, *Energy Fuels*, 16, 1571-1575.

Fish, R.H., Komlenic, J.J., (1984), Molecular characterization and profile identification of vanadyl compounds in heavy crude petroleums by liquid chromatography/graphite furnace atomic absorption spectrometry, *Anal. Chem.* 56, 510-517.

Fish, R.H., Komlenic, J.J., Wines, B.K. (1984) Characterization and comparison of vanadyl and nickel compounds in heavy crude petroleum and asphaltenes by reverse phase and size-exclusion liquid chromatography/graphite furnace atomic absorption spectrometry, *Anal. Chem.* 56, 2452-2460.

Gondal, M.A., Hussain, Y., Baig, M.A., (2006), Detection of heavy metal in Arabian crude oils residue using laser induced breakdown spectroscopy, *Talanta*, 69, 1072-1078.

Gonzarez, M.C., Rodriguez, A.R., Gonzalez. V., (1987), Determination of vanadium, nickel, iron, copper, and lead in petroleum fraction by atomic absorption spectrophotometry with a graphite furnace, *Mirochem. J.* 57, 94-106.

Goreli, M., Vale, V., Silva, M.M., Damin, I.C.F., Filho, P.J. S., Welz. B., (2008), Determination of volatile and non-volatile nickel and vanadium compounds in crude oil using electrothermal atomic absorption spectrometry after oil fractionation into saturates, aromatics, resins and asphaltenes ,*Talanta*, 74, 1385-1391.

Guidr, J.M., Snddon, J., (2002), Fate of vanadium determined by nitrous oxide-acetylene flame atomic absorption spectrometry in unburned and burned Venezuelen crude oil, *Microchem. J.*, 73, 363-366.

Hammond, J.L., Lee, Y., Noble, C.O., Beck, J.N., Proffitt, C.E., Sneddon, J., (1998), Determination of cadmium, lead and nickel by simultaneous multi element flame atomic absorption spectrometry in burned and unburned Venezuelan crude oils, *Talanta*, 47, 261-66.

Istas - Flores, C.A., Buenrostro-Gonzalez, E., Lira-Galeana, C., (2005), Comparisons between open column chromatography and HPLC. SARA fractionations in petroleum, *Energy Fuels*, 19, 2080-2088.

Jakubowski, N., Feldmann, J. Stuewere, D., (1992), Analytical improvement of pneumatic nebulization in ICP-MS by desolvation, *Spectrochim. Acta*, 47B, 107-118.

Kaminski, T.J., Fogler, H.S., Wolf, N., Wattana, P., Mairal, A., (2000), Classification of asphaltenes via fractionation and the effect of heteroatom content on dissolution kinetics, *Energy Fuels*, 14, 25-30.

Kelly, R.W., Long, S.E., Mann, J.L., (2003), Determination of mercury in SRM crude oils and refined products by isotope dilution cold vapour ICP-MS using closed system combustion, *Anal. Bioanal. Chem.*, 376, 753-758.

Khuhawar, M.Y., Arain, G.M., (2006), Liquid chromatographic determination of vanadium in petroleum oils and mineral ore samples using 2-acetylpyridine-4-phenyl-3-thiosemicarbazone as derivatizing reagent, *Talanta*, 68, 535-541.

Khuhawar, M.Y., Lanjwani, S.N., (1996), Simultaneous high performance liquid chromatographic determination of vanadium, nickel, iron and copper in crude petroleum oils using bis(acetylpivalmethane)ethylenediimine as complexing reagent, *Talanta*, 43, 767-770.

Khuhawar, M.Y., Lanjwani, S.N., Khaskhely, G.Q., (1995), High performance liquid chromatographic determination of vanadium in crude petroleum oils using bis(salicylaldehyde)tetramethylethylenediimine, *J. Charomatogr. A*, 689, 39-43.

Khuhawar, M.Y., Memon, A.A., Bhanger, M.I., (1997), Capillary gas chromatographic determination of vanadium in crude petroleum oil using fluorinated ketoamines as derivatizing reagents, *J. Chromatogr. A.*, 766, 159-164.

Korn, M.G.A., Andrade, J.B., de Jesus, D.S., Valfrado, A., Lemos, M.L.S.F., Bandeira, W.N.L., dos Santos, M.A., Bezerra, F.A.C., Amorim, A.S., Ferreira, S.L.C., (2006), Separation and preconcentration procedures for the determination of lead using spectrometric techniques: A review, *Talanta*, 69, 16-24.

Kowalewask Z., Ruszezynska, E., (2005), Cu determination in crude oil distillation, products by atomic absorption and inductively coupled plasma mass spectrometry after analyte transfer to aqueous solution, *Spectrochim. Acta*, 60B, 351-359.

Kumar, U., Dorzey, J.G., Caruzo, J.A., (1994), Metalloporphyrins speciation by liquid chromatography and inductively coupled plasma mass spectrometry, *J. Chromatogr. Sci.*, 32, 282-285.

Kumat, S.J., Gangadharan, S., (1999), Determination of trace elements in naphtha by inductively coupled plasma mass spectrometry using water-in-oil emulsion, *J. Anal. At. Spectrom.*, 14, 967-971.

Langmyhr, F.J., Adalen, U., (1980), Direct atomic absorption spectrometric determination of copper, nickel and vanadium in coal and petroleum coke , *Anal. Chim. Acta*, 115. 365-368.

Las Ebdon, E.H.V., Warren, G.R., Stephen, J.R., (1994), Analysis of geoporphyrins by high temperature inductively coupled plasma mass spectrometry and high performance liquid chromatography inductively coupled plasma mass spectrometry, *J. Anal. At. Spectrom.* 9, 939-943.

Lepri, F.G., Welz, B., Borges, D.L.G., Silva, A.F., Vale, M.G.R., Heitmann, U., (2006), Speciation analysis of volatile and nonvolatile vanadium compounds in Brazilian crude oils using high resolution continuum source graphite furnace atomic absorption spectrometry, *Anal. Chim. Acta*, 558, 195-200.

Lord, C.J., (1991), Determination of trace metals in crude oil by inductively coupled plasma mass spectrometry with micro-emulsion sample introduction, *Anal. Chem.*, 63, 1594-1599.

Luz, M.S., Oliveira, P.V., (2011), Simultaneous determination of Cr, Fe, Ni and V in crude oil by emulsion sampling graphite furnace atomic absorption spectrometry, *Anal. Methods*, 3, 1280-1283.

Magyar, B., Lieneman, P., Vonmont, H., (1986), Some effects of aerosol drying and oxygen feeding on the analytical performance of an inductively coupled nitrogen-argon plasma, *Spectrochim. Acta*, 41B, 27-38.

Mansfield, C.T., Berman, B.N., Thomas, J.V., Mehrotna, A.K., McCann, J.M., (1995), Lubricants : Petroleum and coal, *Anal. Chem.*, 67, 333R-337R.

Marqueza, N., Ysambertta, F., De La Cruz, C. (1999), Three analytical methods to isolate and characterize vanadium and nickel porphyrins from heavy crude oils, *Anal. Chim. Acta*, 395, 343-349.

Meeravali, N.N., Kumar, S.J., (2001), The utility of a W-Ir permanent chemical modifier for the determination of Ni and V in emulsified fuel oils and naphtha by transverse heated electrothermal atomic absorption spectrometer, *J. Anal. At. Spectrom.* , 6, 527-532.

Mello, P.A., Pereira, J.S.F., Moraes, D.P., Dressler, V.L., Flores, E.M.M., Knapp, G., (2009), Nickel, vanadium and sulphur determination by inductively coupled plasma optical emission spectrometry in crude oil distillation residues after microwave, induced combustion, *J. Anal. At. Spectrom.* 24, 911-916.

Microwave assisted acid digestion of sediments, sludges, soils and oils, (1994), US-EPA, A 3051, http://www.epa.gov/osw/hazard/testmethods.

Milner, O.J., Glass, J.R., Kirchner, J.P., Yurick, A.N., (1952), Determination of trace metals in crudes and other petroleum oils, *Anal. Chem.*, 24, 1728-1732.

Mirza, M.A., Kandhro, A.J., Khuhawar, M.Y., Arain, R., (2009), MEKC determination of vanadium from mineral ore and crude petroleum oil samples using precapillary chelation with bis(salicylaldehyde)tetra-methylethylenediimine, *J. Sep. Sci.*, 32, 3169-3177.

Munoz, P.A.A., Correia, P.R.M., Nascimento, A.N., Silva, C.S., Oliveira P.V., Angnes, L., (2007), Electroanalysis of crude oil and petroleum based fuel for trace metals: Evaluation of different micro-wave assisted sample decompositions and stripping techniques. *Energy and Fuel,* 21, 295-302.

Munoz, R.A.A., Correia, P.R.M., Nascimento, A.N., Silva, C.S., Oliveira, P.V., Angnes, L., (2007), Electroanalysis of crude oil and petroleum based fuels for trace metals, Evaluation of different microwave-assisted sample decomposition and stripping techniques, *Energy and Fuel*, 21, 295-302.

Murillo, M., Chirinos, J., (1994), Use of emulsion systems for the determination of sulphur, nickel and vanadium in heavy crude oils samples by inductively coupled plasma atomic emission spectrometry, *J. Anal. At. Spectrom.* 9, 237-240.

Nakamoto, Y., Ishimaru, T., Endo, N., Matsusaki, K., (2004), Determination of vanadium in heavy oils by atomic absorption spectrometry using graphite furnace coated with tungsten, *Anal. Sci.* 20, 739-741.

Odermatt, J., Curiale, J., (1991), Organically bound metals and biomarkers in the Monterey formation of the Santa Maria basin, California, *Chem. Geol.* 91, 99-113.

Oliveira, E., (2003), Sample preparation for atomic spectroscopy: evolution and future trends ,*J. Braz. Chem. Soc.*, 14, 1-17.

Osibanjo, O., Kakulu, S.E., Ajayi, S.Q., (1984), Analytical application of inorganic salt standards and mixed-solvent systems to trace-metal determination in petroleum crudes by atomic absorption spectrophotometry, *Analyst*, 109, 109-127.

Ozcan, M., Akman, S., (2005), Determination of Cu, Co and Pb in gasoline by electrothermal atomic absorption spectrometry using aqueous standard addition in gasoline–ethanol–water three-component system , *Spectrochim Acta*, Part B 60, 399-402.

Pelizzetti, E., Pramauro, E., (1985), Analytical applications of organized molecular assemblies, *Anal. Chim. Acta*, 169, 1-29.

Pena, M.E., Manjarrez, A., Campero, A., (1996), Distribution of vanadyl porphyrins in a Mexican offshore heavy crude oil, *Fuel processing Technol.*, 46, 171-182.

Periria, J.S.F., Moraes, D.P., Antes, F.G., Dieht, L.D., Santos, F.P., Guimaraes, R.C.L., Fonseca, T.C.O., Dressler, V.L., Flores, E.M.M., (2010), Determination of metals and metalloids in light and heavy crude oil by ICP-MS after digestion by microwave-induced combustion, *Microchem. J.*, 96, 4-11.

Peters, K.E., Walters, C.C., Moldowan, J.M., 2nd ed., *The biomarker guide, Volume 2: Biomarkers and isotopes in the petroleum exploration and earth history*, Cambridge University Press, New York, 2004.

Platteau, O., Carrillo, M., (1995), Determination of metallic elements in crude oil-water emulsion by flame AAS, *Fuel*, 74, 74-76.

Quadros, D.P.C., Chaves, E.S., Lepri, F.G., Borges, D.L.G., Welz, B., Becker-Ross, H., Curtius, A.J., (2010), Evaluation of Brazilian and Venezuelan, Crude oil samples by means of the simultaneous determination of Ni and V as their total and non-volatile fractions using high resolution continuum source graphite furnace atomic absorption spectrometry, *Energy Fuels*, 24, 5907-5911.

Quimby, B.D, Dryden, P.C. Sullivan, J.J. (1991), A selective detection of volatile nickel, vanadium and iron porphyrins in crude oils by gas chromatography atomic emission spectroscopy, *J. High Res. Chromatogr.*, 14, 110-116.

Ridyk, B.M., Alturki, Y.I.A., Pillinger, C.T., Eglington, G., (1975), Petroporphyrins as indicators of geothermal maturation ,*Nature* 256, 563-565.

Rudzinski, W.E., Aminabhavi, T.M., (2000), A review on extraction and identification of crude oil and related products using supercritical fluid technology , *Energy Fuels*, 14, 464.

Sant Ana, F.W., Santelli, R.E., Cassella, A.R., Cassella, R.J., (2007), Optimization of an open focused microwave oven digestion procedure for determination of metal in diesel oil by inductively coupled plasma optical emission spectrometry, *J. Hazardous Materials*, 149, 67-74.

Santella, R.E., Bezerra, M.A., Freire, A.S., Oliveira, E.P., de Carvalho, M.F.B., (2008), Non-volatile vanadium determination in petroleum condensate, diesel and gasoline prepared as detergent emulsion using GFAAS, *Fuel*, 87, 1617-1622.

Sebor, G., Long, I., Kolihova, D., Wasser, O., (1982) Effect of the type of organometallic iron and copper compounds on the determination of both metals in petroleum samples by flame atomic absorption spectroscopy, *Analyst*, 107, 1350-1355.

Seidl, P.R., Chrisman, E.C.A.N., Carvalin, C.C.V., Leal, K.Z., de Menezes, S.M.C., (2004), NMR analysis of asphaltenes separated from vacuum residues by selected solvents, *J. Dispersion Sci. Technol.* 25, 349-353.

Seidl, P.R., Chrisman, E.C.A.N., Siwa, R.C., de Menezes, S.M.C., Teixeira, M.A.G., (2004), Critical variables for the characterization of asphaltenes extracted from vacuum residues, *Pet. Sci. Technol.*, 22, 961-971.

Sharma, B.K., Sarowha, S.L.S. Bhagat, S.D., Tiwari, R.K., Gupta, S.K., Venkataramani, O.S., (1998), Hydrocarbon group type analysis of petroleum heavy fractions using the TLC-FID technique, *J. Anal. Chem.*, 360, 539-544.

Silva, M.M., Damin, I.C.F.; Vale, M.G.R., Welz, B., (2007), Feasibility of using solid sampling graphite furnace atomic absorption spectrometry for speciation analysis of volatile and non-volatile compounds of nickel and vanadium in crude oil. *Talanta*, 71, 349-359.

Speight, J.G., (2001), *Handbook of Petroleum Analysis*, John Wiley and Sons Inc. New Jersey, 519.

Stiger, H.P.M., de Haan, R., Guicherit, C.P.A,. Dekkers, M.L., (2000), Determination of cadmium, zinc, copper, chromium and arsenic in crude oil cargoos, *Environ. Pollut.* 107, 451-464.

Tan, S.H., Horlick, H., (1986), Background spectral features in inductively coupled plasma/mass spectrometry, *Appl. Spectrosc.* 40, 445-460.

Tao, H., Murakami, T., Tominaga, M., Miyazaki, A., (1998), Mercury speciation in neutral gas condensate by gas chromatography - inductively coupled plasma mass spectrometry, *J. Anal. At. Spectrom*. 13, 1085-1093.

Teserovsky, E., Arpadjan, E., (1991), Behaviour of various organic solvents and analytes in electrothermal atomic absorption spectrometry, Anal. Atom. Spectrom. 6, 487-491.

Thomaidis, N.S., Piperaki, E.A., (1996), Comparison of chemical modifiers for the determination of vanadium in water and oil samples by electrothermal atomization atomic absorption spectrometry, *Analyst*. 121, 111-117.

Treibs, A., (1936), Chlorophyll and minerals in organic Häminderivate, *Angew Chemie*, 49, 682-686.

Trindade, J.M., Marques, A.L., Lopes, G.S., Margues, E.P., Zhang, J., (2006), Arsenic determination in gasoline by hydride generation atomic absorption spectroscopy combined with a factorial experimental design approach, *Fuel*, 85, 2155-2161.

Turunen, M., Peraniemi, S., Ahlgren, M., Westerholm, H., (1995), Determination of trace elements in heavy oil samples by graphite furnace and cold vapour atomic absorption spectrometry after acid digestion, *Anal. Chim. Acta*, 311, 85-91.

Udoh, A.P., Thomas, S.A., Ekanem., (1992), Application of P-xylenesulphonic acid as ashing reagent in the determination of trace metals in crude oils, *Talanta,* 39, 1591-1595.

Vale, M.G.R., Damin, I.C.F., Klassen, A., Silva, M.M., Welz, B., Silva, A.F., Lepri, F.G., Borges, D.L.G., Heitmann., (2004), Method development for the determination of nickel in petroleum using line-source and high resolution continuum-source graphite furnace atomic absorption spectrometry, *Microchem. J.*, 77, 131-140.

Vale, M.G.R., Silva, M.M., Damim, I.C.F., Sanches Filho, P.J., Welz. B., (2008), Determination of volatile and non-volatile nickel and vanadium compounds in crude oil using electrothermal atomic absorption spectrometry after oil fraction into saturates, aromatics, resins and asphaltenes, *Talanta*, 74, 1385-1391.

Vilhunen, J.K., Bohlen, A., Schmeling, M., Klockenkamper, R., Klockow, D., (1997), Total reflection X-ray fluorescence analyses of samples from oil refining and chemical industries, *Spectrochim Acta*, B, 52, 953-959.

Wilhelm, S.M., Bloom, N., (2000), Mercury in petroleum (Review), *Fuel processing Technology*, 63, 1-27.

Wilhelm, S.M., Liang, L., Kirchgesser, D., (2006), Identification and properties of mercury species in crude oil, *Energy Fuels* 20, 180-186.

Wondimu, W., Goessler, W., Irgolie, K.J., (2000), Microwave digestion of residual fuel oil for the determination of trace elements by inductively coupled plasma mass spectrometry. *Fresenius J. Anal. Chem.*, 367, 35-42.

Xu, H., Que, G., Yu, D., Lu, J.R., (2005),Characterization of petroporphyrins using ultraviolet-visible spectroscopy and laser desorption ionization time of flight mass spectrometry, *Energy and Fuel*, 19, 517-524.

Yan, Z., Hou, X., Jones, B. T., (2003),Determination of wear metals in engine oil by mild acid digestion and energy dispersive X-ray fluorescence spectrometry using solid phase extraction disks, *Talanta*, 59, 673-680.

Zang, Y.D. Uden, P.C., (1994), High temperature gas chromatography - atomic emission detection of m`etalloporphyrins in crude oils , *J. High Resolut. Chromatogr.*, 17, 217-222.

Part 3

Crude Oil Biology

8

Crude Oil by EPR

Marilene Turini Piccinato, Carmen Luisa Barbosa Guedes
and Eduardo Di Mauro
Universidade Estadual de Londrina (UEL) /Laboratório de Fluorescência e Ressonância Paramagnética Eletrônica (LAFLURPE)
Brazil

1. Introduction

Our goal was based on environment questions. Environmental accidents involving crude and by-product oils have motivated laboratory research to evaluate photodegradation in monitored environments, as well as the characterization of crude and by-product oils. EPR was the spectroscopic technique used as analysis tool.

2. What is EPR?

Electron paramagnetic resonance (EPR) or Electron spin resonance (ESR) is a high resolution spectroscopy that consists in energy absorption of microwave, for electron spin, in the presence of a magnetic field (Ikeya, 1993). As the name itself suggests, EPR is applied in samples containing some paramagnetic species or used as an investigative method, to verify the presence of some paramagnetic species. Paramagnetism is characteristic of species with a total magnetic moment different from zero.

Paramagnetism of organic molecules arises almost entirely from unpaired electron spins (Gerson & Huber, 2003). The spin quantum number (S) is the sum of the corresponding numbers, 1/2 of the unpaired electrons. The two possible configurations for an unpaired electron in the presence of an external magnetic field (spin up and spin down) have different energies, which are represented by energy level diagrams (Fig. 1). In the absence of an applied magnetic field, the two spin states are of equal energy (Bunce, 1987).

Energy showed in different spin states in the presence of an external magnetic field is known as "Zeeman effect" and depends on $\vec{H}$ and the magnetic moment ($g\beta M_S$) of the electron (Ikeya, 1993). The Zeeman energy is given by

$$E_z = g\beta H M_s \quad (1)$$

According to the above equation, the energy levels to an unpaired electron have energies equal to

$$E_{z+} = +\frac{1}{2} g\beta H \quad (2)$$

and

$$E_{z-} = -\frac{1}{2} g\beta H \tag{3}$$

corresponding to spin up ($M_s = +\frac{1}{2}$) and spin down ($M_s = -\frac{1}{2}$) respectively.

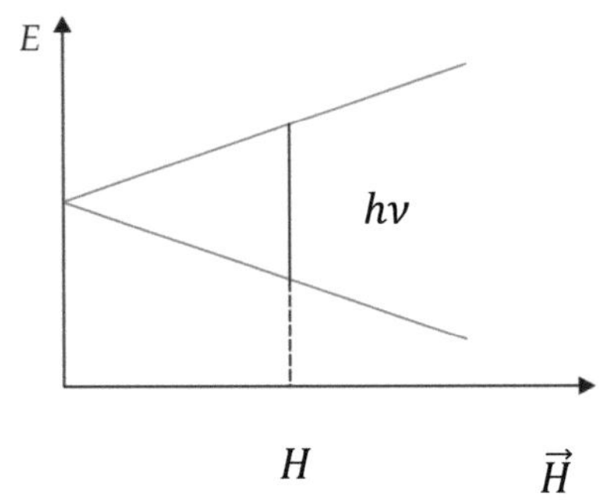

Fig. 1. Energy levels of a single electron in presence of an external magnetic field.

If an oscillating field of radiation in the microwave range acts on an unpaired electron in the presence of a magnetic field, transitions between two energy states are possible. The transition between energy levels will only occur when the following resonance condition is satisfied:

$$h\nu = g\beta H \tag{4}$$

This happens when the incident radiation is equal to the separation between the Zeeman energy levels. $h\nu$ is the energy of the absorbed photon, β is a constant for the electron, the Bohr magneton, H is the external magnetic field, and g (g-factor) is a constant characteristic of spin system (approximately 2.0 for organic free radicals) (Janzen, 1969). The g-factor is sensitive to the chemical neighborhood of the unpaired electron.

In EPR spectroscopy it is common to record the spectrum as first derivative curve (Fig. 2a), as opposed to the direct absorption curve (Fig. 2b), which is the conventional presentation in high-resolution NMR (Bunce, 1987).

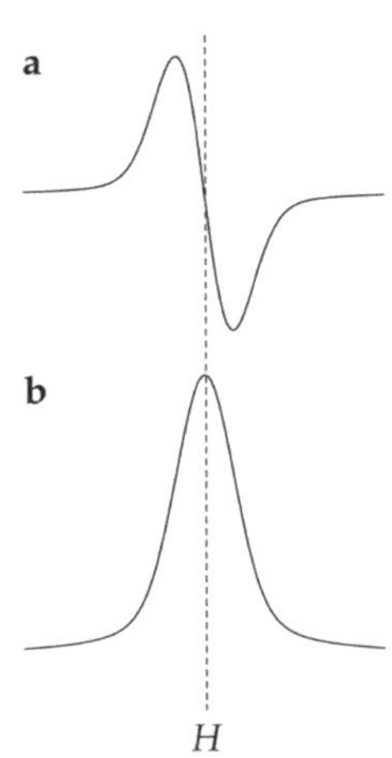

Fig. 2. **a** EPR spectrum as a first derivative curve; **b** Absorption curve.

In addition to the interaction of the unpaired electron with the external magnetic field, interaction can also occur with the nuclei of atoms. If the nucleus of the paramagnetic ion has a magnetic moment, this will interact with the electronic moment, resulting in hyperfine structure in the EPR spectrum (Orton, 1968).

The interaction of the unpaired electron with the nucleus splits the electron energy levels, generating a structure called spectral hyperfine structure or hf splitting (Poole, 1967). Each "M_s state" being split into a closely spaced group of $(2I + 1)$ levels (Orton, 1968). I is the nuclear spin quantum number. The way in which these give rise to hyperfine splitting of the resonance lines is illustrated in Fig. 3.

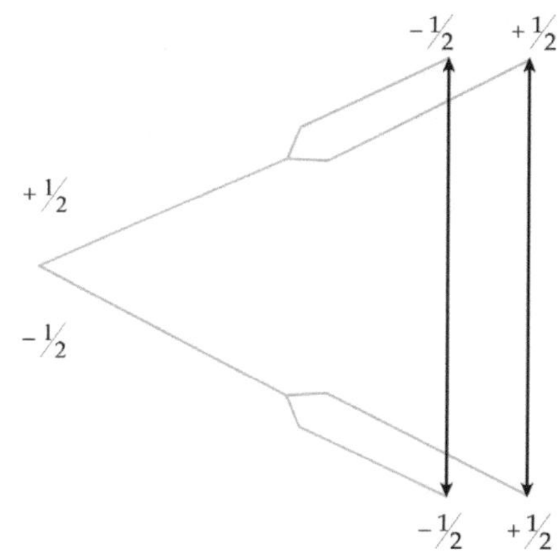

Fig. 3. Schematic diagram of the hf splitting for unpaired electron interaction with a nucleus of nuclear spin $I = 1/2$.

Transitions are allowed under the following selection rules: $\Delta M = \pm 1$, for electron spin levels splitting, and $\Delta m = 0$, for nuclear spins. For the sample taken, the allowed transitions are indicated by arrows in Fig. 3. Each one of these transitions gives rise to a resonance line in the EPR spectrum (Fig. 4). The spacing between the observed lines, usually in gauss, provides the hyperfine coupling constant (A). The spacing between lines is always symmetrically disposed about the center of the spectrum.

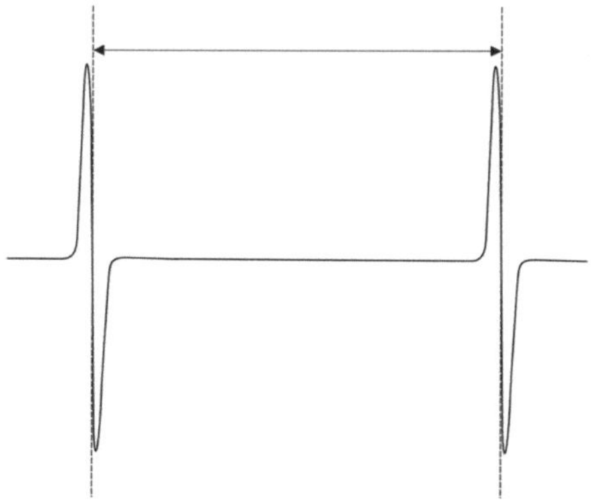

Fig. 4. Resonance lines for unpaired electron interaction with a nucleus of nuclear spin $I = 1/2$ and indication for the hyperfine coupling constant (A).

In crude and by-product oils, the paramagnetic species presence allows that the RPE technique assists to elucidate the complex chemical composition of these systems. Petroleum and related materials such as heavy oils, asphalt, pitch coal tar, tar sands, kerogen, and oil shale have been studied by EPR (Ikeya, 1993).

3. Crude oil by EPR

The EPR spectra of crude oils show signals of two different paramagnetic centers, namely, the vanadyl group VO^{2+} and free radical (Guedes et al., 2001, 2003). These are overlapped in the same magnetic field range (Fig. 5a), being the very intense central line associated with organic free radical (Montanari et al., 1998; Scott & Montanari, 1998; Yen et al, 1962) (Fig. 5b).

The free radical gives rise to a single line corresponding to the transition between the spin $+1/2$ and $-1/2$. This line is interpreted as resulting from the superposition of the signals of the different species of free radicals with very close values of g-factor in crude oil asphaltenes (Guedes et al., 2001, 2003).

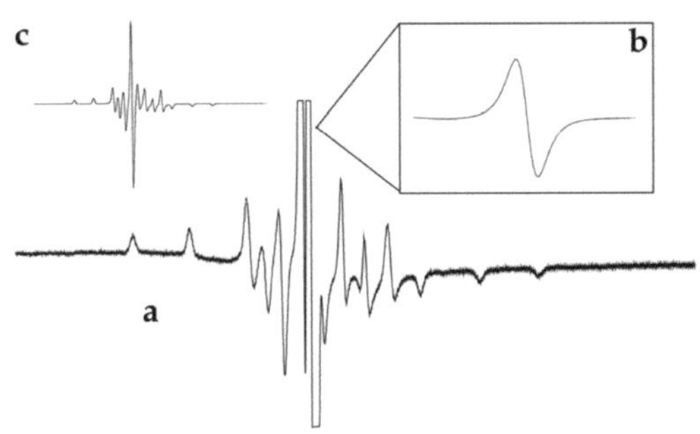

Fig. 5. **a** EPR spectrum for Brazilian crude oil; **b** Signal of the free radical; **c** Simulated spectrum for VO^{2+} (Guedes et al., 2001).

To try to understand the free radical line, Arabian crude oil (Arabian Light Crude Oil) and Colombian crude oil (Cusiana Crude oil) were studied by EPR in X- (9 GHz), Q- (34 GHz), and W- bands (34 GHz). The spectra obtained at different frequencies are shown in Fig. 6 and Fig. 7.

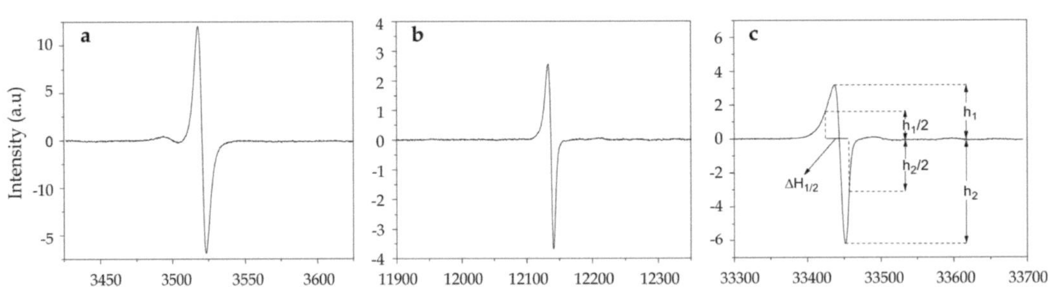

Fig. 6. Free radical EPR spectra of Arabian crude oil at room temperature obtained in: **a** X-band; **b** Q- band; **c** W- band; $\Delta H_{1/2}$ is the half height separation of the EPR derivative peak and ΔH_{pp} is the separation of the EPR derivative peak (Di Mauro et al., 2005).

Asymmetrical lines of the free radical were observed in all EPR spectra (Figs. 6 and 7). However, asymmetry was more pronounced in the spectra obtained in the W- band (Figs. 6c and 7c). The asymmetry in the line is due to the superposition of all the possible orientations of the paramagnetic species in the system and to the contributions of different chemical species that interact with the unpaired electron.

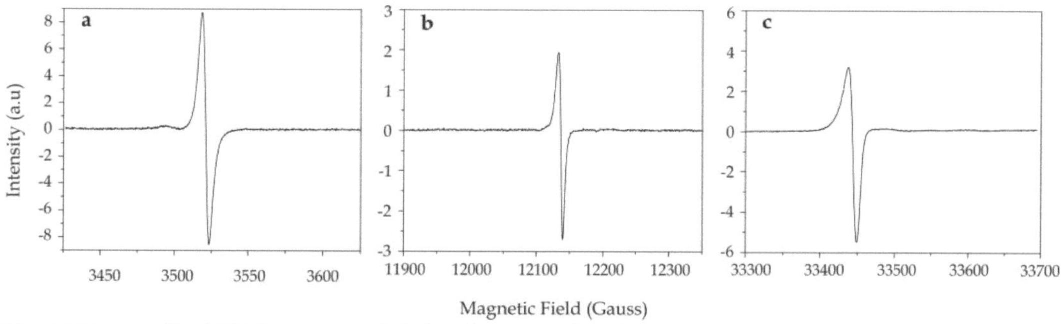

Fig. 7. Free radical EPR spectra of Colombian crude oil at room temperature obtained in: **a** X- band; **b** Q- band; **c** W- band (Di Mauro et al., 2005).

The values of the line width ΔH in the spectra increased linearly with the microwave frequency utilized in EPR experiments (Fig. 8). The ΔH values are obtained directly from the EPR spectrum, according to the representation shown in Fig. 6c.

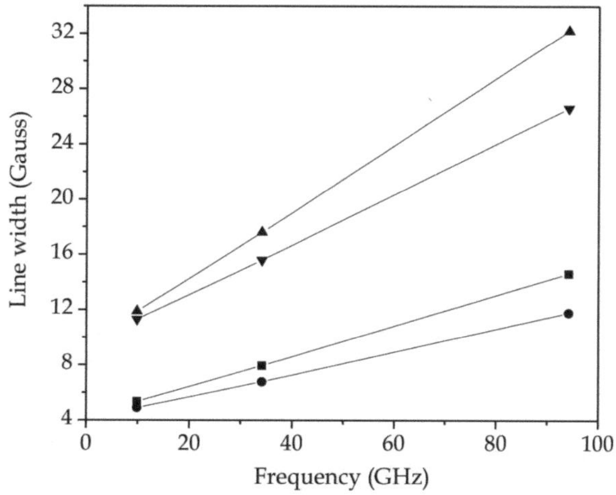

Fig. 8. Line width ΔH of the free radical signal versus microwave frequency of EPR spectra recorded in the X-, Q- and W– bands at room temperature.■, ΔH_{pp} (Arabian petroleum);●, ΔH_{pp} (Colombian petroleum);▲, $\Delta H_{1/2}$ (Arabian petroleum);▼, $\Delta H_{1/2}$ (Colombian petroleum).

The increase in values of the line width ΔH could be due either, to the superposition of all the possible orientations of the paramagnetic species with anisotropic g-factor in the system and/or to the contribution of different chemical species with a different g-factor to the free radical. If the line is the result of a single chemical species, the first cause would be entirely responsible for the broadening of the line with the variation of the microwave frequency and we would be able to mathematically simulate the signal, thus acknowledging that it corresponds to a single species. However, a mathematical simulation of the free radical signal for the EPR spectra in three bands (X-, Q-, and W- bands) with a set of parameters corresponding to a single species does not coincide exactly with the experimental signal, signaling that the hyperfine interaction of the unpaired electron with neighborhood correspond to more than one species of radical in the molecular structure of the crude oil asphaltenes (Di Mauro et al., 2005).

The vanadyl compounds (VO^{2+}) produce EPR signals less intense (Fig. 3c), with anisotropic g-factor and hf splitting. For the vanadium in the presence of an external magnetic field, the interaction of electron spin ($S = 1/2$) with the nucleus V^{51} ($I = 7/2$) has 16 possible states distributed between the two values S ($M_s = +\frac{1}{2}$ and $M_s = -\frac{1}{2}$), as shown in diagram (Fig. 9).

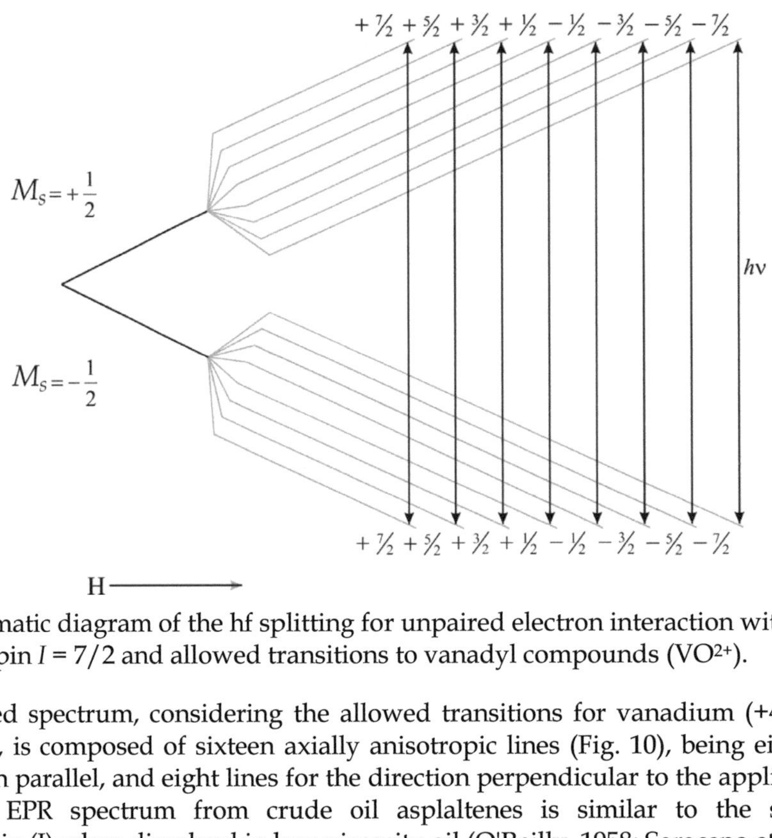

Fig. 9. Schematic diagram of the hf splitting for unpaired electron interaction with a nucleus of nuclear spin $I = 7/2$ and allowed transitions to vanadyl compounds (VO^{2+}).

The expected spectrum, considering the allowed transitions for vanadium (+4) in natural asphaltenes, is composed of sixteen axially anisotropic lines (Fig. 10), being eight lines for the direction parallel, and eight lines for the direction perpendicular to the applied magnetic field. This EPR spectrum from crude oil asplaltenes is similar to the spectrum of etioporphyrin (I) when dissolved in low-viscosity oil (O'Reilly, 1958; Saraceno et al., 1961).

The spectrum of vanadyl can be interpreted in terms of the following spin Hamiltonian with axial symmetry (O'Reilly, 1958):

$$\mathcal{H} = g_{\parallel}\beta H_z S_z + g_{\perp}\beta(H_x S_x + H_y S_y) + A_{\parallel} I_z S_z + A_{\perp}(I_x S_x + I_y S_y) \qquad (5)$$

When the molecule is rotating about with a correlation time much shorter than the reciprocal of the spread of the spectrum in frequency, Eq. (5) is reduced to an ''isotropic'' Hamiltonian, with $g_0 = (1/3)(g_{\parallel} + 2g_{\perp})$(O'Reilly, 1958).

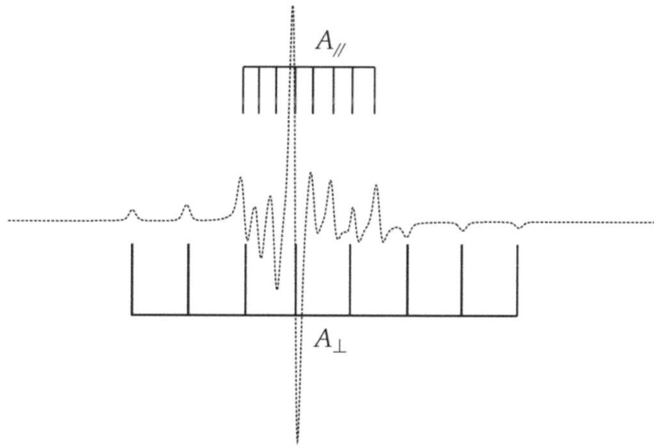

Fig. 10. Simulated spectra of VO^{2+} in crude oil and indication of lines corresponding to the parallel and perpendicular directions in relation to the applied magnetic field.

EPR spectra with hyperfine interaction assist in the identification of the porphyrin or non-porphyrin vanadium complexes in crude oil (Saraceno et al., 1961; Espinosa et al., 2001). The experimental parameter used in this identification is Δg_0 (chemical shift), which expresses chemical changes. This is calculated from the spectroscopic factors $g_0\{\Delta g_0 = (2.0023 - g_0) \times 10^3\}$ (Dickson & Petrakis, 1974). The different values obtained experimentally for the Δg_0 parameter represent structural differences in the distribution of the ligands around the VO^{2+} in complexes.

4. Degradation of Crude oil by EPR

Since laboratory monitoring has shown photochemical degradation of crude oil, an efficient process for oil oxidation and removal must also occur in the environment (Guedes, 1998). Several scientific investigations have been carried out in an attempt to reduce the actual damages caused by-products that attack nature. In some studies, sunlight and artificial light were used on crude oil in order to verify the changes caused by all sunlight spectrum and some specific wavelengths. Monitoring the paramagnetic species in crude oil by EPR is possible to follow the changes in the molecular structure of the oils, once it reflects changes in the unpaired electron neighborhood.

4.1 Degradation by artificial light

Crude oils were irradiated with Hg lamp and He-Ne laser. Two types of oil were used, with three different samples. One sample is from Campos Basin - RJ. The other two were obtained from Arabian oil: part of the oil was used without treatment (total Arabian oil), and part was distilled at 260 °C, (partial Arabian oil). The oil samples were subjected to irradiation under commercial mercury lamp (street lighting), of 450 W, which had its protective cover removed. Oil samples were also subjected to irradiation under He-Ne laser with an output of 15 mW and monochromatic emission of 632 nm. EPR measurements were performed on the BRUKER ESP 300E Series equipment, operating in X- band (9 GHz) at room temperature.

Hg lamp					He-Ne laser				
Brazilian crude oil									
Time	$A_\perp$ 10^{-4} cm^{-1}	$A_\parallel$ 10^{-4} cm^{-1}	g_0	Δg_0	Time	$A_\perp$ 10^{-4} cm^{-1}	$A_\parallel$ 10^{-4} cm^{-1}	g_0	Δg_0
0 h	54.0	156.0	1.980	22.8	0 h	54.0	156.0	1.979	22.8
2 h	54.0	156.0	1.986	17.0	8.5 h	53.5	155.5	1.963	39.0
9 h	54.0	157.0	1.990	12.6					
Arabian crude oil									
0 h	54.0	156.0	1.980	22.8	0 h	54.0	156.0	1.979	22.8
2 h	54.0	156.0	1.985	17.0	8.5 h	53.5	155.0	1.963	39.0
9 h	54.0	156.0	1.990	12.6					
Partial Arabian oil									
0 h	54.0	157.0	1.980	22.8	0 h	54.0	157.0	1.979	22.8
2 h	53.5	157.0	1.986	17.0	8.5 h	53.5	155.5	1.963	39.0
9 h	54.0	157.5	1.990	12.6					

Table 1. EPR parameters of VO^{2+} in irradiated oil.

The values obtained to $\Delta g_0 \cong 22.8$ for non-irradiated oils reveal the presence of vanadyl porphyrins in the Brazilian and Arabian crude oils. All non-irradiated oil samples had the same value of g_0 (Table 1). After irradiation, the g-factor values changed due to changes in the areas around paramagnetic VO^{2+} species. When these oils are irradiated by UV-visible, values such as $\Delta g_0 \cong 12.6$ are shown, corresponding to vanadium non-porphyrin complexes containing sulfur as a binder. When oils are irradiated at 632 nm, these values increase significantly (Δg_0= 39.0), indicating the presence of vanadyl non-porphyrin complexes containing oxygen as a binder (Guedes et al., 2001).

Both irradiations cause destruction of porphyrins. The radiation on 632 nm is responsible for Δg_0 values further apart from those obtained for vanadyl porphyrin. The UV-visible irradiation causes a decrease in the Δg_0 values with exposure to time. According to Table 1, the first hours of exposure are the most significant for the destruction of vanadyl porphyrin.

The oils analyzed have different values for the g-factor and line width, corresponding to the free radical EPR signal (Table 2). The signal of free radical in Brazilian oil has a value of de g = 2.0046 and in the Arabian oil of g = 2.0053, due to the high percentage of aromatics in Brazilian oil. This aromaticity explains also the narrow line that corresponds to the signal of the radical in the Brazilian oil. On the other hand, partial Arabian oil (distilled at 260 °C) is more viscous; therefore it originates a broadening of the line of the radical due to the dipolar interaction of the spins. The values of g-factor suggest the presence of free radical in carbon and nitrogen. The same mechanisms indicate that the percentage of sulfur radical in Arabian oil is higher than that of Brazilian oil.

With irradiation, UV-visible radiation increases the values of the g-factor and the line width of the free radical in Brazilian oil EPR spectra (Table 2), caused by the destruction of aromatics. Irradiation at 632 nm of the Arabian oil causes a reduction of the value of g-factor because of the changes on the molecular structure of the photosensitive species in the region.

Hg lamp			He-Ne laser		
Brazilian crude oil					
Time	ΔH_{pp} (Gauss)	g-factor	Time	ΔH_{pp} (Gauss)	g-factor
0 h	5.04	2.0045	0 h	5.04	2.0046
2 h	5.04	2.0047	8.5 h	5.85	2.0044
9 h	5.80	2.0053			
Arabian crude oil					
0 h	5.70	2.0053	0 h	5.70	2.0053
2 h	5.70	2.0053	8.5 h	6.62	2.0047
9 h	5.91	2.0055			
Partial Arabian oil					
0 h	6.78	2.0053	0 h	6.78	2.0052
2 h	7.38	2.0052	8.5 h	6.75	2.0043
9 h	7.34	2.0055			

Table 2. EPR parameters of free radical in irradiated oil.

4.2 Degradation by sunlight

In countries where the incidence of solar light is significance, the process of photochemical weathering is an important mechanism for the removal of foreign substance from the environment (Nicodem et al., 1998). The effects of photochemical oxidation of petroleum films over water were studied by Nicodem et al (Nicodem et al., 2001).

The photochemical weathering of Brazilian oil (Campos Basin in the state of Rio de Janeiro), Arabian oil (Arabian light crude oil), and Colombian oil (Cusiana crude oil), as a film over seawater, was monitored by EPR. In all experiments, 5 ml of petroleum was placed floating over 20 ml of seawater. The resulting oil film was 0.8-mm thick. Petri dishes with Pyrex lids were used. This Pyrex® transmits 75% at 295 nm and 85% at 300 nm, with little sunlight attenuation (Nicodem et al., 1998). Its use is common practice for samples with considerable absorption in the UVA and visible portions of the solar spectrum (El Anba-Lurot et al., 1995; Lartiges & Garrigues, 1995; Nicodem et al., 1998). Crude oil absorbs sunlight in the ultraviolet, visible and near infrared, as reported by Nicodem et al. (1997). Samples were irradiated by exposure to sunlight on the laboratory's building roof on cloud less days from 9:00 AM to 3:00 PM. For every irradiated sample was used a non-irradiated control, which was handled in the same way except that a black cover plate was used to eliminate irradiation. After irradiation, the two phases were separated by centrifugation and crude oils were submitted to EPR experiments at X- band (9.5 GHz), at room temperature. The WINEPR SimFonia Version 1.25 software of Bruker® was used in the simulation option for the determination of paramagnetic species parameters.

No variations in g-factor were observed in non-irradiated samples for any of the paramagnetic species (Table 3). The g-factor determined for the free radical signal in Brazilian oil (Campos Basin in the state of Rio de Janeiro) was 2.0045±0.0001 (Table 3), suggesting the presence of phenoxy radicals, i.e. radicals partially localized in aromatic systems due to the oxygen. There was no variation in the g-factor values for the free radical, whereas the line width (ΔH_{pp}) showed a significant decrease (Fig. 11).

Specimen	g-factor	Sample	0 h	2 h	5 h	20 h	40 h	60 h	100 h
VO^{2+}	$g_{\parallel}$	NI[a]	1.9675	1.9682	1.9685	1.9685	1.9685	1.9685	1.9685
	$g_{\parallel}$	I[b]	1.9675	1.9690	1.9690	1.9690	1.9700	1.9700	1.9705
	$g_{\perp}$	NI	1.9873	1.9870	1.9870	1.9872	1.9875	1.9875	1.9875
	$g_{\perp}$	I	1.9873	1.9873	1.9877	1.9880	1.9885	1.9889	1.9889
	g_0	NI	1.9807	1.9807	1.9808	1.9810	1.9812	1.9812	1.9812
	g_0	I	1.9807	1.9812	1.9815	1.9817	1.9823	1.9826	1.9828
Free radical	g	NI	2.0045	2.0045	2.0045	2.0045	2.0045	2.0045	2.0045
	g	I	2.0045	2.0045	2.0045	2.0045	2.0045	2.0045	2.0045

[a]Non-irradiated.
[b]Irradiated.

Table 3. EPR g values for the paramagnetic species of Brazilian crude oil at room temperature.

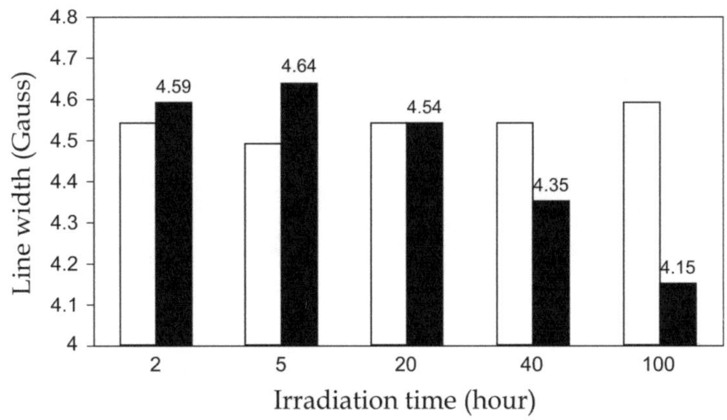

Fig. 11. Variation of the line width of the free radical versus sunlight irradiation time. Filled columns represent the irradiated samples and blank columns the non-irradiated samples.

In asphaltenes the hyperfine interaction is generally between the electron spin delocalized in an aromatic k orbital and the nuclear magnetic moments of H attached to the aromatic C. The line width broadening of the free radical cannot be attributed unequivocally to the unresolved hyperfine structure of the EPR spectrum. In petroleum asphaltenes, the effects of the aromaticity and the different degrees of substitution on the line width and the line shape probably overlap, and different number of spins can also contribute to the line width by dipolar interaction (Scotti & Montanari, 1998).

Solar irradiation caused an increase in the line width of the signal corresponding to the free radical within the first 5 h (Fig. 11). Since we know that the photodegradation of this Brazilian oil under solar light begins from singlet oxygen and continues with the formation of free radicals and the destruction of aromatic components of the oil by a photochemical effect (Nicodem et al., 1998), we can say that the widening of the line was due to the increase of the concentration of free radicals and the decrease of aromaticity in asphaltenes. After 20 h of irradiation, narrowing of the line was detected. At the end of 100 h of irradiation the line width was reduced by 10.6%, indicating rearrangement among radicals present in the structure and probably the partial destruction of the asphaltenic fraction of the oil.

An increase in the g_0 values for VO^{2+} was observed in irradiated samples (Table 3). The chemical shift obtained for Brazilian oil was Δg_0 = 21.6, indicating that this oil contains vanadyl in the porphyrin and non-porphyrin structures. Based on literature data (Dickson & Petrakis, 1974), it is possible to suggest that vanadyl in Brazilian oil has VO(N4), VO(NS3), VO(N2S2) and VO(N3S2) as possible environments.

After 100 h of solar irradiation the variation in the Δg_0 value to 19.5 (Fig. 12) must be attributed to the preferential destruction of the vanadyl porphyrin complexes due to the decrease in the Δg_0 value (Dickson & Petrakis, 1974). The uncertainty in the determination of the Δg_0 value is ±0.1.

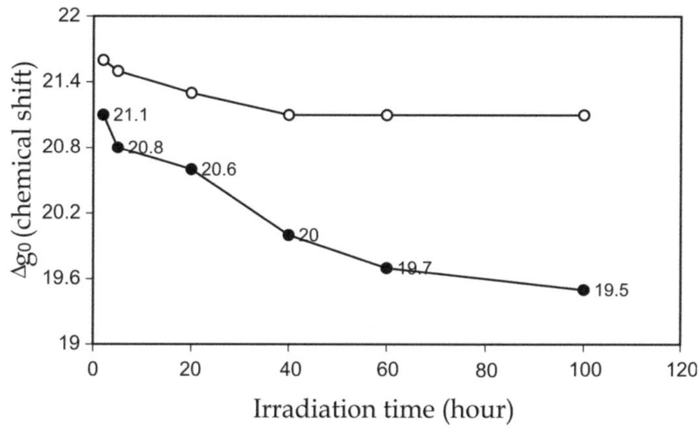

Fig. 12. Variation of the Δg_0 (chemical shift) versus sunlight irradiation time for Brazilian crude oil. Full circles represent the irradiated samples and blank circles represent the non-irradiated samples.

The g-factor determined for the free radical signal in Arabian and Colombian petroleum was 2.0033±0.0001 and 2.0030±0.0001, respectively. One possible interpretation for the g-factor values observed corresponds to neutral radicals of carbon or nitrogen (Yen et al., 1962). Thus, Arabian crude oil should have a lower percentage of aromatic carbon than Colombian crude oil, in which the percentage of heteroatoms should be higher. However, it is also possible that a difference in the distribution of the anisotropy in both the g-factor and the hyperfine coupling constants of the two samples produces a difference in the spectral shape, which could cause a small variation in the g-factor feature (Di Mauro et al., 2005). It is interesting to observe that this variation could result in increased localization of the electron on the heteroatom. No changes were observed in the free radical g-factor in crude oil by exposure of the samples (irradiated and non-irradiated) to sunlight.

In agreement with Scotti and Montanari (Scotti & Montanari, 1998), the g-factor obtained by EPR for free radicals were found to be lower when the aromatic carbon fraction was larger, in the asphaltene of several crude oils registered by NMR. The fact that Arabian oil (g = 2.0033) presents a heteroatom weight percentage of 2.79%, smaller than Colombian oil (19.77% wt), and the fact that nitrogen and sulfur are located mainly in the aromatic systems in crude oil indicate that this oil is less aromatic than the Colombian oil (g = 2.0030). Another important fact obtained by EPR is that Colombian oil presents 19.6% wt of nitrogen and since this heteroatom is related mainly with porphyrinic and non-porphyrinic systems in petroleum asphaltenes, it is possible to affirm that the oil of Colombian origin has a larger asphaltene fraction than Arabian oil.

The line width ΔH_{pp} of the free radical signal was 5.2 ± 0.1 Gauss for Arabian crude oil and 5.3±0.1 Gauss for Colombian crude oil.

Solar irradiation did not alter the line width of the signal corresponding to the free radical within a few hours of exposure of the oil films. A decrease in the ΔH_{pp} values was observed for both oils in samples irradiated for 100 hours (Fig. 13 and 14). The reduction in line width from 5.1 to 4.6 Gauss (9.8%) in Arabian oil and from 5.4 to 4.4 Gauss (18.5%) in Colombian oil indicates photochemical degradation of the crude oils under solar light.

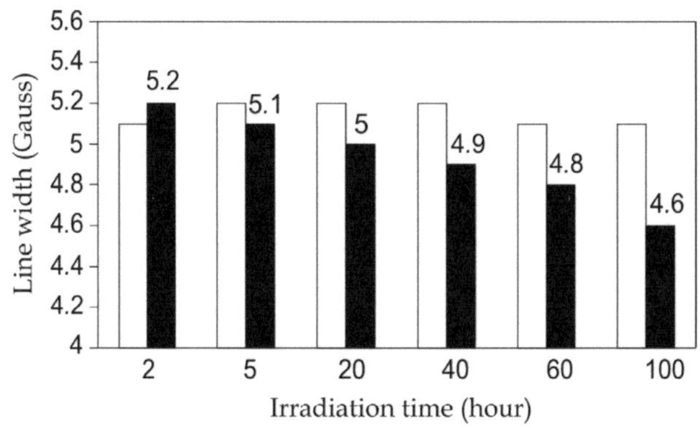

Fig. 13. Variation of the line width of the free radical versus sunlight irradiation time for Arabian crude oil. Filled columns represent the irradiated samples and blank columns represent the non-irradiated samples.

The spin relative counts of the free radical signal in crude oils, using Varian strong pitch signal as a spin counter standard, indicated a reduction of 12% in this paramagnetic species in irradiated Arabian oil, while in irradiated Colombian oil this corresponded to 35% after 100 hours under solar light. In the non-irradiated samples (control) the spin counts revealed an increase of 16% and 9% in radicals in Arabian and Colombian oils, respectively, indicating that the photochemical process is capable of degrading the aromatic components present in crude oils.

The degradation of crude oils can be observed by a reduction in the amount of free radicals related to spin counts and to line width narrowing of the EPR signal of these paramagnetic species, which react with atmospheric oxygen and can be extinguished during the exposure of crude oils to solar light. There was no significant variation in the ΔH_{pp} values for the free radicals in the non-irradiated samples. The line shape parameter was determined in the EPR spectra and showed no modification in relation to the value obtained before irradiation.

It is possible to affirm that the narrowing of the EPR line corresponding to free radicals in the irradiated oils was due to the rearrangement among the radicals present. The electromagnetic source, in the case of the solar light, with chemical modification properties in relation to some substances, can break links that result in the generation of free radicals, which upon suffering rearrangements or recombinations can produce other chemical species different from the precursory compounds (Guedes et al., 2006).

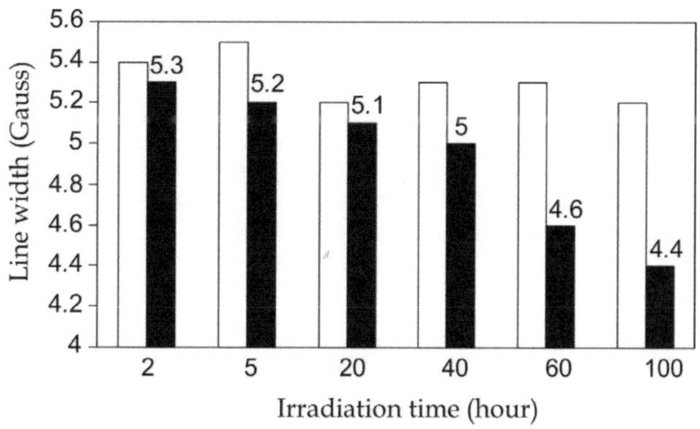

Fig. 14. Variation of the line width of the free radical versus sunlight irradiation time for Colombian crude oil. Filled columns represent the irradiated samples and blank columns represent the non-irradiated samples.

The oils studied in the present investigation were significantly affected by the action of sunlight under tropical conditions. Observations revealed that solar irradiation reduces the aromaticity of crude oil, degrading porphyrin complexes and at least partially destroying asphaltene fraction of oil. The EPR technique proved to be useful in the characterization of the molecular structure of asphalttenes in crude oil and also revealed changes of the photochemical nature in the oil under the effect of sunlight.

5. By-product oil by EPR

The interest in studying by-product oil is due to the fact this low-viscosity, when compared to the crude oil, allows high mobility of free radical in its environment. The by-product oil investigated by EPR was Marine diesel (fluid catalytic diesel, bunker, ship fuel). The marine diesel spectrum (Fig. 15) consist of signal from radicals with a typical hf splitting of protons, exhibiting a septet of lines with intensities proportional to 1, 6, 15, 20, 15, 6, 1. These correspond to different ways of form spins +3, +2, +1, 0, −1, −2, −3 due to the interaction of six equivalent and strong coupled protons [$A(^1H)$]. Each of seven hfs lines is split into four lines (quartet) due the three weakly coupled protons [$A'(^1H)$] with intensities proportional to 1, 3, 3, 1, and corresponding to the different ways of form spins $+3/2$, $+1/2$, $-1/2$, $-3/2$.

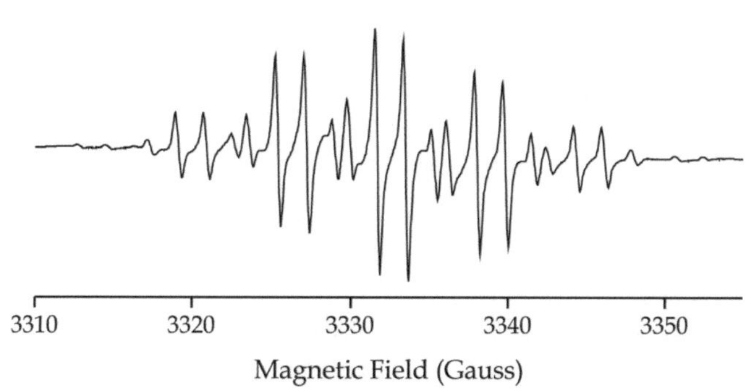

Fig. 15. RPE spectrum of marine diesel (Di Mauro et al., 2007).

The analysis suggesting first- and second-order hf splitting, with spin configurations described above, provides the layout of an energy diagram (Fig. 16).

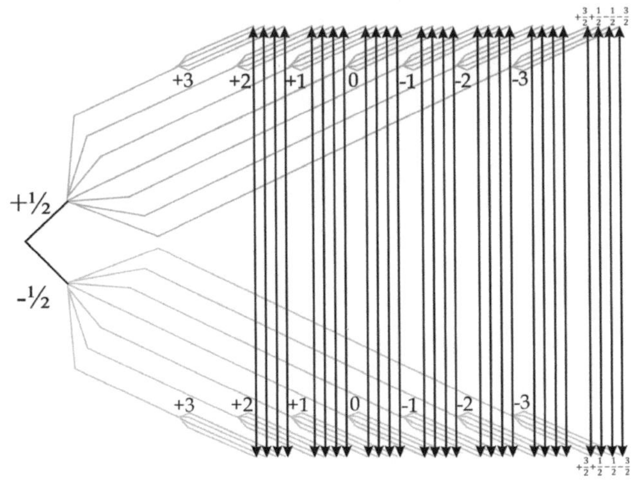

Fig. 16. Energy diagram to free radical in marine diesel (bunker).

Considering the selection rules, the allowed transitions are indicated by 28 vertical arrows representing each one of the spectral lines.

The radical rotates at a shorter correlation time than the reciprocal of spectrum in frequency. Therefore, in Fig. 15 the spectrum can be interpreted in terms of the following "isotropic" spin Hamiltonian:

$$\mathcal{H} = g\beta HS + AIS + A'IS \tag{6}$$

where $S = 1/2, I = 1/2$.

The WINEPR SimFonia Version 1.25 software of Bruker® was used in the simulation option for determining g, A, A' and ΔH_{pp} (peak-to-peak line with) of the free radical species. The parameter values found were g = 2.0028±0.0005, A = 6.31±0.01 G for 6 equivalent protons, A' = 1.80±0.01 G for three equivalent protons, and ΔH_{pp}= 0.38±0.02 G.

Therefore, with the parameters above and no superposition among spectrum lines, 28 lines are expected with intensities 1, 3, 3, 1, 6, 18, 18, 6, 15, 45, 45, 15, 20, 60, 60, 20, 15, 45, 45, 15, 6, 18, 18, 6, 1, 3, 3, 1 (Fig. 17a). The theoretically expected intensity, when compared with the intensities of spectrum lines (Fig. 17b), shows results in far agreement, except for the outer lines with low intensity.

The septet-quartet signal has been observed in petroleum-rich mudstone and carbonates (Ikeya & Furusawa, 1989). A similar signal observed in flints has been assigned to stable perinaphthenyl radicals (Chandra et al., 1988). Some authors (Ikeya, 1993; Uesugi, 2001) presented splitting similar to the pattern mentioned above assigning to *t*-butyl molecules. However, according to Forbes et al. (Forbes et al., 1991) the hf coupling for ^1H in *t*-butyl is 22.6 G, which does not fit the spectrum of marine diesel. Sogo et al. (Sogo et al., 1957) on the other hand, determined hf parameters for perinaphthene that fit fairly well to the spectrum

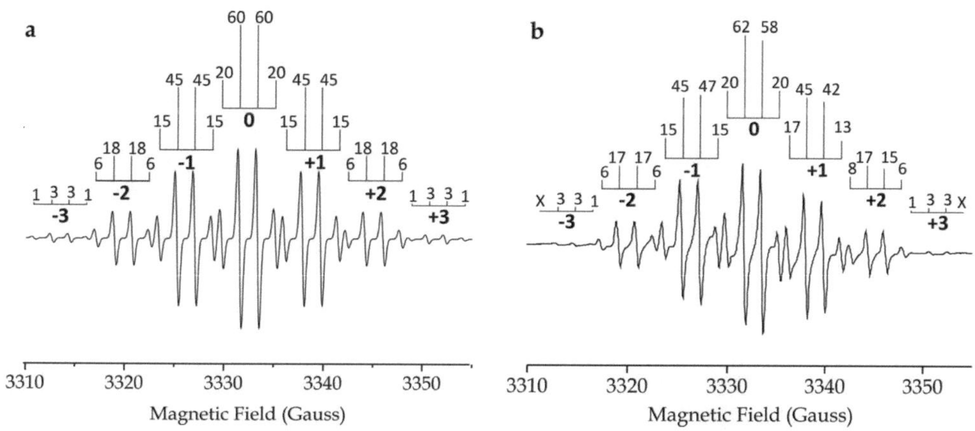

Fig. 17. **a** Simulation of septet-quartet EPR spectrum with g = 2.0028±0.0005, A = 6.31±0.01 G, A' = 1.80±0.01 G and ΔH_{pp} = 0.38±0.02 G. **b** EPR spectrum of marine diesel in X- band at room temperature, showing the hf separation into seven lines due to interaction of six equivalents strongly coupled protons and each of the seven lines is resolved into four lines due the three weakly coupled protons. The line intensities are indicated (Di Mauro et al., 2007).

of marine diesel. Besides, according to Gerson and Huber (Gerson & Huber, 2003), perinaphthenyl can be detected in pyrolysis products of petrol fractions. Another chemical evidence favorable to perinaphthenyl is its persistence (Sogo et al, 1957) when compared to the no very persistent *t*-butyl.

The results obtained for the hfs interaction of free radicals in marine diesel and the discussion regarding the organic molecule models indicate that perinaphthenyl radicals (Fig. 18) are the probably responsible for the septet-quartet EPR spectrum of this oil by-product (Di Mauro et al., 2007).

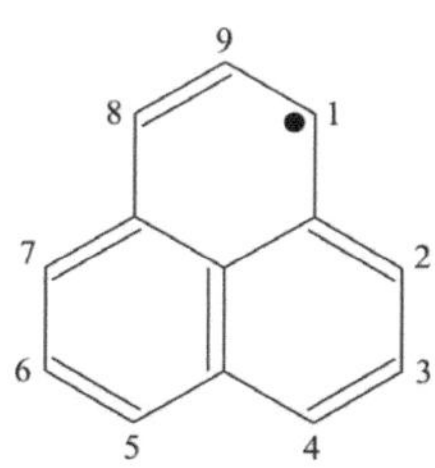

Fig. 18. Structural representation to perinaphthenyl radical indicating, 1 to 9, hydrogen atoms responsible for the hyperfine splitting observed in marine diesel spectrum.

According to Fig. 18, the hf splitting arise from the hyperfine interaction of the unpaired electron with the hydrogen atoms around the molecule. The six hydrogen atoms in positions 1, 2, 4, 5, 7, 8 are responsible for first-order hyperfine interaction and the other three atoms in positions 3, 6 and 9, the second-order hyperfine interaction.

It was verified that EPR signal attributed to the perinaphthenyl radical in marine diesel decreases in intensity and finally disappears with time, depending on the time that samples were exposed to air; this suggested that the radical undergoes a chemical reaction, probably with oxygen in air, since phenalenyl is sufficiently persistent in dilute deoxygenated solutions (Gerson, 1966; Hicks, 2007). Senglet et al (Senglet et al, 1990) observed weak phenalenyl radical spectra after six months of storage to fuel samples. Another possibility is that the perinaphtenyl radicals form a dimer (Fig. 19) becoming diamagnetic and, consequently, exhibiting no EPR signal. In the liquid state, usually all but the most stable free radicals rearrange or polymerize (Lewis & Singer, 1969). Studies (Gerson, 1966; Reid, 1958) indicate that the phenalenyl radical and its derivatives show self-association and formation of a diamagnetic dimer.

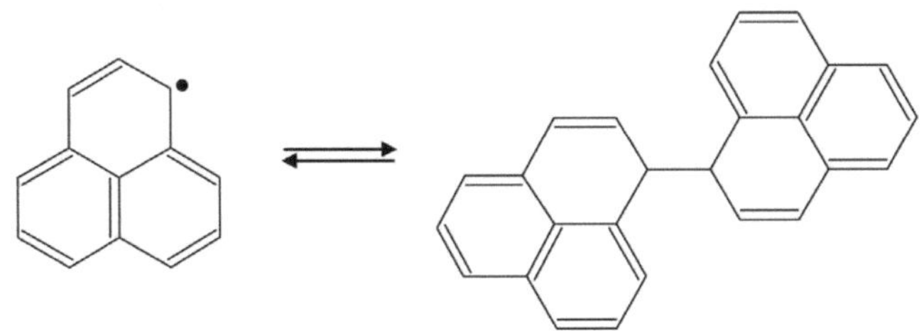

Fig. 19. Spontaneous self-associations of phenalenyl radicals forming the σ-dimer.

More recently, quantitative EPR studies (Zaitev et al, 2006; Zheng et al, 2003) confirmed that the phenalenyl dimerization occurs reversibly in carbon tetrachloride, toluene and dichloromethane, resulting in a complete signal loss at low temperatures due to dimer formation. Given that the phenalenyl radical generally exists in equilibrium with its diamagnetic dimer (Gerson, 1966) and taking into account the high mobility of this radical in marine diesel (Di Mauro et al, 2007), dimerization even at room temperature should be considered.

Concerning this possibility, marine diesel sample that exhibited no hyperfine resolved lines (Fig. 20a) was investigated by EPR in the temperature range from 170 to 400 K.

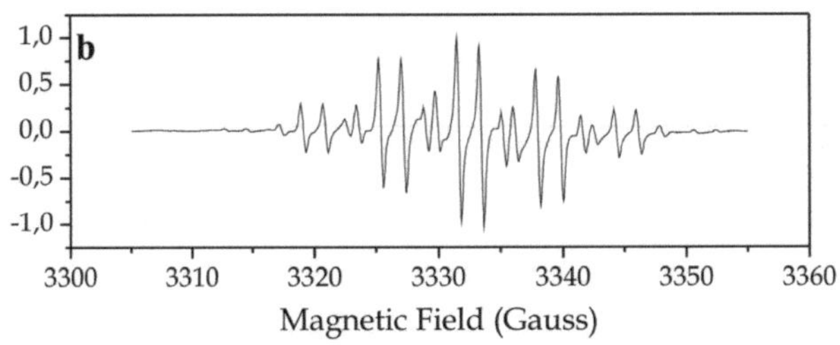

Fig. 20. Comparison between EPR spectrum of marine diesel. **a** EPR spectrum of marine diesel (older sample) at 9.37 GHz at room temperature. **b** EPR spectrum of marine diesel (fresh sample) (Piccinato et al., 2009).

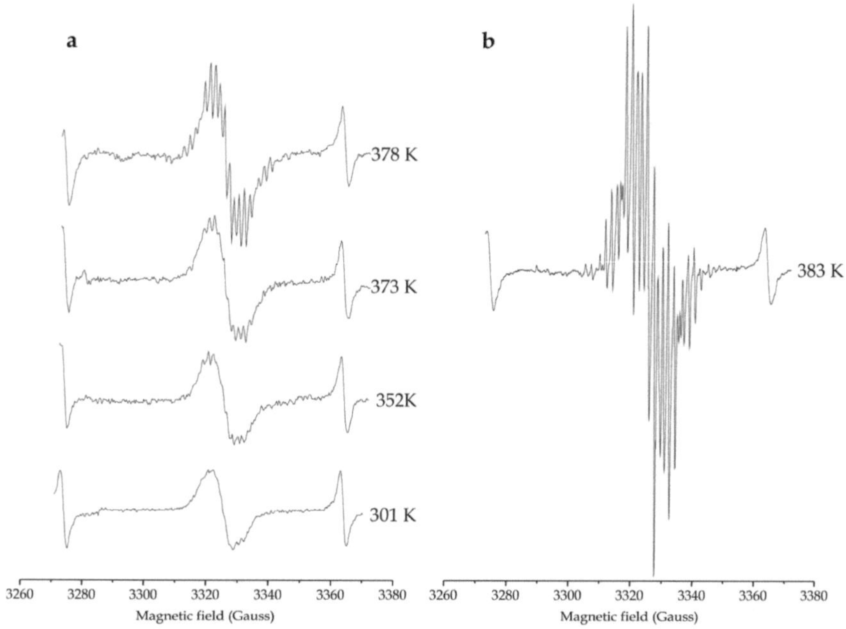

Fig. 21. **a** EPR spectra of marine diesel (older sample) at 9.37 GHz in the temperature range from 301 to 378 K. **b** Resolved hfs lines at 383 K.

The free radical EPR spectrum in marine diesel revealed a progressive appearance of the typical hfs of protons with heating (Fig. 21a). The hfs lines were superposed over the single line with a peak-to-peak line width of about 9 G present in all spectra (Fig. 21a), whose intensity also increased with temperature. Up to 378 K, it was impossible to determine the interaction of a free electron with hydrogen atoms protons. At 383 K, the spectrum became very intensive (Fig. 21b) exhibiting resolved lines.

This spectrum was analyzed to determine the types of free radicals manifested in this experiment. In order to investigate only resolved hfs lines, a single unresolved line (Fig. 22a) was subtracted from the spectrum (Fig. 22b). The spectrum resulting the subtraction (Fig. 22c) can be interpreted in terms of the isotropic spin Hamiltonian (eq. 6).

The First attempt in the simulation, with WINEPR SimFonia Version 1.25 software of Bruker®, was to consider a septet-quartet RPE spectrum attributed to the perinaphthenyl radical ($C_{13}H_9^{\bullet}$). However, this interpretation was not sufficient to reproduce the spectrum presented in Fig. 22c, indicating the superposition with other groups of less intensive lines which could be due to phenalenyl radicals with different number of splitting protons (Zaitev et al., 2006).

The investigation of the remaining lines in the EPR spectrum after subtraction of the first group of lines simulated (Fig. 23a) revealed the need to add a second group of lines due to the interaction of five equivalent and strongly coupled protons (sextet) and the interaction of three weakly coupled protons (quartet) (Fig. 23b). The chemical structure corresponding to this interaction is presented in Fig. 24b. The sum of these two simulated groups (Fig. 23a and b) reproduced all spectrum lines but not their intensities. The intensity problem was solved by the addition of a third group of lines due to the interaction of four equivalent and

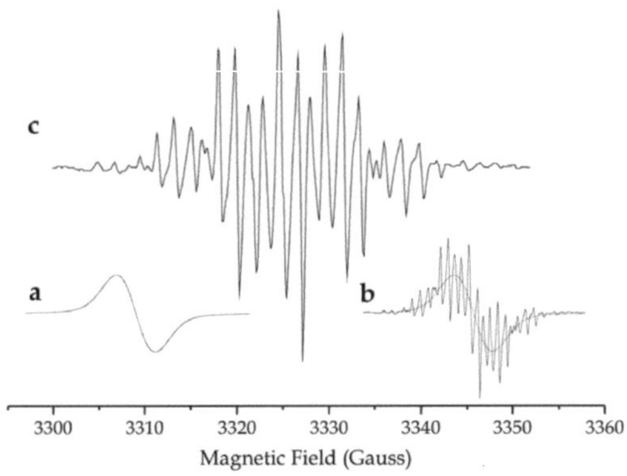

Fig. 22. Spectra subtraction for analysis of EPR hyperfine lines: **a** Unresolved line simulated by the software WINEPR SimFonia; **b** Overlap of the simulated spectrum and marine diesel spectrum at 383 K for subtraction of the unresolved line; **c** Result of the spectra subtraction.

strongly coupled protons (quintet) and the interaction of three weakly coupled protons (quartet) (Fig. 23c) whose structure is presented in Fig. 24c.

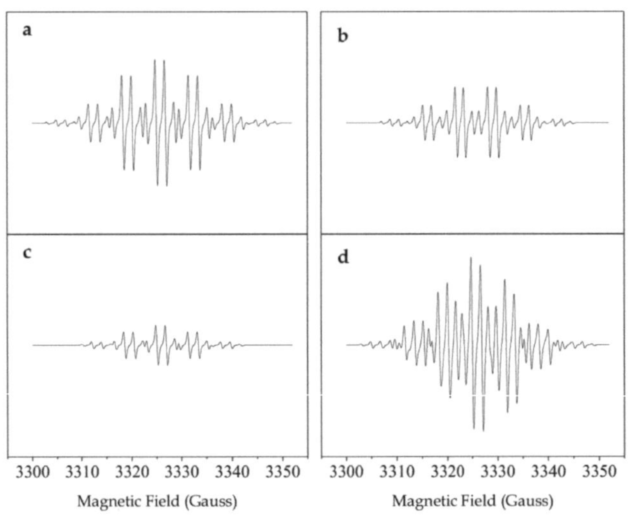

Fig. 23. **a** Simulation of the septet-quartet EPR spectrum. **b** Simulation of the sextet-quartet EPR spectrum. **c** Simulation of the quintet-quartet spectrum. **d** superposition of the septet-quartet, sextet-quartet and quintet-quartet with weight percentages of the 53.5, 30.0, and 16.5%, respectively.

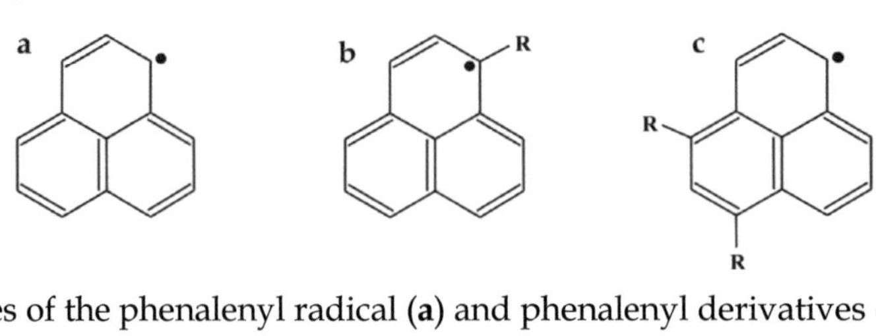

Fig. 24. Structures of the phenalenyl radical (**a**) and phenalenyl derivatives (**b** and **c**).

The superposition of three groups of lines generates a set of lines shown in Fig. 23d. The theoretical model with three groups of lines overlaps with the experimental spectrum (Fig. 25). The weight percentages was 53.5, 30.0 and 16.5% for the first, second and third groups, respectively. The hyperfine parameters (*A* and *A*´) and weight percentages in the intensity of lines, used in the simulation of the three groups of lines, are presented in Table 4.

Simulated spectrum	***A* (G)**	***A*´ (G)**	**Intensity (%)**
septet-quartet	6.41±0.03	1.82±0.02	53.3
sextet-quartet	6.21±0.03	1.64±0.02	30.0
quintet-quartet	6.16±0.03	1.83±0.02	16.5

Table 4. Hyperfine parameters and weight percentages in intensity of the lines used in the simulation of lines groups.

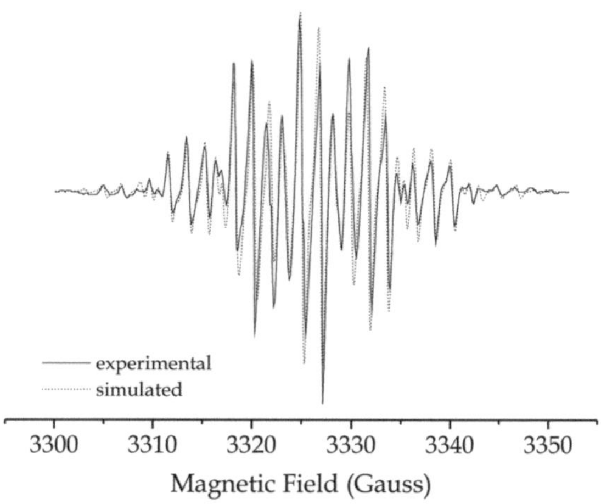

Fig. 25. Superposition of theoretical model, with three groups of lines (dotted line), and experimental spectrum (solid line).

The proposed model of three overlapped paramagnetic species accurately reproduced the experimental lines. Three paramagnetic species, phenalenyl plus two of its derivatives, found after heating, indicate that the system (older marine diesel) somehow preserved the phenalenyl structure.

Like the paramagnetic species were overlapped in the same spectrum, apart from the difficulty of obtaining a high-resolution spectrum that allowed the observation of the splitting due to the functional group protons, it was impossible to identify the functional group that substituted the hydrogen atoms.

Spectra to paramagnetic species with different functional groups that substituted one hydrogen atom are presented in literature (Lewis & Singer, 1969; Rabold et al, 1965; Wain et al, 2006). The calculated hfs coupling constants for the second lines group is in strong agreement with the values reported by Rabold et al (Rabold et al, 1965) to the hydroxiperinaphthenyl radical.

Yamada and Toyoda (Yamada & Toyoda, 1973) observed the formation of the 4,6-dimethylperinaphthenyl radical when acenaphthylene dissolved in an inert solvent was heat up to 440 °C, resulting in the EPR spectrum splitting due to the methyl protons. The two methyl groups occupying two α-positions produce a spectrum similar to that of the third group (Fig. 23c). Despite the observation of this radical only at high temperatures, it appears upon heating and as an intermediate compound in the reaction.

Despite the fact that it is impossible to describe the exact chemical transformations and the mechanisms involved in the appearance of magnetic species in marine diesel due to complexity of this oil by-product, the information provided by EPR spectroscopy, especially the hfs coupling, allowed to monitor the modifications and to suggest the type of free radical species formed in this oil during heating. Thus, with organic molecule models, perinaphthenyl radical are thermally recuperated by breaking the linkage formed in the dimer, and hydroxyperinaphthenyl and 4,5-dimethylperinaphthenyl radicals are the most likely phenalenyl derivatives yielded upon heating the marine diesel (Piccinato et al., 2009).

6. Conclusion

The EPR technique proved to be useful in the characterization of the molecular structure of asphaltenes in crude and by-product oil. The EPR parameters are able to reveal changes of the photochemical nature in the oil under the effect of sunlight or in some specific wavelength. This spectroscopy technique can be further explored and, consequently, applied in the monitoring of petroleum weathering and its by-products in environmental matrixes whenever accidents occur or routine operations result in oil spill. The period of permanence of toxic and refractory aromatic components of petroleum in aquatic systems can be evaluated not only through quantitative chromatographic methods, but also using nondestructive and more economical qualitative methods.

7. References

Bunce, N. J. (1987). Introduction to the Interpretation of Electron Spin Resonance Spectra of Organic Radicals. *Journal of Chemical Education*, Vol. 64, No. 11, (November, 1987), pp. 907-914, ISSN 0021-9584

Chandra, H., Symons, M. C. R. & Griffiths, D. R. (1988). Stable Perinaphthenyl Radicals in Flints. *Nature*, Vol. 332, No. 6164, (April 1988), pp. 526-527, ISSN 0028-0836

Di Mauro, E., Guedes, C. L. B. & Nascimento, O. R. (2005). Multifrequency (X-band to W-band) CW EPR of the Organic Free Radical in Petroleum Asphaltene. *Applied Magnetic Resonance*, Vol. 29, No. 4, (December 2005), pp. 569-575, ISSN 0937-9347

Di Mauro, E., Guedes, C. L. B. & Piccinato, M. T. (2007). EPR of Marine Diesel. *Applied Magnetic Resonance*, Vol. 32, No. 3, (November 2007), pp. 303-309, ISSN 0937-9347

Dickson, F. E. & Petrakis, L. (1974). Application of Electron Spin Resonance and Electronic Spectroscopy to the Characterization of Vanadium Species in Petroleum Fractions. *Analytical Chemistry*, Vol. 46, No. 8, (July 1974), pp. 1129–1130, ISSN 0306-7319

El Anba-Lurot, F., Guiliano, M., Doumenq, P. & Mille, G. (1995). Photo Oxidation of 3,3´- and 4,4´-dimethylbiphenylsin Natural Sea Water. *International Journal of Environmental Analytical Chemistry*, Vol. 61, No. 1, pp. 27–34, ISSN 0306-7319

Espinosa, M., Campero, A. & Salcedo, R. (2001). Electron Spin Resonance and Electronic Structure of Vanadyl-porphyrin in Heavy Crude Oils. *Inorganic Chemistry*, Vol. 40, No. 18, (July 2001) pp. 4543–4549, ISSN 0020-1669

Gerson, F. & Huber, W. (2003). *Electron Spin Resonance Spectroscopy of Organic Radicals*. Wiley-VCH, ISBN 978-3-527-30275-8, Weinheim, Germany

Gerson, F. (1966). Notiz über das ESR.-Spektrum des Phenalenyl-Radikals. *Helvetic Chimica Acta*. Vol. 49, No. 5, pp. 1463-1467, ISSN 1522-2675

Guedes, C. L. B., Di Mauro, E., Antunes, V. & Mangrich, A. S. (2003). Photochemical Weathering Study of Brazilian Petroleum by EPR Spectroscopy. *Marine Chemistry*, Vol. 84, No. 1, (December 2003), pp. 105-112, ISSN 0304-4203

Guedes, C. L. B. (1998). Intemperismo Fotoquímico de Petróleo Sobre Água do Mar: Estudo do Processo Natural e Efeito da Adição da Tetrafenilporfina. PhD Thesis, Instituto de Química. Universidade Federal do Rio de Janeiro, Brazil, unpublished

Guedes, C. L. B., Di Mauro, E., Campos, A., Mazzochin, L. F., Bragagnolo, G. M., Melo, F. A. & Piccinato, M. T. (2006). EPR and Fluorescence Spectroscopy in the Photodegradation Study of Arabian and Colombian Crude Oils. *International Journal of Photoenergy*, Vol. 2006, No. 1, (June 2006), 48462, pp. 1-6, ISSN 1110-662X

Guedes, C. L. B., Di Mauro, E., Mangrich, A. S., Ramoni, M. & Antunes, V. (2001). Study of the Photodegradation of Oil by Electronic Paramagnetic Resonance, In: *Série Ciência-Técnica-Petróleo, Seção Química, CD-R 3 145-154*, Available from< http://www2.uel.br/grupo-pesquisa/meioambiente/fotopetro/arquivos/artigos/0011.pdf>

Hicks, R. G. (2007). What's New in Stable Radical Chemistry? *Organic & Biomolecular Chemistry*, Vol. 5, No. 9, pp. 1321-1338, ISSN 1477-0520

Ikeya, M. & Furusawa, M. (1989). A Portable Spectrometer for ESR Spectrometry, Dosimetry and Dating. *International Journal of Radiation Applications and Instrumentation. Part A. Applied Radiation and Isotope*, Vol. 40, No. 10-12, pp. 845-850, ISSN 0883-2889

Ikeya, M. (1993). *News Applications of Electron Spin Resonance*, World Scientific, ISBN 13 978-9810211998, Singapore

Janzen, E. G. (1969). Substituent Effects on Electron Spin Resonance Spectra and Stability of Free Radicals. *Accounts of Chemical Research*, Vol. 2, No. 9, (September, 1969), pp. 279-288, ISSN 0001-4842

Lartiges, S. B. & Garrigues, P. P. (1995). Degradation Kinetics of Organophosphorus and Organonitrogen Pesticides in Different Waters under Various Environmental Conditions. *Environmental Science & Technology*, Vol. 29, No. 5, (May 1995), pp. 1246–1254, ISSN 0013-936X

Lewis, I. C. & Singer, L. S. (1969). Carbonization of Aromatic Hydrocarbons. *Preprints of Papers - American Chemical Society, Division of Fuel Chemistry*. Vol. 13, No. 4, pp. 86-100

Montanari, L., Clericuzio, M., Del Piero, G. & Scott, R. (1998). Asphaltene Radicals and Their Interaction with Molecular Oxygen: An EPR Probe of Their Molecular Characteristics and Tendency to Aggregate. *Applied Magnetic Resonance*, Vol. 14, No. 1, (February 1998), pp. 81-100, ISSN 0937-9347

Nicodem, D. E., Fernandes, M. C. Z., Guedes, C. L. B. & Correia, R. J. (1997). Photochemical Processes and the Environmental Impact of Petroleum Spills. *Biogeochemistry*, Vol. 39, No. 2, (November 1997), pp. 121–138, ISSN 0168-2563.

Nicodem, D. E., Guedes, C. L. B. & Correia, R. J. (1998). Photochemistry of Petroleum I. Systematic Study of a Brazilian Intermediate Crude oil. *Marine Chemistry*, Vol. 63, No. 1-2, (December 1998), pp. 93–104, ISSN 0304-4203

Nicodem, D. E., Guedes, C. L. B., Fernandes, M. C. Z., Severino, D., Correa, R. J., Coutinho, M. C. & Silva, J. (2001). Photochemistry of Petroleum. *Progress in Reaction Kinetics and Mechanism*, Vol. 26, No. 2-3, pp. 219–238, ISSN 1468-6783

O'Reilly, D. E. (1958). Paramagnetic Resonance of Vanadyl Etioporphyrin I. *The Journal of Chemical Physics*, Vol. 29, No. 5, (November 1958), pp. 1188–1189, ISSN 0021-9606

Orton, J. W. (1968). *Electron Paramagnetic Resonance*, Iliffe Books Ltd, ISBN 13 9780592050416, London, United Kingdom

Piccinato, M. T., Guedes, C. L. B. & Di Mauro, E. (2009). EPR Characterization of Organic Free Radicals in Marine Diesel. *Applied Magnetic Resonance*, Vol. 35, No. 3, (April 2009), pp. 379-388, ISSN 0937-9347

Poole, C. P. (1967). *Electron Spin Resonance – A Comprehensive Treatise on Experimental Techniques*, John Wiley & Sons Inc, ISBN 13 978-0470693865, New York, United State of America

Reid D. H. (1958). Stable π-Electron Systems and New Aromatic Structures. *Tetrahedron*, Vol. 3, No. 3, pp. 339-352, ISSN 0040-4020

Saraceno A. J., Fanale D. T. & Coggeshall, N. D. (1961). An Electron Paramagnetic Resonance Investigation of Vanadium in Petroleum Oils. *Analytical Chemistry*, Vol. 33, No 4, (April 1961), pp. 500-505, ISSN 0003-2700

Scott, R. & Montanari L. (1998). Molecular Structure and Intermolecular Interaction of Asphaltenes by FI-IR, NMR, EPR, In: *Structures and Dynamics of Asphaltenes*, Oliver C. Mullins and Eric Y. Sheu, pp. 79–113, Plenum Press, ISBN 0306459302, New York, United State of America

Senglet, N., Faure, D., Des Courières, T., Bernasconi, C. & Guilard, R. (1990). E.S.R. Characterization of Phenalenyl Radicals in Various Fuel Samples. *Fuel*, Vol. 69, No. 2, (February 1990), pp. 203-206, ISSN 0016-2361

Sogo, P. B., Nakazaki, M. & Calvin, M. (1957). Free Radical from Perinaphthene. *Journal of Chemical Physics*, Vol. 26, No. 5, (May, 1957) pp. 1343-1345, ISSN 0021-9606

Uesugi, A. & Ikeya, M. (2001). Electron Spin Resonance Measurement of Organic Radicals in Petroleum Source Rock Containing Transition Metal Ions. *Japanese Journal of Applied Physics Part 1-Regular Papers Short Notes & Review Papers*. Vol. 40, No. 4A, (April 2001), p. 2251-2254, ISSN 0021-4922

Yen, T. F., Erdman, J. G. & Saraceno, A. J. (1962). Investigation of the Nature of Free Radicals in Petroleum Asphaltenes and Related Substances by Electron Spin Resonance. *Analytical Chemistry*, Vol. 34, No. 6, (May 1962), pp. 694–700, ISSN 0003-2700

Zaitev, V., Rosokha, S. V., Head-Gordon, M. & Kochi, J. K. (2006). Steric Modulations in the Reversible Dimerizations of Phenalenyl Radicals via Unusually Weak Carbon-Centered π- and σ-Bods. *Journal of Organic Chemistry*, Vol. 71, No. 2, pp. 520-526, ISSN 0022-3263

Zheng, S., Lan, J., Khan, S. I. & Rubin, Y. (2003). Synthesis, Characterization, and Coordination Chemistry of the 2-Azaphenalenyl Radical. *Journal of American Chemical Society*, Vol. 125, No. 19, (April 2003), pp. 5786-5791, ISSN 0002-7863

Rabold, G. P., Bar-Eli, K. H., Reid, E. & Weiss, K. (1965). Photochemically Generated Free Radicals. I. The Perinaphthenone System. *The Journal of Chemical Physics*, Vol. 42, No. 7, (April 1965), pp. 2438-2447, ISSN 0021-9606

Wain, A. J., Drouin, L. & Compton, R. G. (2006). Voltammetric Reduction of Perinaphthenone in Aqueous and Non-Aqueous Media: An Electrochemical ESR Investigation. *Journal of Electroanalytical Chemistry*, Vol. 589, No. 1, (April 2006), pp. 128-138, ISSN 1572-6657

Yamada, Y. & Toyoda, S. (1973). An Electron Spin Resonance Study of the Carbonization of Acenaphthylene. *Bulletin of the Chemical Society of Japan*, Vol. 46, No. 11, pp. 3571-3573, ISSN 1348 0634

9

Effects of Crude Oil Contaminated Water on the Environment

Noyo Edema
Department of Botany, Delta State University, Abraka, Delta State Nigeria

1. Introduction

With the development of oil industry, the general environment and in particular wetland ecosystem has become extremely vulnerable to damaging effects of oil pollution. Contamination of aquatic environment by crude oil and petroleum products constitute an additional source of stress to aquatic organisms (Omoregie *et al.*, 1997) and is of importance to the wetland environment. Oil contaminated water resulted in water becoming unsuitable for the growth of macrophytes (Edema, 2006) only scanty data are available for levels of chemical pollution of aquatic plants since most studies of biota are concentrated in fish (FAO, 1993).

2. The water environment

Water quality is one important factor of an aquatic environment. Water analysis consists of an assessment of the condition of water in relation to set goals. For example, water samples with decreased electrical conductivity measurement indicate a good measure of purity (Hoagland, 1972). During spillage, water supply becomes critical.

Toxic pollutants in water refer to a whole array of chemical which are leached into ground water or which are discharged directly into rivers. Contamination of aquatic environment by crude oil and petroleum products constitute an additional source of stress to aquatic organisms (Omoregie *et al.*, 1997) and is of importance to wetland environment.

Water pollutants can also include excessive amounts of heavy metals, radioactive isotopes, faecal coliform bacteria, phosphorus, nitrogen, sodium and other useful (even necessary) elements as well as certain pathogenic bacteria and viruses (Botkin and Keller, 1998).

The water environment experiences many dynamic changes induced by various natural events such as the spillage of toxic chemicals that may have significant impact on aquatic life (Camougis, 1981).

Even in Roman times, heavy metals from mining and pathogens from cities caused serious though local, water contamination (FAO, 1993). Some of the major factors associated with accelerating pace of fresh water pollution is accidental damage of pipes and tankers, major leaks and local spills. These cause varying degrees of aquatic toxicity and material damage. Industrial accidents involving spillage of long lasting pollutants such as persistent organic substances have the most serious effects on water quality. Many of these substances become concentrated in living tissue because organisms have no means of excreting them. They accumulate and are pass on at successively greater concentration of predators higher up the food chain.

This chapter is primarily focused on the biological impact of the exploration activities of oil companies in our water environment.

3. Crude oil

Crude oil is a colloidal mixture of huge number of hydrocarbon and non-hydrocarbon (Cadwallaer, 1993). The source material for nearly all petroleum products is crude oil. Spill, leaks and other releases of gasoline, diesel, fuels, heating oils and other petroleum products often result in the contamination of soil and water. Hydrocarbon form over 90 percent petroleum oil are grouped according to their chemical structures such as straight, branched and cyclic alkanes and aromatics. The non-hydrocarbon components of petroleum include (O_2, N, S ---) and some metals related porphyrin oxygen containing compounds e.g. naphthenic acid, carboxylic acid, esters, ketones, phenols etc (Odu, 1981). Oil pollution occurs when oil is introduced into the environment directly or indirectly by men's impacts resulting in unfavorable change in such a way that safety and welfare of any living organisms is endangered. Crude oil if spilled into the water spreads over a wide area forming a slick and oil in water immediately begins to undergo a variety of physical, chemical and biological changes including evaporation of high volatile fractions, dissolution of water-soluble fractions, photochemical oxidation, drill, emulsification, microbial degradation and sedimentation (Muller, 1987). Crude oil is a complex mixture of hydrocarbon and organic compounds of sulphur, nitrogen, oxygen and a certain quantity of water which varies in composition from place to place (Anoliefo, 1991). Crude oil is produced from decay of plants and animals over millions of years. It is also referred to as mineral oil. Crude oil, which is a mixture of hydrocarbons and inorganic compounds is drilled through rocks. Crude oil discharged on the sea surface undergoes physical, chemical and biological alteration. Rapid physical and chemical processes include spreading and movement by wind and currents, injection into the air, evaporation of volatile components, dispersion of small droplets into water, dissolution and chemical oxidation (Nelson Smith, 1972). Concurrent with these are relevant biological processes. They include degradation by micro-organisms and uptake by larger organisms followed by metabolism and storage or discharge (Nelson-Smith, 1972).

4. Water soluble fraction

Water and oil are usually considered to be non-miscible. However, crude oil contains a very small soluble portion referred to as the water soluble fraction (WSF) Kavanu, 1964. The soluble constituents are dispersed particulate oil, dissolved hydrocarbons and soluble contaminants such as metallic ions (Kauss and Hutchinson, 1975). Non-hydrocarbon components of crude oil include polar components containing nitrogen, sulphur and oxygen (Westlake, 1982). Oxygen containing compounds include esters and ketones, while nitrogen containing compounds include pyrimidine and quinoline (Obire, 1985).

The concentration of hydrocarbon and non-hydrocarbon components in crude oil from different sources differ greatly (Brunnock *et al.*, 1968). The components of crude oil that go into solution make up the WSF. They are taken up by living cells and metabolized. This is ecologically important because in the event of an oil spill or effluent discharge from engine oil vehicles or where there is deliberate discharge of petroleum products into aquatic habitats, these hydrocarbons are absorbed by living organisms, with serious effects on the ecosystem (Michael, 1977). The lower the molecular weight of the constituent hydrocarbon of crude oil, the higher is its concentration in the water-soluble fraction (Bohon and Clausen, 1951). Anderson *et al.* (1974) analyzed the water soluble extract of South Loisiana and Kuwait crude oils and reported that the WSF contained 20 aromatic compounds ranging from benzene to dimethyl phenothrenes and up to 14 saturated hydrocarbons ranging from C_{14} to parafins.

Adverse biological effects have been attributed to dissolved low molecular weight hydrocarbon particularly aromatics such as toluene. Anderson *et al.* (1974) and Winter (1976) considered naphthalene as a more important source of crude oil toxicity than low molecular weight aromatics. According to another source, the low boiling point unsaturated hydrocarbon such as benzene, toluene, xylene and naphthalene, are the most toxic components in crude oils, the toxicity can be said to be a function of the presence of these substances (Nelson-Smith, 1972).

When there is delay in clean up action for any reason, after spillage has occurred, the water soluble components of crude oil seep into the aquatic ecosystem. The components of crude oil that go into solution make up the WSF. Concave (1979) reported that pure hydrocarbon yield 4.2mgl^{-1} of WSF. Baker (1970a) observed that water soluble fraction (WSF) is produced during a long period of oil water contact.

5. Preparation of WSF

The Water Soluble Fraction was prepared according to the method of Anderson *et al.* (1974). A sample of crude oil (500ml) was slowly mixed in equal volume of deionized water in a 2 liter screw-cap conical flask. A Gallenkamp table top magnetic stirrer supplied with 7/1cm magnetic bar was used for mixing. Stirring was done for 20hrs and at room temperature (27^{0}C ± 2^{0}C). After mixing, the oil water mixture was allowed to stand overnight in a separating funnel. The lower phase was collected and used as the WSF. It was referred to as 100% or full- strength WSF. The stock WSF was diluted with water to give 50% and 25% strength WSF which were stored in crew-cap bottles prior to use. The WSF samples were applied at three levels, 25%, 50% and 100%.

6. Composition of WSF

Ionic components: These include the cations and anions.

Cation

Cations are positively charged ions. The four major cations of the total ionic salinity of water for all practical purposes are Ca^{++}, Mg^{++}, Na^{+} and K^{+} (Wetzel, 2001). These elements are required by plants in large amounts (Hopkins, 1999). Water soluble fraction of crude oil has been found to contain the following cations Na^{+}, Ca^{++}, Mg^{++}, Fe^{++}, Fe^{+++} , NH_4^{+}, K^{+} (Edema, 2006)

Anions

Anions are negatively charged ions (Botkin and Keller, 1998). Anions, such as Cl^-, NO_3^- and SO_4^{2-} are soluble and are present in living plants largely as ions in solution (Hopkins, 1999). The major anions that constitute the ionic salinity of water for all practical purposes are Cl^-, SO_4^{2-}, HCO_3^- and CO^-_3 (Wetzel, 2001). Water soluble fraction of crude oil was found to contain, Cl^-, SO^{2-}_4, NO_3^-, PO_4^{2-} and HCO_3^- (Edema, 2006).

Heavy metals

The contamination of the aquatic system with heavy metals has been on the increase since the last century due to industrial activities (Ali and Mai, 2007). Heavy metals are taken up as cations. Among the heavy metals detected in WSF are Pb, Cu, Zn, Cd, Ni, Cr, and V (Edema, 2006). This is in agreement with the statement of Kauss and Hutchinson (1975) that the WSF of crude oil contains metallic ions among other soluble contaminants. Botkin and keller (1998) stated that Pb, Cr and V are among metals that pose hazard to living organisms. Heavy metals are non-biodegradable and are toxic under certain condition (Rana, 2005).

7. Physical composition

The physical components found to be present in WSF of crude oil include hydrogen ion concentration (pH), chemical oxygen demand (COD), total dissolved solids (TDs) and electrical conductivity (EC) (Edema, 2006). Chemical substances in WSF are capable of changing the hydrogen ion concentration (pH) of the medium (Neff and Anderson, 1981). Elevation of pH values after the introduction of macrophytes to WSF of crude oil were recorded by Edema *et al.* (2008). The pH values recorded were within the maximum permissible level (pH 6.5 - 9.8 values) (WHO, 1995). The values for electrical conductivity, total dissolved solids, chemical oxygen demand were significantly found to increase after exposure to plants. High EC, TDS and COD are signs of pollution. This means that more ions were available after the introduction of test macrophytes. Increase in EC indicates that most inorganic elements exist in abundance (Kadiri, 2006). Although the total dissolves solids (TDS) values were found to increase after exposure to plants were still within the highest desirable limit of WHO (500mg/l) and Talling and Talling (1965), classification Scheme of African water within conductivities of between 6,000 - 16,000μScm^{-1}. With continuous spillage, the values may rise and exceed the WHO values.

8. Salinity

The concentration of the 4 major cations Ca^{++}, Mg^{++}, Na^{++} and K^+ and 4 major anions, HCO_3^-, CO_3^-, SO_4^{2-} and Cl^- usually constitute the total ionic salinity of water for all practical purposes (Wetzel, 2001). Concentrations of ionized components of other elements such as N, P and Fe and numerous minor elements are of immerse biological importance but are usually minor contributors to total salinity (Wetzel, 2001). The sum of all ionic concentrations is the basis for salinity measurement (Covich, 1993). Total dissolved solids and ionic conductivity of water, are generally used measurement (Covich, 1993).

The values of ions increased with increase in concentration of WSF prior to used. This is in agreement with the report of McOliver (1981) that when there is oil spillage more salts are released into river. Thus, the amount of salts contained in aquatic ecosystem increased. These increases could be due to leakage of the cells brought about by salt (ionic) stress and associated oxidative damage (Burdon *et al.*, 1996). Salt stress refers to an excess of ions and is not limited to Na^+ and Cl^- ion (Hopkins, 1999). According to Hernandez *et al.* (1985) oxidative stress is influenced by environmental factors, metal ion deficiency and toxicity. The sum of ions in the WSF of crude oil studied had higher values than river waters of Africa as reported by Wetzel (2001).

WSF, %	Sum of all ionic contents (as the total salinity)		
	Before	**After**	**Difference**
25	123.33	245.50	121.17
50	151.25	263.70	112.43
100	206.26	385.80	197.54

Table Ia. Sum of all ionic contents of Amukpe well- head WSF before and after exposure to *Pistia stratiotes.* Source: Edema and Okoloko (2008)

WSF, %	Sum (EC+ TDS)		
	Before	**After**	**Difference**
25	145.83	191.10	45.27
50	167.62	263.40	95.79
100	220.35	374.56	154.21

Table Ib. Sum of EC and TDS (as total salinity) of the WSF of Amukpe well-head crude oil before and after exposure to *Pisitia stratiotes.* Source: Edema and Okoloko (2008)

WSF, %	Sum of all ionic concentration		
	Before	**After**	**Difference**
25	12.95	262.05	249.02
50	22.90	338.21	315.31
100	46.75	371.69	324.94

Table IIa. Sum of all ionic concentration of the WSF of Ogini well-head crude oil before and after exposure to *Azolla sp.*

WSF, %	Sum of (EC + TDS)		
	Before	After	Difference
25	14.60	287.28	264.68
50	21.40	340.38	318.94
100	36.02	455.00	418.98

Source: Edema (2009)

Table IIb. Sum of EC +TDS (as total salinity) of the WSF of Ogini well-head crude oil before and after exposure to *Azolla sp.*

9. Produced water

Almost all offshore oil produces large quantities of contaminated water that can have significant environmental effects if not handled appropriately (Will, 2000). Oil and gas reservoirs have natural water layer (called formation water). Because, it is denser is found under the hydrocarbons. Oil reservoirs frequently contain large volume of water. To achieve maximum oil recovery, additional water is usually injected into surface. The formation and injected water are eventually produced along with hydrocarbons. The product of the formation and injected water is referred to as produced water.

There is more in produced water than water and oil. Neff and Anderson (1981) described produced water for ocean discharge as containing up to 48ppm of petroleum. This is because it had usually been in contact with oil in the reservoir rock. There were also elevated concentration of barium, beryllium, cadmium, chromium, copper, Iron, lead, nickel, silver and zinc and "small amount of the neutral radionucleids radium 226 and radium 228 and non-volatile dissolved organic material of unknown composition". Due to rapid mixing with seawater, most physical - chemical features of produced water (low dissolved oxygen and pH, elevated salinity and metals) do not pose any hazard to water, elevated concentration of hydrocarbon may be detected in surface sediments up to about 1,000 from the discharge, that contains aromatic hydrocarbons and metals. These aromatic hydrocarbons and metals in produced water were reported by Neff and Anderson (1981) to be toxic to organisms.

10. Biological effects of crude oil

Baker (1970) reported that oil pollution effects vary according to the type and amount of oil involved, the degree of weathering, the time of the year, the plant species concerned and the age of the plant. Cowell (1977) include the physical and chemical properties of the oil as well as the quantity of the water being polluted. The water soluble fraction of crude oils has been found to reduce the growth rate of biomas turnover of some marcophytes (Gunlack and Hayes, 1977). Kauss and Hutchinson (1975) found that aquatic macrophytes population was reduced in the presence of water-soluble petroleum components. The inhibitory effects of petroleum components are known to be dependent upon the concentration of the crude oil as well as that of water soluble components (Shew, 1977).

Aquatic macrophytes e.g. water fern (*Azolla africana*) have been used to remove heavy metals from solution (Talor and Sela, 1992). Heavy metals (nickel, mercury, cadmium) are potential carcinogens from drinking water. These elements may become 4,000 - 20,000 times more concentrated in plants, than in water (Brix and Shierup, 1989., Horan, 1990., Mason, 1993) as a result of bioaccumulation. Duckweeds (*Spirodela polyrrhiza* and *Lemna* minor) have thus been used for the removal of excessive nutrients from polluted water (Culley and Epps, 1993). *Pistia stratiotes* has been similarly employed for the removal of nutrient from polluted bodies (Aoi *et al.*, 1996). The plants absorb and incorporate the dissolved materials into their structure.

Studies with various species of organisms have demonstrated that the developmental stages of organisms are often more sensitive to toxicant than older stages (Alexander, 1977). Oil with aromatic content of 33 percent reduced the growth of maize plants to 31 percent.

Flowers and Haji bagheri (2001) reported that the effect of ion on the growth of leaves is determined by the ability of plants to accommodate the ions within compartments of the leaves cells where they will not do damage. If the ions are accommodated in the vacuole and concentration rises in the leaf apoplast then there will be osmotic effect on the leaf growth. Okoloko and Bewley (1982) reported the enhancement of protein synthesis in moss (*Tortula ruralis*) gametophytes exposed to 5mM aqueous SO_2. Higher levels were toxic. Some components, particularly compounds, are toxic to aquatic animals and plants (Odiete, 1999). They are acutely lethal and chronically lethal in sublethal concentration of part per billion (ppb). However, plants and animals vary widely in their sensitivity (Clark, 1982).

The soluble fraction of crude petroleum depress phytoplankton, photosynthesis, respiration and growth, and also kill or cause developmental abnormalities of young metallic ions present in the WSF may inhibit root growth (Winter *et al.* 1976).

Factors connected with phytotoxicity of oil as already mentioned, are the properties of the oil, the quantity of the oil applied and the environmental conditions. Other factors include the species of plants and the parts of the plant affected. Also of importance are the thickness of cuticle, number and structure of stomata, chloroplast physiology, CO_2 fixation pattern and photosynthetic electron transport system. Relatively low levels of pollution can cause rapid depression in the rate of net photosynthesis (Fitter and Hay, 1987).

The adverse effects of petroleum and its components on growth have earlier been recorded by Gill *et al.,* 1992. Oil contaminated water soluble fraction resulted in water becoming unsuitable for growth of aquatic macrophytes. Anoliefo (1991) reported that slight pollution with petroleum products or WSF makes carbonyl compounds available in the soil, hence the observed rapid increase in the growth of melon plants. Baker (1970b) reported that they inhibit metabolic processes. Edema and Okoloko (1997) showed increase in inhibition with increase in concentration of WSF. Studies on the effects of the water soluble fraction of Escravos hight and Odidi well oil on *Allium cepa* showed a "fertilizer effect" at 12.5% WSF level (Edema and Okoloko, 1997). Only higher levels of WSF are toxic. Also, growth enhancement and early flowering of *L. esculentus* at 0.25 x 10^{-3}ml/g concentration of crude oil treated soil was reported by Edema and Etioyibo (1999).

Edema (2010b) reported leaf bud formation in *Allium cepa* at 25% WSF and root initiation at 50% and 100% WSF treatments. Total inhibition of root was recorded for produced water at

100% which shows that produced water was more stressful to the plant than WSF of the same crude oil. Edema (2010b) also reported increase in catalase activity for WSF and decrease in catalase activity for produced water. Increased level of catalase activity is an indication of increased production of free radicals occasion by exposure of plants to crude oil. While decrease in activities of detoxification mechanisms of hydrogen peroxide can generate severe cell damage due to increase production of toxic oxygen radicals (Hernandez *et al;* 1995).

Salts influence the activities of aquatic plants resulting in the death of aquatic plants. Salt stress has been reported by Concave (1979) to reverse the condition that could make essential nutrients available to plants thereby resulting in mitoclondria damage (Cowell, 1977). The presence of ions in plants may block the oxidation of pyruvate for energy production (Anoliefo, 1991).

High salinity could be toxic to bacteria and may inhibit their activities. Besides, salinity may also affect the cellular infiltration pressure, leading to plasmolysis or even cell breakup (Ji *et al;* 2009).

Ajao *et al.* (1981) reported that undiluted formation water was toxic to millet at 1 - 100mg/l (1 - 100ppm) level. Edema (2010) also reported that produced water (PW) of crude oil was more toxic to *Allium cepa* than the water soluble fraction (WSF) of crude oil. Catalase activity could not be measured for *Allium cepa* exposed to produced water because of inhibition of growth (Edema, 2010).

The exposure of plants to metals results in the synthesis of phytochelatins, a metal binding polypeptide (Stern, 2000). Phytocheletins are known to sequester and detoxify metals through the formation of metals. Phytochelatin complexes (Stern, 2000; Clemens *et al;* 1999, Ha *et al.*, 1999). Plant nutrient elements such as Ca, Mn, Fe and Zn for example may enter through Ca channels or by means of broad-range metal transporter previously identified as Fe transporter (Clemens *et al.*, 1999). Edema (2006) reported that the levels of Fe in WSF before and after exposure to *Pistia* were far above the highest desirable limit for drinking water and still within the maximum permissible limit of 1.0mg/l of WHO value.

Edema and Asagba (2007) reported reduction in temperature and DO after exposure to the different levels of WSF. Reduction in the DO means reduction in the chemical oxygen demand and biological oxygen demand. An aquatic ecosystem with inadequate oxygen supply is considered polluted for organisms that require dissolved oxygen above the existing level (Botkin and Keller, 1998). Decrease in THC at 25% and 50% WSF after exposure to *Pistia* species was also reported by Edema (2007). This shows uptake or metabolism of THC by *Pistia stratiotes.* Odokuma and Dickson (2003) reported marked decrease in the percentage of THC in all the treatments applied in the bioremediation of crude oil in the tropical rainforest (using indigenous hydrocarbon utilizing bacteria) soil of the Niger-Delta except in the control treatment.

Thus, the introduction of WSF of crude oil into the aquatic system can increase the ionic, heavy metals and physical characteristics of aquatic ecosystem. And with continuous spillage there would be a build up of these ions in the aquatic environment. The need for government to implement measures to safeguard and reduce the effect of oil contaminated water on the environment cannot be overemphasized.

11. References

Alexander, V. (1977). Preliminary results of studies on toxicity and effects of petroleum hydrocarbons and marine phytoplankton : In: Wolf, D. (ed). Symposium of fate and effects of petroleum hydrocarbons in marine ecosystem and organisms. pergamon Press. New York. 646p.

Anderson, J.N.,Neff, J.M.,Cox,, B.A.,Tatan, H.E., Hightower, G.M.1974. Characteristics of dispersions and water soluble extracts of crude oils and their toxicity to estuarine crustaceans and fish. *Marine Biology 27*: 75-88.

Anoliefo, G.O. (1991). Forcados Blend Crude oil Effects on Respiratory Metabolism, Mineral Element Composition and Growth of *Citrutlus vulgaris*. Sehrad. Ph.D Thesis.

Ali, A.D. and Mai, G.R. (2007). Concentrations of some heavy metals in water and three macrophytes from disused tin-mined ponds in Jos, Nigeria. *Nigerian Journal of Botany. 21*:51-58.

Aoi, T. Hayashi, T and Balley, D. (1996). Nutrient removal by water lettuce (*Pistia stratiotes*). *Water Science and Technology. 34*:7-8.

Baker, J.M. (1970a). The effect of oil on plants. *Environmental Pollution. 1*: 27-34.

Baker, J.M. (1970b). Growth Stimulation Following Oil Pollution. In: Cowell, E.B. (ed). The Ecological Effects of Oil Pollution and Littoral Communities. Applied Science Publishers, London. Pp 72-77.

Botkin, D.B. and Keller, E.A. (1989). Environmental Science. Second Edition. John Wiley and Sons, Inc. New York. 637pp.

Bohon, R. L. and Clausen, W.F. (1951). Solubility of aromatic hydrocarbon in water. *Journal of American Chemical Society. 73*:1511 - 1578

Brix, H. and Shierup, H.H. (1989). The use of aquatic macrophytes in water pollution control. *Botany.* 75:100-107.

Brunnock, J.V., Duckworth, D.F. and Stephens, G.C. (1968). Analysis of Bench Pollutants. In: Hepple, P. (ed). Scientific Aspects of Pollution of the Sea by Oil. Institute of Pettroleum Printing Press. London. Pp 12-27.

Burdon, R. H., O'Kane, D. Fadzillah, N. Gill, V., Boyd, P. A. and Finch, R.R. (1996). Oxidative stress and responses in *Arabidopsis thallana* and *Oryza sativa* subjected to chilling salinity stress. *Biochemical Society Transactions. 24*:470-472.

Cadwellaer, S. (1993). Encyclopaedia of Environmental Science and Engineering. 4th ed.

Camougis, G. (1981). Environmental Biology for Engineering. A Guide to Environmental Assessment. Academic Press. New York. 26p.

Concave, E. (1979). The Environmental Impact of Refinery Effluent. Report prepared by Concawe. Water Pollution Control Species Task Force, No. 8

Cowell, E.B. (1977). The Ecological Effect of Oil Pollution on Littoral Community. Applied Science Publisher Ltd. England. Pp 88-99.

Covich, A.P. (1993). Water and Ecosystem. In: Gleick, P.M. (Ed). Water in Crisis, A Guide to the World's Fresh Water Resources, Oxford University Press. New York. Pp. 40-45.

Clark, R.B. (1982). Biological effects of oil pollution. *Water, Science and Technology.* 14:1185-1194

Clemens, S.K.M., E. J., Neumanna, D. and Schroede J.I. (1999). Tolerance to toxic metals by a gene family of phytochelatin synthesis from plants and yeast. *EMBO Journal 18*:3325-3333.

Culley, D.D. and Epps. E. A. (1993). Use of duckweed for waste water treatment and animal feed. *Journal of the Water Pollution Control Federation*. *45*:335-347.

Edema, N.E. and Okoloko, G.E. (1997): Effects of the water soluble fraction of Escravos light and Odidi well crude oils on root growth, mitotic cell division and chromosome morphology of *Allium cepa L. Bulletin of Science Association of Nigeria. 21,* 15-18

Edema, N.E. and Etioyibo, E.L. (1999). The effect of Amukpe Flowstation crude oil on germination, height fresh and dry weight of *Lycopersicon esculentum* and *Abelmoschus esculentus* species. *Transaction of the Nigerian Society for Biological Conservation (NSBC). 6*:1-4

Edema, N. E. (2006). Ionic and Physical Characteristics of the Water Soluble Fraction of Crude Oil and the Effects and Physiology of Aquatic Macrophytes. Ph.D. Thesis. University of Benin, Benin City.

Edema, N.E. and Asagba, S.O. (2007). Influence of nutrient supplementation on crude oil induced toxicity of okra (*Abelmoschus esculentus*). *Nigerian Journal of Science and Environment*. *6*:58-65.

Edema, E.N., G.E. Okoloko (2008). Composition of the water-soluble fraction (WSF) of Amukpe well head crude oil before and after exposure to *Pistia stratiotes* L. *Research Journal of Applied Science 3*: 143-146.

Edema, E.N., Okoloko, G.E. and Agbogidi, O.M, (2008). Physical Ionic characteristics in water soluble fraction (WSF)of Olomoro well-head crude oil before and after exposure to *Azolla Africana* Dev. *African Journal of Biotechnology. 7:* 035-040.

Edema, N.E. (2009). Total salinity of the water soluble fraction (WSF) of Ogini well-head crude oil before and after exposure to *Azolla Africana* Devs. *Nigerian Journal of Botany. 22*(2):239-246

Edema, N.E. (2010). Comparative assessment of produced water (PW) and water soluble fraction (WSF) of crude oil on the growth and catalase activity of *Allium cepa L. Journal of Applied Biosciences. 30*:1866-1872

FAO, (1993). Inland Fisheries of Africa Committee Report of the 4th Session of the Working Party on Pollution and Fisheries. Accra. Ghana, 18-22 October. Food and Agricultural Organization of the United Nations Report No. 502.

Fitter, A.H. and Hay, R.K.M. (1987). Environmental Physiology of Plants. Second Edition. Academic Press. New York. Pp 284-295.

Flowers, T.J. and Hajibagheri, M.A. (2001). Salinity tolerance in *Hordeum vulgare*: Iron concentrations in root cells of cultivars differing in salt tolerance. *Plant and Soil. 231:1-9.*

Gill, L.S., Nyawame, H.C.K. and Ehikhametor, A. (1992). Effects of crude oil on the growth and anatomic features of *Chromolaema odoranta. Journal of Chromolaema Newsletter. 5*:1-9.

Gunlack, E. R. and Hayas, M.O. (1977). The Urguiola oil spill: case history and discussion of methods of control and clean up. *Marine Pollution Bulletin. 8*:132-136.

Ha, S.H., Smith, A.P., Howden, R., Dietrich, W.M., Bugg, S., Oconnell, M.J., Coldsbrongh, P.B. and Cobbett, C.S. (1999). Phytochelatin synthesis genes from Arabidopsis. *Plant cell. 11*:1152-1164.

Hernandez, J.A. Olmos, E. Corpas, T., Sevilla, F. and Rio, L.A. (1995). Salt induced oxidative stress in chloroplast of pea plants. *Plant Science. 105*:151-167.

Hoagland, D.R. (1972). Mineral Nutrition of Plant. Principles and Practices. John Wiley and Sons. New York. 412p

Hopkins, W.G. (1999). Introduction to Plant Physiology. Second Ed. John Wiley and Sons, Inc. New York. 512p

Horan, N.J. (1990). Biological Wastewater Treatment System. John Wiley and Sons. Chichester. 190pp.

Kadiri, M.O. (2006). Phytoplankton flora and physico-chemical attributes of some water in the Eastern Niger-Delta Area of Nigeria. *Nigerian Journal of Botany. 19*:188-200.

Kauss, P.B. and Hutchinson, T.C. (1975). The effects of water soluble oil components on the growth of *chlorella vulgans Beinzerinck. Environmental Pollution. 9*:157-174

Kavanu, J.L. (1964). Water and water soluble interaction. Holden Day Publishers, San Francisco. Pg.101.

Mason, C.F. (1993). Biology of Freshwater Pollution. Second Edition. John Wiley and Sons. New York. 351p.

McOliver, (1981). The Nigeria and Industry. Importance and Role in Economic Development. Lagos. Nigeria, 330p.

Michael, A.P. (1977). Ecological Effects of Petroleum in Marine System. In: Wofe, D. (ed) Symposium on Fate of Petroleum Hydrocarbons in Marine Ecosystem and Organism. New York. 646p.

Neff, J. M. and Anderson, J. N. (1981). Response of Marine Animal to Petroleum and Specific Petrols in Hydrocarbons. Applied Science Publisher Ltd. London. Pp 1 -170.

Nelson - Smith, A. (1972). Oil Pollution and Marine Ecology. Paul Clack Scientific Brok Ltd. London. 420p.

Obire, O. (1985). Studies on the Development of Bacteria Inocula to rid the Aquatic Environment of Spilled Petroleum Hydrocarbons. Ph. D. Thesis. University of Benin, Benin - City. 88p.

Odiete, W.O. (1999). Environmental Physiology of Animals and Pollution. Published by Diversified Resources Ltd. Nigeria. Pp 157-258

Odu, C.T. (1981). Degradation and Weathering of crude oil under Tropical Condition. Proceeding of an International Seminar on Petroleum Industry and the Nigerian Environment. Thomopoulas Environmental Pollution Consultants Incorporation with the Petroleum Inspectorate of NNPC. Pp 143-153

Odokuma, L.O. and Dickson, A.A. (2003). Bioremediation of crude oil polluted tropical rain forest soil. *Global Journal of Environmental Science.* 2(1): 29-40

Okoloko, G.E. and Bewley, J.D. (1982). Potentiation of sulphur dioxide induced inhibition of protein synthesis by desiccation. New Phytology. 91:169-176

Omoregie, E., Ufodike B.C, O and Onwuliri, C.O. E. (1997). Effects of water soluble fractions of oil on carbohydrate reserves of *Oreochromis niloficus (L). Journal of Aquatic Science.* 12:1-7.

Rana, S.V.S. (2005). Essentials of Ecology and Environmental Science. 2nd Edition. Practice Hall of India Private Limited. New Delhi, 488pp.

Shew, D.G. (1977). Hydrocarbons in water column. In: Wolfe, D.A. (ed). Fate and Effects of Petroleum Hydrocarbons in Marine Ecosystems. Pergamon Press. New York. Pp 8-12.

Stern, K.R. (2000). Introductory Plant Biology. Eight edition. London. 272p.

Talling, J.F. and Talling, I.B. (1965). Three chemical composition of African lake waters. Int. *Revueges Hydrobiol. 50*:421-563

Taylor, D.D., Green N.P.O., Stout G.W. (1998). Biological Science,3rd Edition. Cambridge University Press, UK. Pg. 121.

Tel-or, E and Sela, M. (1992). Bioremediation of Heavy Metals and Radioactive Waste. Third U.S.A. EPLA-Israel Workshop on Bioremediation. Ministry of Environment. Jerusalem, Israel.

WHO, (1995). Guidelines for Drinking Water Quality. World Health Organization.

Westlake, D.W.D. (1982). Microbial Activities and Changes in the Chemical and Physical Properties of Oil. Conference on Microbial Enhancement of Oil Recovery.

Wetzel, R.G. (2001). Limnology: Lakes and River Ecosystem, 3rd Edition, Academic Press, San Diego, U.S.A. Pp 110 - 115.

Wills, M.A. (2000). Ekologicheskaya Vahkta Sakhalina (Sakhalin Environment Watch) Ph.D. Thesis M. Inst. Petroleum.

Winter, K., Daniel, R.O. Batterlon, J.C. and Van Ballon, C. (1976). Water Soluble components of four fine oils chemical characterization and effects on growth of water plants. *Marine Biology. 36*:269-276.

Part 4

Natural Components in Crude Oil

10

Factors Affecting the Stability of Crude Oil Emulsions

Manar El-Sayed Abdel-Raouf
Petroleum Application Department
Egyptian Petroleum Research Institute
Egypt

1. Introduction

Crude oils are typically water in crude oil (w/o) emulsions, which are often very stable. Among the indigenous natural surfactants contained in the crude oils, asphaltenes and resins are known to play an important role in the formation and stability of w/o emulsions. Asphaltenes are defined as the fraction of the crude oil precipitating in pentane, hexane, or heptane, but soluble in toluene or benzene. Asphaltenes are the most polar and heaviest compounds in the crude oil. They are composed of several poly nuclear aromatic sheets surrounded by hydrocarbon tails, and form particles whose molar masses are included between 500 and 20,000 g mol^{-1}. They contain many functional groups, including some acids and bases. Resins are molecules defined as being soluble in light alkanes (pentane, hexane, or heptane), but insoluble in liquid propane. They consist mainly of naphthenic aromatic hydrocarbons; generally aromatic ring systems with alicyclic chains. Resins are effective as dispersants of asphaltenes in crude oil. It was postulated that asphaltenes stabilize w/o emulsions in two steps. First, disk-like asphaltene molecules aggregate into particles or micelles, which are interfacially active. Then, these entities upon adsorbing at the w/o interface aggregate through physical interactions and form an interfacial network.

The introduced chapter deals with different factors that affecting the stability of crude oil emulsions and also those factors causing asphaltenes precipitation.

1.1 Types of emulsions

Emulsions have long been of great practical interest due to their widespread occurrence in everyday life. They may be found in important areas such as food, cosmetics, pulp and paper, pharmaceutical and agricultural industry [1, 2]. Petroleum emulsions may not be as familiar but have a similar long-standing, widespread, and important occurrence in industry, where they are typically undesirable and can result in high pumping costs, pipeline corrosions, reduced throughput and special handling equipment. Emulsions may be encountered at all stages in the petroleum recovery and processing industry (drilling fluid, production, process plant, and transportation emulsions [3, 4].

Emulsions are defined as the colloidal systems in which fine droplets of one liquid are dispersed in another liquid where the two liquids otherwise being mutually immiscible. Oil and water produce emulsion by stirring; however, the emulsion starts to break down immediately after stirring is stopped.

Depending upon the nature of the dispersed phase, the emulsions are classified as, O/W emulsion or oil droplets in water and W/O emulsion or water droplets in oil. Recently, developments of W/O/W type emulsion or water dispersed within oil droplets of O/W type emulsion and O/W/O type, Figure 1.

i. **Oil-in-water emulsions (O/W):** The emulsion in which oil is present as the dispersed phase and water as the dispersion medium (continuous phase) is called an oil-in-water emulsion.
ii. **Water-in-oil emulsion (W/O):** The emulsion in which water forms the dispersed phase, and the oil acts as the dispersion medium is called a water-in-oil emulsion.

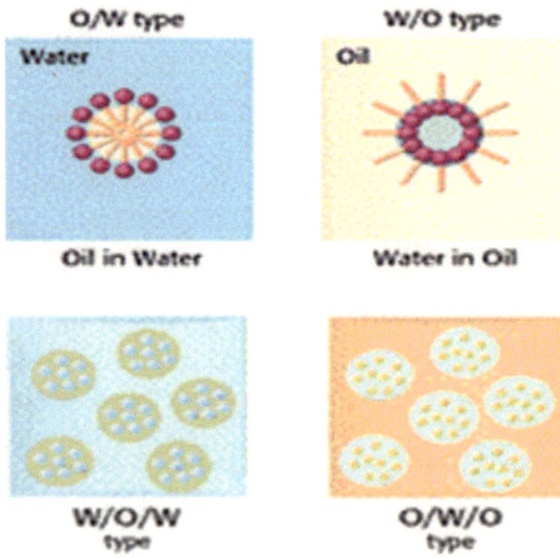

Fig. 1. Types of emulsions.

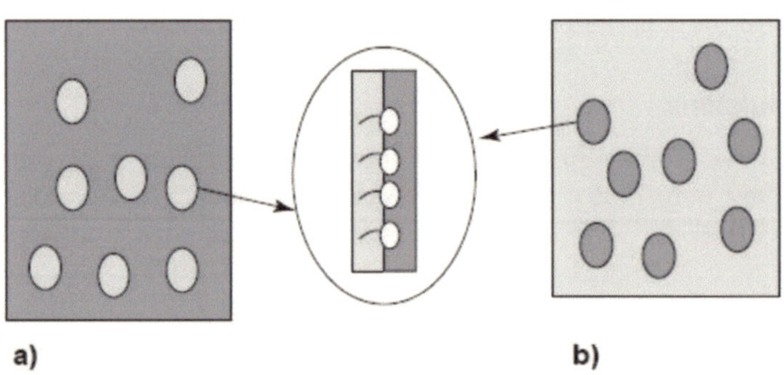

Fig. 2. (5) Schematic representation of emulsion structures. a) O/W emulsion; b) W/O emulsion. Encircled: enlarged view of a surfactant monolayer sitting at the oil-water interface.

1.2 Properties of emulsions (5-7)

The emulsions satisfy the following criteria:

i. Emulsions show all the characteristic properties of colloidal solution such as Brownian movement, Tyndall effect, electrophoresis etc.
ii. These are coagulated by the addition of electrolytes containing polyvalent metal ions indicating the negative charge on the globules.
iii. The size of the dispersed particles in emulsions in larger than those in the sols. It ranges from 1000 Å to 10,000 Å. However, the size is smaller than the particles in suspensions.
iv. Emulsions can be converted into two separate liquids by heating, centrifuging, freezing etc. This process is also known as demulsification.

Many advances have been made in the field of emulsions in recent years. Emulsion stability depends on presence of adsorbed structures on the interface between the two liquid phases. Emulsion behavior is largely controlled by the properties of the adsorbed layers that stabilize the oil-water surfaces (7-9). The knowledge of surface tension alone is not sufficient to understand emulsion properties, and surface rheology plays an important role in a variety of dynamic processes. When a surface active substance is added to water or oil, it spontaneously adsorbs at the surface, and decreases the surface tension γ (7). In the case of small surfactant molecules, a monolayer is formed, with the polar parts of the surface-active molecules in contact with water, and the hydrophobic parts in contact with oil (Fig. 2). The complexity of petroleum emulsions comes from the oil composition in terms of surface-active molecules contained in the crude, such as low molecular weight fatty acids, naphthenic acids and asphaltenes. These molecules can interact and reorganize at oil/water interfaces. These effects are very important in the case of heavy oils because this type of crude contains a large amount of asphaltenes and surface-active compounds (10-14). In petroleum industry, water-in-oil (w/o) or oil-in-water (o/w) emulsions can lead to enormous financial losses if not treated correctly. Knowing the particular system and the possible stability mechanisms is thus a necessity for proper processing and flow assurance. Thus, there are desirable or undesirable emulsions as shown in Table (1).

1.3 Emulsion formation

There are three main criteria that are necessary for formation of crude oil emulsion (16):

1. Two immiscible liquids must be brought in contact;
2. Surface active component must present as the emulsifying agent;
3. Sufficient mixing or agitating effect must be provided in order to disperse one liquid into another as droplets.

During emulsion formation, the deformation of droplet is opposed by the pressure gradient between the external (convex) and the internal (concave) side of an interface. The pressure gradient or velocity gradient required for emulsion formation is mostly supplied by agitation. The large excess of energy required to produce emulsion of small droplets can only be supplied by very intense agitation, which needs much energy (17-21).

Occurrence	Usual type [a]
Undesirable emulsions	
Well-head emulsions	W/O
Fuel oil emulsions (marine)	W/O
Oil Sand Floatation process	W/O or O/W
Oil spill mousse emulsions	W/O
Tanker bilge emulsions	O/W
Desirable emulsions	
Heavy oil pipe line emulsions	O/W
Oil sand floatation process slurry	O/W
Emulsion drilling fluid, oil-emulsion mud	O/W
Emulsion drilling fluid, oil-base mud	W/O
Asphalt emulsion	O/W
Enhance oil recovery in situ emulsions	O/W

Table 1. Desirable and Undesirable petroleum emulsions [(10)].

A suitable surface active component or surfactant can be added to the system in order to reduce the agitation energy needed to produce a certain droplet size. The formation of surfactant film around the droplet facilitates the process of emulsification and a reduction in agitation energy by factor of 10 or more can be achieved. A method requiring much less mechanical energy uses phase inversion. For instance, if ultimately a W/O emulsion is desired, then an O/W emulsion is first prepared by the addition of mechanical energy. Then the oil content is progressively increased. At some volume fraction above 60-70%, the emulsion will suddenly invert and produce a W/O emulsion of smaller water droplet sizes than were the oil droplets in the original O/W emulsions [(22)].

Emulsions of crude oil and water can be encountered at many stages during drilling, producing, transporting and processing of crude oils and in many locations such as in hydrocarbon reservoirs, well bores, surface facilities, transportation systems and refineries [(23, 24)].

1.4 Emulsion breakdown

In general there are three coupled sub-processes that will influence the rate of breakdown processes in emulsions. These are aggregation (Flocculation), coalescence and phase separation [(15)], Figure 3. They will be discussed in some details.

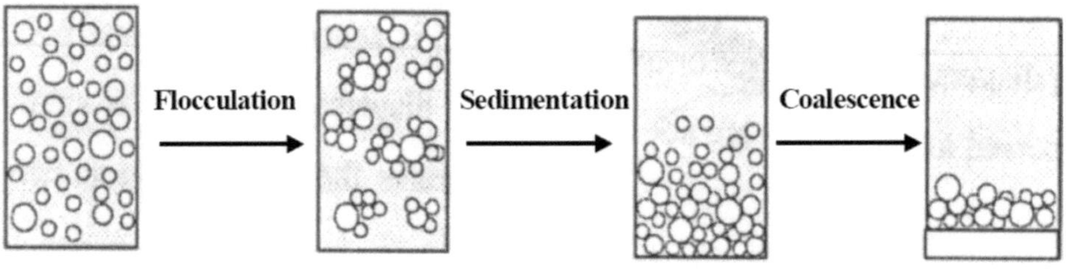

Fig. 3. Emulsion breakdown [(15)].

Flocculation [15]

It is the process in which emulsion drops aggregate, without rupture of the stabilizing layer at the interface. Flocculation of emulsions may occur under conditions when the van der Waals attractive energy exceeds the repulsive energy and can be weak or strong, depending on the strength of inter-drop forces.

The driving forces for flocculation can be:

1. Body forces, such as gravity and centrifugation causing creaming or sedimentation, depending on whether the mass density of the drops is smaller or greater than that of the continuous phase.
2. Brownian forces or
3. Thermo-capillary migration (temperature gradients) may dominate the gravitational body force for very small droplets, less than 1 µm.

Coalescence [15]**:**

It is an irreversible process in which two or more emulsion drops fuse together to form a single larger drop where the interface is ruptured. As already mentioned, for large drops approaching each other (no background electric field), the interfaces interact and begin to deform. A plane parallel thin film is formed, which rate of thinning may be the main factor determining the overall stability of the emulsion. The film thinning mechanism is strongly dependent on bulk properties (etc. viscosity) in addition to surface forces. The interaction of the two drops across the film leads to the appearance of an additional disjoining pressure inside the film, Figure 4.

Phase separation

The processes of flocculation and coalescence are followed by phase separation, i.e. emulsion breakdown.

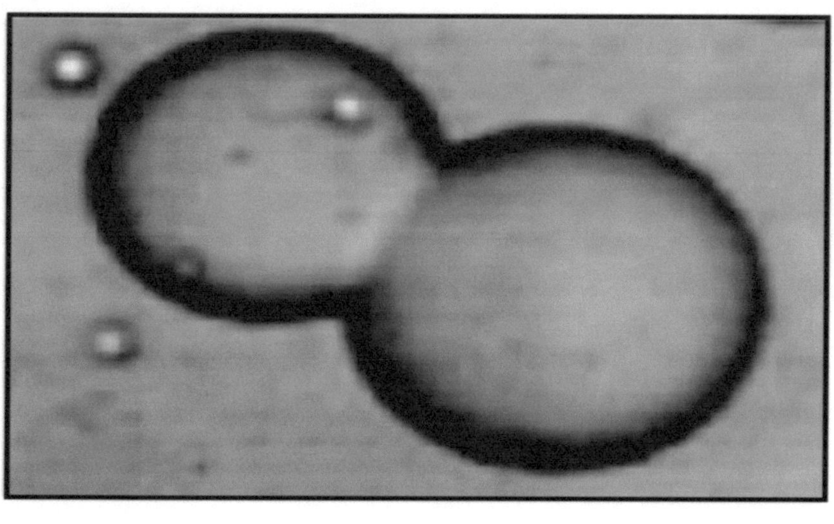

Fig. 4. Adhesion between two emulsion droplets [10].

1.5 Stabilization of the emulsion

There are many factors that usually favour emulsion stability such as low interfacial tension, high viscosity of the bulk phase and relatively small volumes of dispersed phase. A narrow droplet distribution of droplets with small sizes is also advantageous, since polydisperse dispersions will result in a growth of large droplets on the expense of smaller ones. The potent stabilization of the emulsion is achieved by stabilization of the interface (25- 27).

1.5.1 Steric stabilization of the interface

The presence of solids at interfaces may give rise to repulsive surface forces which *thermodynamically* stabilize the emulsion. As concluded (28-30), many of the properties of solids in stabilizing emulsion interfaces can be attributed to the very large free energy of adsorption for particles of intermediate wettability (partially wetted by both oil and water phases). This irreversible adsorption leads to extreme stability for certain emulsions and is in contrast to the behavior of surfactant molecules which are usually in rapid dynamic equilibrium between the oil: water interface and the bulk phases. According to the asphaltene stabilization mechanism, coalescence requires the solid particles to be removed from the drop-drop contact region. Free energy considerations suggest that lateral displacement of the particles is most likely, since forcing droplets into either phase from the interface require extreme energies (31-38). The asphaltenes stabilization effect for water droplets has already been pictured in Figure 5 where droplet contact is prevented by a physical barrier around the particles.

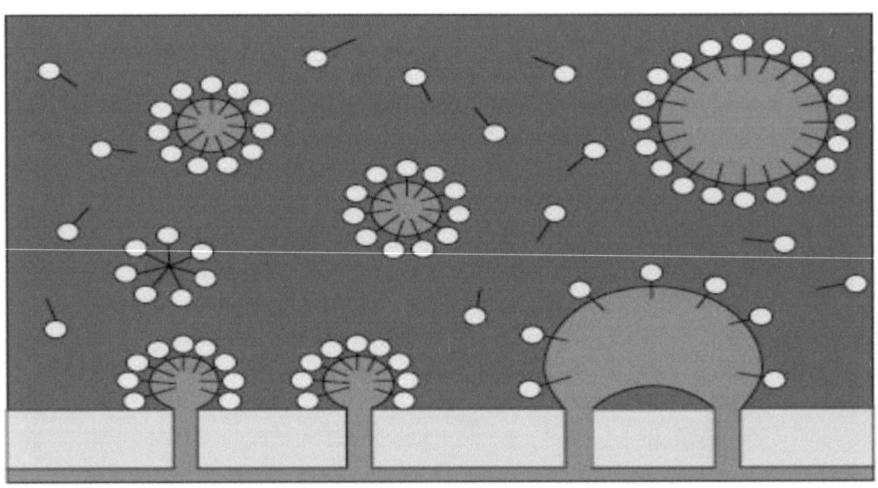

Fig. 5. Steric stabilization of the interfaces (40).

As the stability of w/o emulsions appeared clearly related to the presence of an interfacial network surrounding the water droplets (40-43), many rheological studies of water/asphalted oil interfaces have been made to clarify the mechanisms involved (44-46). The rheological properties of these interfaces have been found to be strongly dependent on the nature of solvent used for dilution, the oil concentration, the asphaltenes and resin concentrations, the resin to asphaltenes ratio, and so on (47-51).

1.5.2 Electric stabilization of the interface

Electrical double layer repulsion or charge stabilization by polymers and surfactants with protruding molecular chains may prevent the droplets to come into contact with each other (52-54). Also, polymers, surfactants or adsorbed particles can create a mechanically strong and elastic interfacial film that act as a barrier against aggregation and coalescence. A film of closed packed particles has considerable mechanical strength, and the most stable emulsions occur when the contact angle is close to 90°, so that the particles will collect at the interface. Particles, which are oil-wet, tend to stabilize w/o emulsions while those that are water-wet tend to stabilize o/w emulsions. In order to stabilize the emulsions the particles should be least one order of magnitude smaller in size than the emulsion droplets and in sufficiently high concentration. Nevertheless, stable w/o emulsions have been generally found to exhibit high interfacial viscosity and/or elasticity modulus. It has been attributed to physical cross-links between the naturally occurring surfactants in crude oil (i.e. asphaltenes particles) adsorbed at the water–oil interface (55, 56).

1.5.3 Composition of crude oil

A good knowledge of petroleum emulsions is necessary for controlling and improving processes at all stages of petroleum production and processing. Many studies have been carried out in the last 40 years and have led to a better understanding of these complex systems (57-60). However there are still many unsolved questions related to the peculiar behavior of these emulsions. The complexity comes mostly from the oil composition, in particular from the surface-active molecules contained in the crude. These molecules cover a large range of chemical structures, molecular weights, and HLB (Hydrophilic-Lipophilic Balance) values; they can interact between themselves and/or reorganize at the water/oil interface. Oil-water emulsions are fine dispersions of oil in water (O/W) or of water in oil (W/O), with drop sizes usually in the micron range (61, 62). In general, emulsions are stabilized by surfactants. In some cases multiple emulsions such as water in oil in water (W/O/W) or oil in water in oil (O/W/O) can be found. Emulsions can be stabilized by other species, provided that they adsorb at the oil-water interface and prevent drop growth and phase separation into the original oil and water phases. After adsorption, the surfaces become visco-elastic and the surface layers provide stability to the emulsion (63- 65). Crude oils contain asphaltenes (high molecular weight polar components) that act as natural emulsifiers. Other crude oil components are also surface active: resins, fatty acids such as naphthenic acids, porphyrins, wax crystals, etc, but most of the time they cannot alone produce stable emulsions (65). However, they can associate to asphaltenes and affect emulsion stability. Resins solubilize asphaltenes in oil, and remove them from the interface, therefore lowering emulsion stability. Waxes co-adsorb at the interface and enhance the stability. Naphthenic and other naturally occurring fatty acids also do not seem able to stabilize emulsions alone. However, they are probably partly responsible for the important dependence of emulsion stability upon water pH. The composition of crude oil is given in Figure 6 and examples of these structures are given in Figure 7 a & b (66).

These components can be separated by simple technique known as SARA analysis (Saturated, Aromatic, Resin and Asphaltenes). Some examples of the resin and asphaltenes that can be separated by using SARA analysis of are given in Figure 7a& b. The SARA analysis process is shown in Figure 8. Particles such as silica, clay, iron oxides, etc. can be

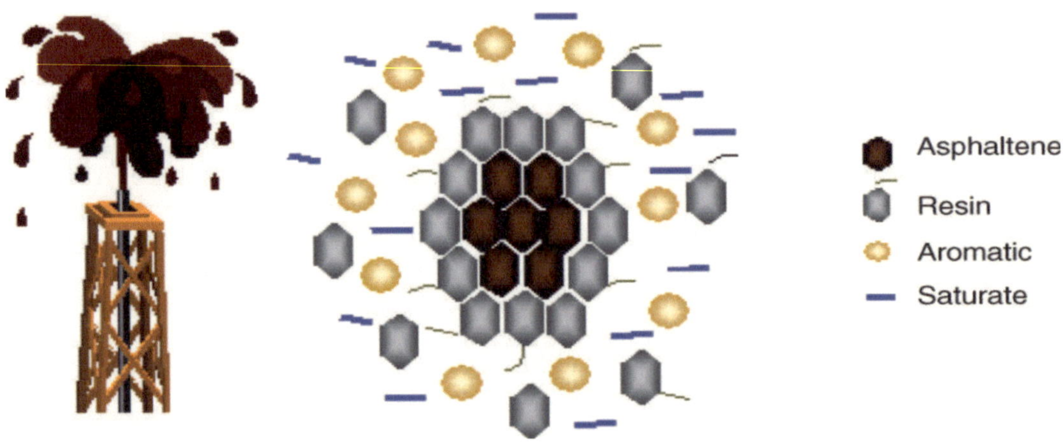

Fig. 6. Composition of crude oil.

present in crude oils. These particles are naturally hydrophilic, but can become oil-wet (hydrophobic) due to long term exposure to the crude in the absence of water. Emulsions with particles and asphaltenes combined can be much more stable than those stabilized by asphaltenes alone, provided that enough asphaltenes are present: all the adsorption sites on the particle surface need to be saturated by asphaltenes (67, 68). These species will be mentioned in some details in the next section.

1.5.4 Asphaltenes

There are many definitions of asphaltenes. Strictly speaking, asphaltenes are the crude oil components that meet some procedural definition (69). A common definition is that asphaltenes are the material that is:

Insoluble in n-pentane (or n-heptane) at a dilution ratio of 40 parts alkane to 1 part crude oil and (2) Re-dissolves in toluene.

Chemically, asphaltenes are polycyclic molecules that are disc shaped, and have a tendency to form stacked aggregates. The tendency of asphaltenes to self-aggregate distinguishes them from other oil constituents. Asphaltene aggregation is the cause of complex non-linear effects in such phenomena as adsorption at solid surfaces, precipitation, fluid's rheology, emulsion stability, etc (70, 71). Asphaltenes are regarded to be polar species, formed by condensed poly aromatic structures, containing alkyl chains, hetero atoms (such as O, S and N) and some metals.

They contain also polar groups (ester, ether, carbonyl) and acidic and basic groups (carboxylic and pyridine functional groups) that can be ionized in a certain range of pH by accepting or donating protons. Typical composition of asphaltenes is provided in Table 2. Asphaltenes average molecular weights range from approximately 800 to 3000 g mol^{-1}. They are molecularly dispersed in aromatic solvents such as toluene, and precipitate in alkanes (ASTM D2007-93, IP 143).

The procedure should also specify the temperature at which the mixing and separation takes place, the amount of time that must elapse before asphaltenes are separated from the oil/alkane mixture, and even the method used to accomplish the separation (filter size,

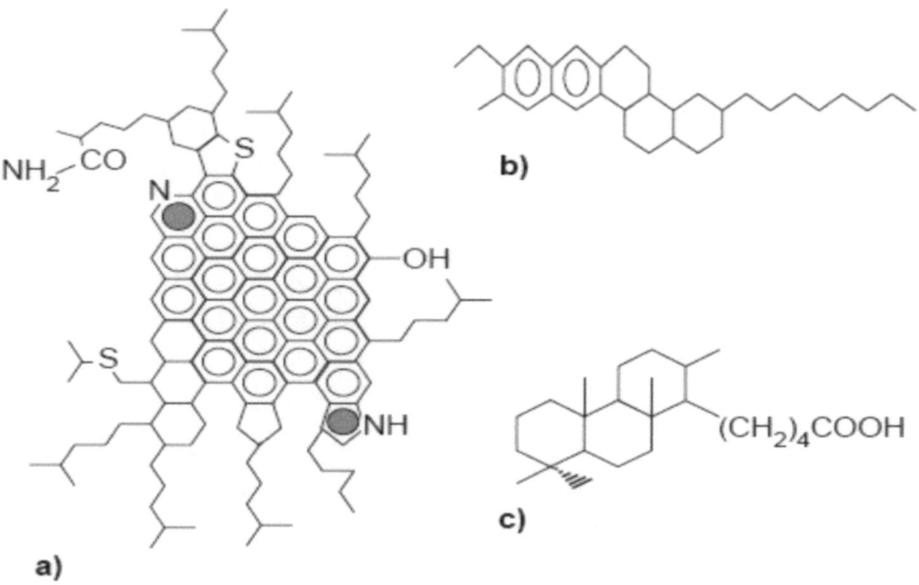

Fig. 7a. Examples of molecular structures in crude oil. a- Asphaltenes, b- Resins, c- naphthenic acids.

filtration rate), since all of these factors can affect the final result. Although there are several standardized procedures, but in reality every lab uses its own procedure. These may vary a little or a lot from the standards. They differ in color and in texture (72).

Material separated with still lower molecular weight alkanes (e.g., propane) would be sticky and more liquid-like than those separated by n-heptane as shown in Figure 9.

Some authors point out that the precipitation techniques may provide an excessively strong interference into the delicate molecular organization of asphaltenes associates (Figure 10), leading to their irreversible transformation, so that the supra-molecular architecture in solutions of the precipitated material may be different from that in native crude. Consequently, studies of aggregation in crude oil solutions may supply valuable information regarding the manner of asphaltenes–asphaltenes interactions in the presence of other crude oil components.

Element (in wt. %)	*Range*	*Typical*
Carbon	78-90	82-84
Hydrogen	6.1-10.3	6.5-7.5
Nitrogen	0.5-3.0	1.0-2.0
Sulfur	1.9-10.8	2.0-6.0
Oxygen	0.7-6.6	0.8-2.0
Vanadium(ppm)	0-1200	100-300
H/C	0.8-1.5	1.0-1.2

Table 2. Range and Typical Values of Elemental Composition of Asphaltenes.

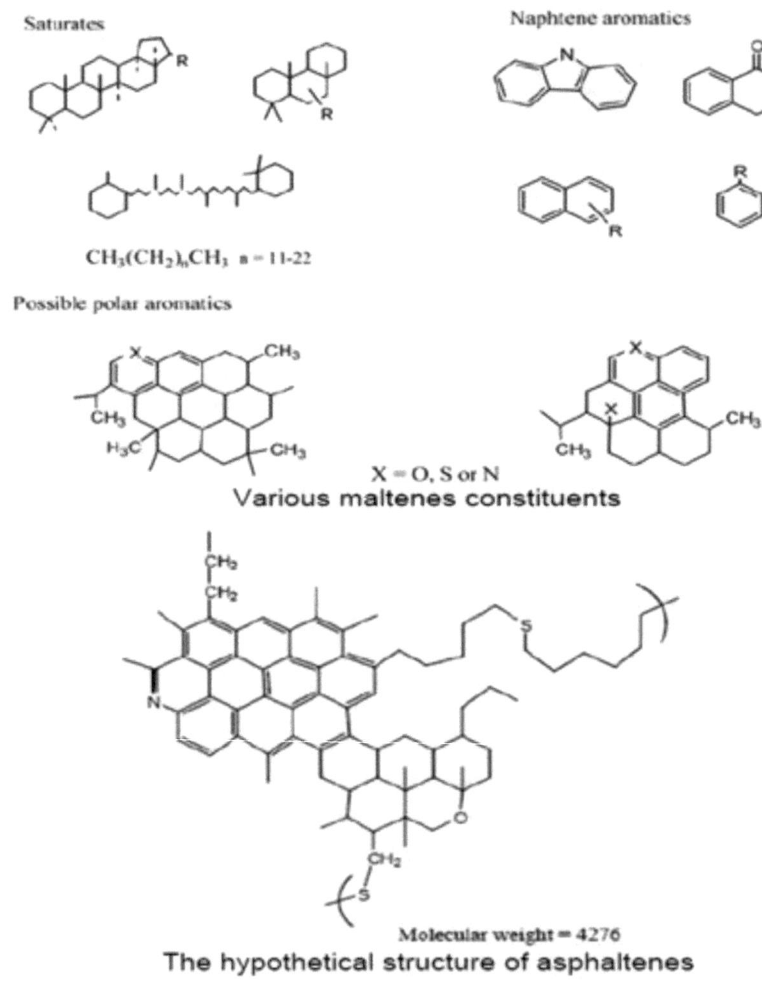

Fig. 7b. Other examples of molecular structures in crude oil.

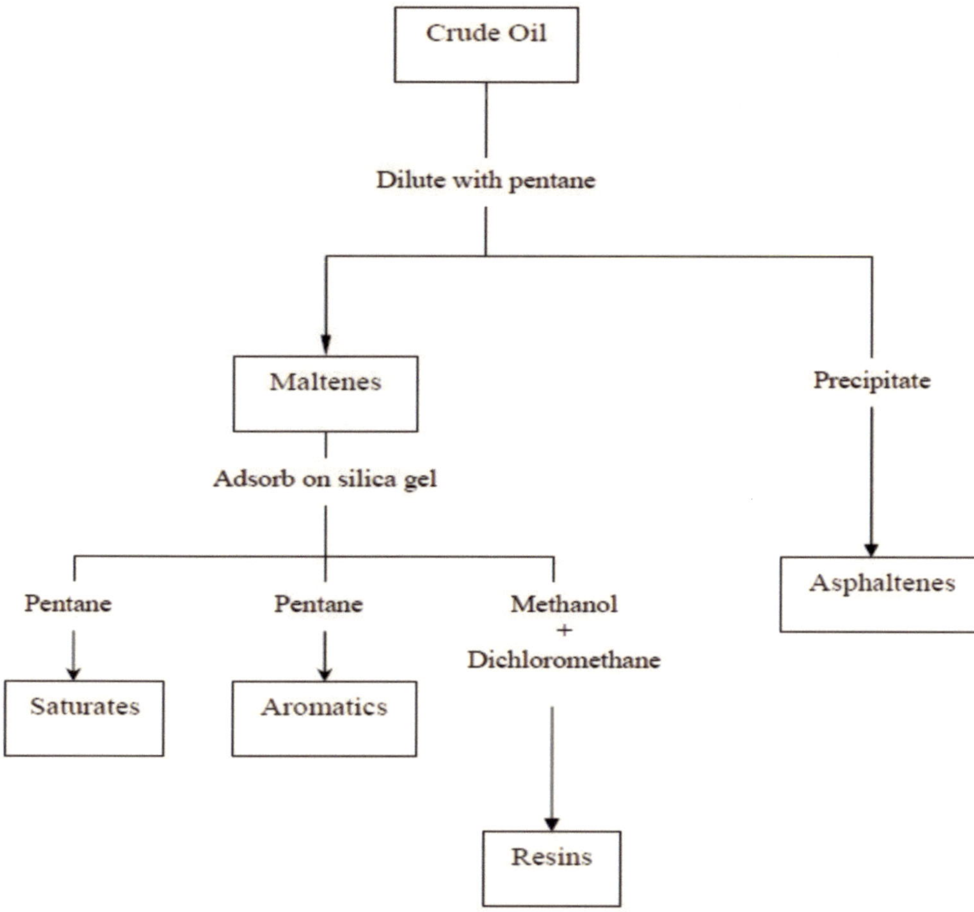

Fig. 8. SARA analysis for separation of crude oil components.

n-C5 asphaltene

n-C7 asphaltene

Fig. 9. asphaltenes separated by different alkanes.

Asphaltenes stabilize the crude oil emulsion by different modes of action. When asphaltenes disperse on the interface, the film formed at a water/ crude oil interface behaves as a skin whose rigidity can be shown by the formation of crinkles at interface when contracting the droplet to a smaller drop size (78). They can also aggregate with resin molecules on the interfaces and prevent droplet coalescence by steric interaction (Figure 11). Some authors suggest that asphaltenes stabilize the emulsion by formation of hydrogen bonding between asphaltenes and water molecules (79-80).

Singh et al. (81) postulated that asphaltenes stabilize w/o emulsions in two steps. First, disk-like asphaltenes molecules aggregate into particles or micelles, which are interfacially active. Then, these entities upon adsorbing at the w/o interface aggregate through physical interactions and form an interfacial network. Different modes of action of asphaltenes are represented in Figure 12.

1.5.5 Resins

Just as the asphaltenes have only a procedural definition, resins also are procedurally defined. There are at least two approaches to defining resins. In one approach the material that precipitates with addition of propane, but not with n-heptane, is considered to constitute the resins.

There is no universal agreement about the propane/n-heptane pair, but the general idea is that resins are soluble in higher molecular weight normal alkanes, but are insoluble in lower molecular weight alkanes. A standard method exists to quantify resins by a completely SARA analysis. Resins can be also defined as the most polar and aromatic species present in

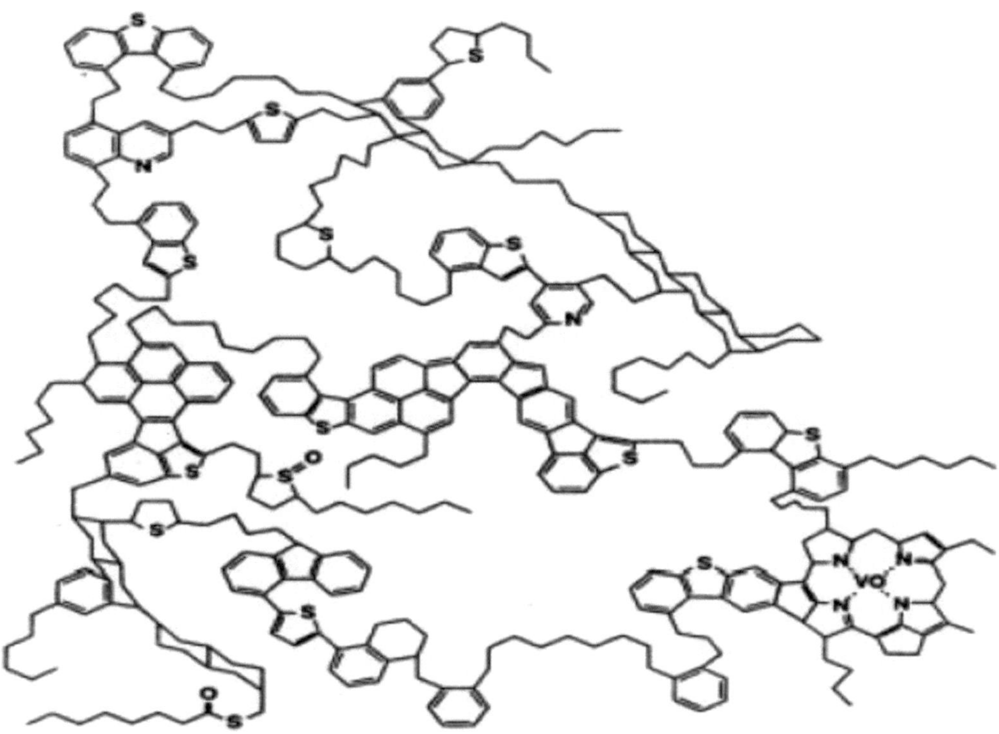

Fig. 10. Asphaltene associates (77).

deasphalted oil and, it has been suggested, contribute to the enhanced solubility of asphaltenes in crude oil by solvating the polar and aromatic portions of the Asphaltenic molecules and aggregates (82-85). The solubility of asphaltenes in crude oil is mediated largely by resin solvation and thus resins play a critical role in precipitation, and emulsion stabilization phenomena (86-88).

Resins are thought to be molecular precursors of the asphaltenes. The polar heads of the resins surround the asphaltenes, while the aliphatic tails extend into the oil, Figure 13. Resins may act to stabilize the dispersion of asphaltene particles and can be converted to asphaltenes by oxidation. Unlike asphaltenes, however, resins are assumed soluble in the petroleum fluid. Pure resins are heavy liquids or sticky (amorphous) solids and are as volatile as the hydrocarbons of the same size. Petroleum fluids with high-resin content are relatively stable. Resins, although quite surface-active, have not been found to stabilize significantly water-in-oil emulsions by themselves in model systems. However, the presence of resins in solution can destabilize emulsions via asphaltenes solvation and/or replacement at the oil/water interface (89-96), Figure 13.

1.5.6 Saturates (97-100)

Saturates are nonpolar and consist of normal alkanes (*n*-paraffins), branched alkanes (iso-paraffins) and cyclo-alkanes (also known as naphthenes). Saturates are the largest single source of hydrocarbon or petroleum waxes, which are generally classified as paraffin wax,

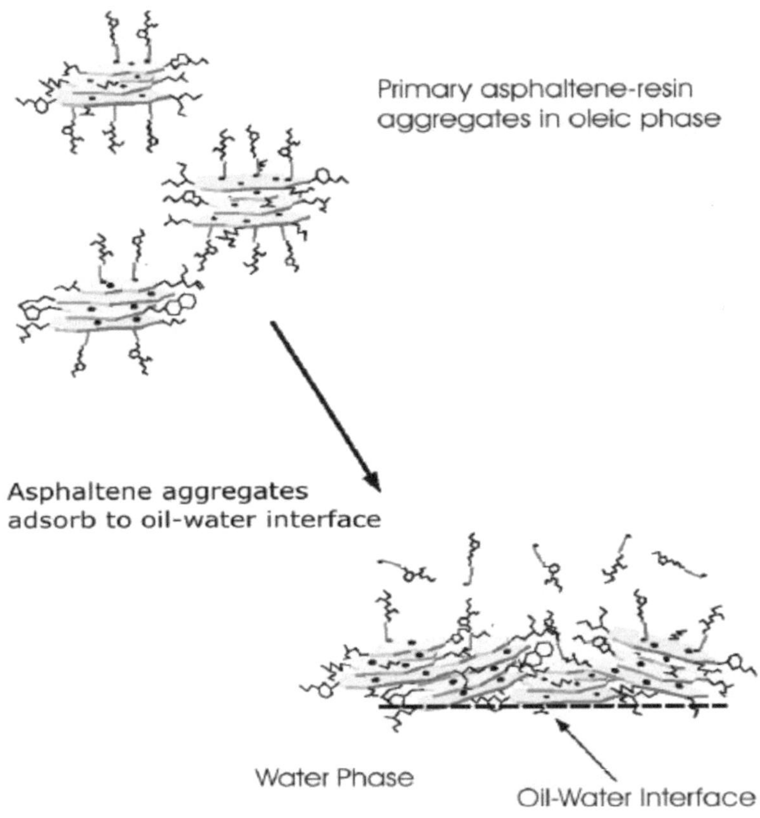

Fig. 11. Proposed stabilizing mechanism for asphaltenes in petroleum by resin molecules [82].

microcrystalline wax, and/or petrolatum (Figure 7b). Of these, the paraffin wax is the major constituent of most solid deposits from crude oils.

1.5.7 Aromatics [101-106]

Aromatics are hydrocarbons, which are chemically and physically very different from the paraffins and naphthenes. They contain one or more ring structures similar to benzene. The atoms are connected by aromatic double bonds (Figure 7b).

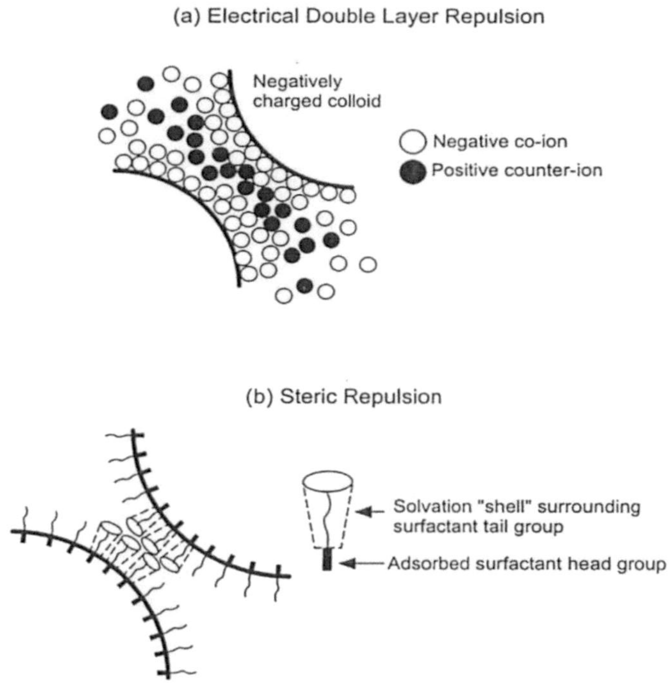

Fig. 12. Different modes of action of asphaltenes in stabilizing crude oil emulsions.

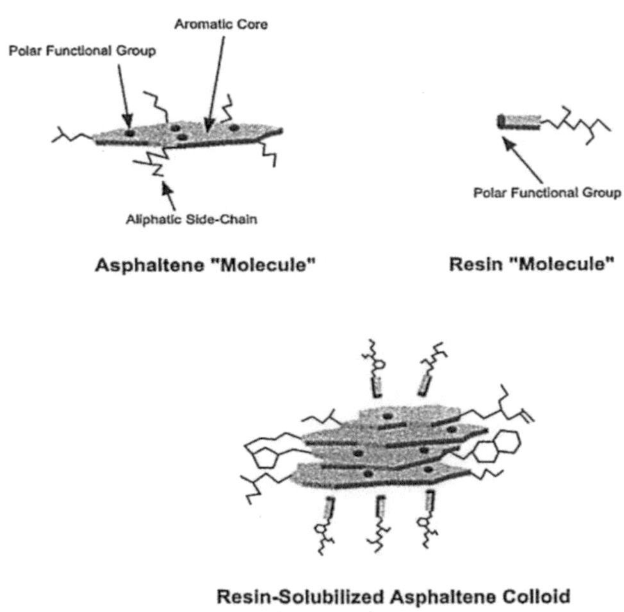

Fig. 13. Solubilization of asphaltenes by resin molecules.

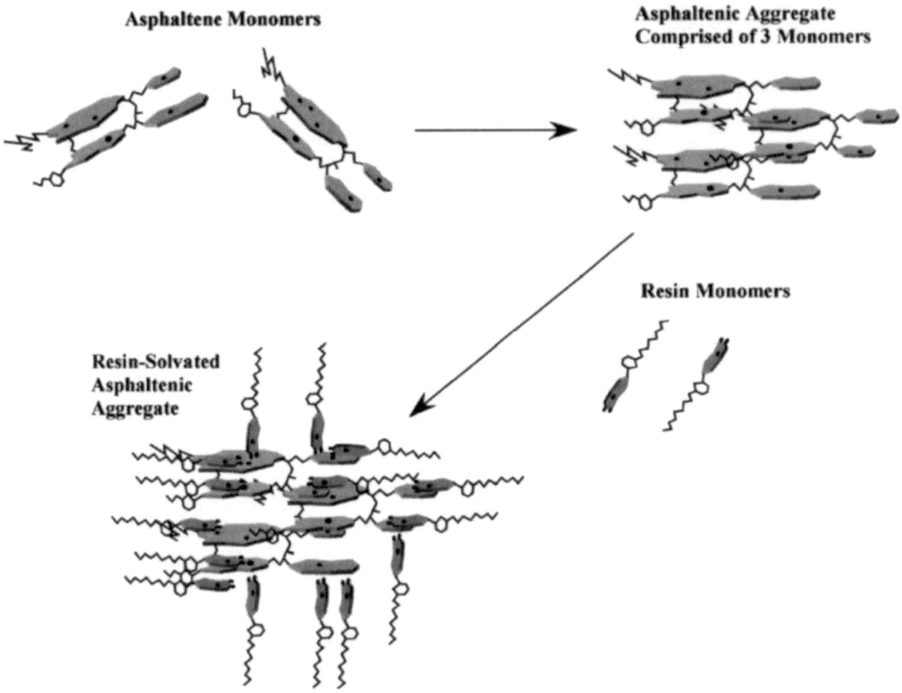

Fig. 14. Schematic illustration of asphaltene aggregates in absence and in presence of resins (82).

2. References

[1] J. L. Grossiord, M. Seiller (1998), Multiple Emulsions: Structure, Properties and Applications. Editions de Sante, Paris.

[2] P. Walstra, (1993), Principles of Emulsion Formation. *Chemical Engineering Science*, 48, 333-349

[3] G.W. Mushrush, J.G. Speight, Petroleum Products: Instability and Incompatibility, Taylor & Francis, Bristol, PA, 1995.

[4] L.L. Schramm, Emulsions: fundamentals and applications in the petroleum Industry, Advances in Chemistry Series 231, ACS, Washington DC, 1992.

[5] D. Langevin, S. Poteau, I. Hénaut and J.F. Argillier, Crude Oil Emulsion Properties and their Application to Heavy Oil Transportation, Oil & Gas Science and Technology, *Rev. IFP*, Vol. 59 (2004), No. 5, pp. 511-521

[6] C.W. Angle, in: J. Sjoblom (Ed.), Chapter 20. Encyclopedic handbook of emulsion Technology, Marcel Dekker, NY, 2001.

[7] C.W. Angle, in: J. Sjoblom (Ed.), Chapter 24. Encyclopedic handbook of emulsion Technology, Marcel Dekker, NY, 2001.

[8] D.A. Storm, S.J. DeCanio, E.Y. Sheu, in: T.F. Yen (Ed.), Asphaltene Particles in Fossil Fuel Exploration, Recovery, Refining, and Production Processes, Plenum, New York, 1994, p. 81.

[9] J.-R. Lin, J.-K. Park, T.F. Yen, in: T.F. Yen (Ed.), Asphaltene Particles in Fossil Fuel Exploration, Recovery, Refining, and Production Processes, Plenum, New York, 1994, p. 91.

[10] R. Cimino, S. Correra, A.D. Bianco, T.P. Lockhart, in: O.C. Mullins (Ed.), Asphaltenes: Fundamentals and Applications, Plenum, New York, 1995, p. 97.

[11] J. S. Buckley, Microscopic Investigation of the Onset of Asphaltene Precipitation, *Fuel Sci. Tech. Internat.* (1996) 14, 55-74.

[12] S. Bin Othman, Study on Emulsion Stability and Chemical Demulsification Characteristics, Bachelor of Chemical Engineering, Faculty of Chemical & Natural Resources Engineering Universiti Malaysia Pahang, May 2009.

[13] Mc Lean, J.D., Kilpatrick, P.K., 1997a. Effects of asphaltenes aggregation in model heptane–toluene mixtures on stability of water-in-oil emulsions. *J. Colloid Interface Sci.* 196, 23–34.

[14] Mc Lean, J.D., Kilpatrick, P.K., 1997b. Effects of asphaltenes solvency on stability of water-in-crude-oil emulsions. *J. Colloid Interface Sci.* 189, 242–253.

[15] M. Sarbar, K. M. Al-Jaziri, 1995. Laboratory investigation of factors affecting the formation and stability of tight oil-in-water emulsions in produced fluids. OAPEC Conference on New Technology 1, 261–268.

[16] M.E. Abdul-Raouf, A. M. Abdul-Raheim and A. A. Abdel-Azim, (2011) Surface Properties and Thermodynamic Parameters of Some Sugar- Based Ethoxylated Amine Surfactants: 1-Synthesis, Characterization, and Demulsification Efficiency, *Journal of Surfactant and Detergent* 14:113–121.

[17] J. Bibette, F. Leal-Calderon, V. Schmitt, and P. Poulin, (2002) Introduction, in: Emulsion science - Basic principles, Springer.

[18] S.M. Farouq Ali, Heavy oil–evermore mobile, (2003) *J. Petrol. Sci. Eng.* 37, 5–9.

[19] J. S. Buckley, J. X. Wang, and J. L. Creek, (2007) "Solubility of the Least Soluble Asphaltenes," Chapter 16 in *Asphaltenes, Heavy Oils and Petroleomics*, eds., Springer 401-437.

[20] R. Cimino, S. Correra, A. Del Bianco, and T.P Lockhart, (1995): "Solubility and Phase Behavior of Asphaltenes in Hydrocarbon Media," *Asphaltenes: Fundamentals and Applications*, E.Y. Sheu and O.C. Mullins (eds.), NY: Plenum Press 97-130.

[21] R. B. Long, (1981) "The Concept of Asphaltenes," *Chemistry of Asphaltenes*, J.W. Bunger and N.C. Li (eds.), ACS, Washington, DC 17-27.

[22] R. B.Boer, K. Leerlooyer, M. Eigner and A. van Bergen, (Feb. 1995) "Screening of Crude Oils for Asphalt Precipitation: Theory, Practice, and the Selection of Inhibitors," *SPE PF* 10, 55-61.

[23] T. Fan and J. Buckley (2002) "Rapid and Accurate SARA Analysis of Medium Gravity Crude Oils," *Energy & Fuels* 16, 1571-1575.

[24] G. Mansoori, (1997). Modeling of asphaltene and other heavy organics depositions. *J. Petrol. Sci. Eng.* 17, 101–111.

[25] J. Murgich, (1996). Molecular recognition and molecular mechanics of micelles of some model asphaltenes and resins. *Energy and Fuels* 10, 68–78.

[26] Hirschberg, L. deJong, B. A. Schipper, and J.G.Meijer, (June 1984) "Influence of Temperature and Pressure on Asphaltene Flocculation," *SPEJ* 24, 283-293.

[27] J. Bibette, F. Leal-Calderon, V.Schmitt, V. and P. Poulin, (2002) Introduction, in: *Emulsion science - Basic principles*, Springer.

[28] P. M. Spiecker, K. L. Gawrys, C. B.Trail, C.B. and P. K. Kilpatrick, (2003) Effects of Petroleum Resins on Asphaltene Aggregation and Water-in-Oil Emulsion Formation. *Colloids and Surfaces A-Physicochemical and Engineering Aspects*, 220, 9-27.

[29] B. P. Binks, (2002) Particles as Surfactants - Similarities and Differences. *Current Opinion in Colloid & Interface Science*, 7, 21-41.

[30] G.A. van Aken and F. D. Zoet, (2000) Coalescence in Highly Concentrated Coarse Emulsions. *Langmuir*, 16, 7131-7138.

[31] S. Acevedo, G. Escobar, L. B. Gutierrez, H. Rivas, and X. Gutierrez, (1993) Interfacial Rheological Studies of Extra- Heavy Crude Oils and Asphaltenes - Role of the Dispersion Effect of Resins in the Adsorption of Asphaltenes at the Interface of Water-In-Crude Oil-Emulsions. *Colloids and Surfaces A-Physicochemical and Engineering Aspects*, 71, 65-71.

[32] J. Sjöblom, (2001) *Encyclopedic Handbook of Emulsion Technology*, Marcel Dekker, New York.

[33] R. F. Lee, (1999) Agents which Promote and Stabilize Water-in- Oil Emulsions. *Spill Science and Technology Bulletin*, 5, 117-126.

[34] M. Jeribi, B. Almir-Assad, D. Langevin, I. Henaut and J. Argillier, (2002) Adsorption Kinetics of Asphaltenes at Liquid Interfaces. *Journal of Colloid and Interface Science*, 256, 268-272.

[35] W. Adamson, (1976) Physical Chemistry of Surfaces. Wiley, New York.

[36] T. Tri, J. Phan, H. Harwell and A. Sabatini, (2010) Effects of Triglyceride Molecular Structure on Optimum Formulation of Surfactant-Oil-Water Systems, *J Surfact Deterg*, 13:189–194

[37] Yeung, A., Dabros, T., Masliyah, J. and Czarnecki, J. (2000) Micropipette: a New Technique in Emulsion Research. *Colloids and Surfaces A-Physicochemical and Engineering Aspects*, 174, 169-181.

[38] J. D. Mclean and P. K. Kilpatrick (1997) Effects of Asphaltene Aggregation in Model Heptane-Toluene Mixtures on Stability of Water-in-Oil Emulsions. *Journal of Colloid and Interface Science*, 196, 23-34.

[39] D. Langevin, S. Poteau, I. Henaut and F. Argillier, (2004), Crude oil emulsion properties and their application to heavy oil transportation, Oil and Gas science and technology- Rev. IFP, vol.59 No.6, pp. 652-655.

[40] J. D. Mclean and P. K. Kilpatrick, (1997) Effects of Asphaltene Solvency on Stability of Water-in-Crude-Oil Emulsions. *Journal of Colloid and Interface Science*, 189, 242-253.

[41] P. Sullivan, and P. K. Kilpatrick, (2002) The Effects of Inorganic Solid Particles on Water and Crude Oil Emulsion Stability. *Industrial & Engineering Chemistry Research*, 41, 3389-3404.

[42] S. Arditty, C. P. Whitby, B. P. Binks, V. Schmitt and F. Leal- (2003) Some General Features of Limited Coalescence in Solid-Stabilized Emulsions. *European Physical Journal*, 11, 273-281.

[43] N. S. Ahmed, A. M. Nassar, N. N. Zaki and K. H. Gharieb,. (1999) Stability and Rheology of Heavy Crude Oil-in-Water Emulsion Stabilized by an Anionic-Nonionic Surfactant Mixture. *Petroleum Science and Technology*, 17, 553-576.

[44] K. Khristov, S. D. Taylor, J. Czarnecki and J. Masliyah, (2000), Thin Liquid Film Technique - Application to Water- Oil-Water Bitumen Emulsion Films. *Colloids and Surfaces A-Physicochemical and Engineering Aspects*, 174, 183-196.

[45] O. V. Gafonova and H. W. Yarranton, (2001) The Stabilization of Water-in-Hydrocarbon Emulsions by Asphaltenes and Resins. *Journal of Colloid and Interface Science*, 241, 469-478.

[46] Hénaut, L. Barré, J. F. Argillier, F. Brucy, and R. Bouchard, (2001) Rheological and Structural Properties of Heavy Crude Oils in Relation with their Asphaltenes Content. *SPE*, 65020.

[47] E. Strassner, (1968) Effect of pH on Interfacial Films and Stability of Crude Oil-Water Emulsions. *SPE*, 1939.

[48] R. Sun and C. Shook, (1996) Inversion of Heavy Crude Oil-in-Brine Emulsions. *Journal of Petroleum Science and Engineering*, 14, 169-182.

[49] A.A.M Elgibaly, I.S. Nashawi, and M.A. Tantawy (1997) Rheological Characterization of Kuwaiti Oil-Lakes Oils and their Emulsions. *SPE* 37259, 493-508.

[50] Sharma, V. K. Saxena, A. Kumar, H.C. Ghildiyal, A. Anurada, N. D. Sharma, B.K. Sharma and B.K. Dinesh, (1998) Pipeline Transportation of Heavy Viscous Crude oil as Water Continuous Emulsion in North Cambay Basin (India). *SPE*, 39537.

[51] R. G. Gillies, R. Sun and C. A. Shook, (2000) Laboratory Investigation of Inversion of Heavy Oil Emulsions. *Canadian Journal of Chemical Engineering*, 78, 757-763.

[52] T. H. Plegue, S. Frank, D. Fruman and J.L Zakin, (1986) Viscosity and Colloidal Properties of Concentrated Crude Oil in Water Emulsions. *Journal of Colloid and Interface Science*, 114, 88-105.

[53] P. R. Garrett, (1993) *Defoaming, Theory and Industrial Applications*, Surfactant Science Series 45, M. Dekker, New-York.

[54] J. Zhang, D. Chen, D. Yan, and X. Yang, (1991) Pipelining of Heavy Crude Oil as Oil-in-Water Emulsions. *SPE* , 21733.

[55] B.M Yaghi and A. Al Bemani, (2002) Heavy Crude Oil Viscosity Reduction for Pipeline Transportation. *Energy Sources*, 24, 93-102.

[56] N.N. Zaki, T. Butz, and D. Kessel, (2001) Rheology, Particle Size Distribution, and Asphaltene Deposition of Viscous Asphaltic Crude Oil-in-Water Emulsions for Pipeline Transportation. *Petroleum Science and Technology*, 19, 425- 435.

[57] N.N., Zaki, N.S. Ahmed and A.M. Nassar, (2000) Sodium Lignin Sulfonate to Stabilize Heavy Crude Oil-in-Water Emulsions for Pipeline Transportation. *Petroleum Science and Technology*, 18, 1175-1193.

[58] T.H., Plegue, J.L., Zakin, S.G. Frank and D.H. Fruman, (1985) Studies of Water Continuous Emulsions of Heavy Crude Oils. *SPE* 15792.

[59] I.A., Layrisse, D.R., Polanco, H., Rivas, E., Jimenez, L. Quintero, L., J. Salazar, M. Rivero, A. Cardenas, M. Chirinos, D. Rojas, D. and H. Marquez,. (1990) Viscous Hydrocarbon- in-Water Emulsions. *US Patent 4 923 483.*

[60] N.N. Zaki, (1997) Surfactant Stabilized Crude Oil-in-Water Emulsions for Pipeline Transportation of Viscous Crude Oils. *Colloids and Surfaces A-Physicochemical and Engineering Aspects*, 125, 19-25.

[61] G. Nunez, M. Briceno, C. Mata, H. Rivas and D. Joseph, (1996) Flow Characteristics of Concentrated Emulsions of Very Viscous Oil in Water. *Journal of Rheology*, 40, 405-423.

[62] D.H. Fruman and J. Briant, (1983) Investigation of the Rheological Characteristics of Heavy Crude Oil-in-Water Emulsions. *International Conference on the Physical Modeling of Multi-Phase Flow*, Coventry, England.

[63] J.L., Salager, M.I. Briceno, and C.L. Brancho, (2001) Heavy Hydrocarbon Emulsions, In *Encyclopedic Handbook of Emulsion Technology*, Sjöblom, J. (ed), Marcel Dekker, New York.
[64] N.S., Ahmed, A.M., Nassar, N.N. Zaki and H.K. Gharieb, (1999) Formation of Fluid Heavy Oil-in-Water Emulsions for Pipeline Transportation. *Fuel*, 78, 593-600.
[65] M.R. Khan, (1996) Rheological Properties of Heavy Oils and Heavy Oil Emulsions. *Energy Sources*, 18, 385-391.
[66] O.V. Gafonova and H.W. Yarranton, (2001) The stabilization of water-in-hydrocarbon emulsions by asphaltenes and resins, J. Coll. Interf. Sci. 241,469–478.
[67] J. Sjoblom, N. Aske, I.H. Auflem, O. Brandal, T.E. Havre, O. Saether, A. Westvik, E.E. Johnsen and H. Kallevik, (2003), Our current understanding of water-in crude oil emulsions. Recent characterization techniques and high pressure performance, *Adv. Col. Interf. Sci.* 100–102, 399–473.
[68] T.H Plegue, S.G Frank, D.H., Fruman and J.L. Zakin (1989) Studies of Water-Continuous Emulsions of Heavy Crude Oils Prepared by Alkali Treatment. *SPE*, 18516.
[69] Jeribi, B. Almir-Assas, D. Langevin, I. Henaut, J.F.Argillier, Adsorption kinetics of asphaltenes at liquid interfaces, *J. Coll. Interf. Sci.* 256 (2002) 268–272.
[70] Yan, M. Gray, J. Masliyah, (2001) On water-in-oil emulsions stabilized by fine solids, *Coll. Surf. A* 193 97–107.
[71] H.W. Yarranton, H. Hussein, J. Masliyah, (2000) Water-in-hydrocarbon emulsions stabilized by asphaltenes at low concentrations, *J. Coll. Interf. Sci.* 228, 52–63.
[72] J.L. Creek, (2005). Freedom of action in the state of asphaltenes: Escape from conventional wisdom. *Energy Fuels* 19(4), 1212–1224.
[73] F. Verzaro, M. Bourrel, O. Garnier, H.G., Zhou and J.F. Argillier, (2002) Heavy Acidic Oil Transportation by Emulsion in Water. *SPE*, 78959.
[74] S. Poteau, (2004) Heavy Oil Transportation as Aqueous Emulsion. *PhD Thesis Report*, Paris 6 university.
[75] V.A. Adewusi and A.O. Ogunsola, (1993) Optimal Formulation of Caustic Systems for Emulsion Transportation and Dehydration of Heavy Oil. *Chemical Engineering Research & Design*, 71, 62-68.
[76] R.C. Shaw, L.L. Schramm, J. Czarnecki, Suspensions in the hot water flotation process for Canadian oil sands, in: L.L. Schramm (Ed.), Suspensions: Fundamentals and Applications in the Petroleum Industry, Advances in Chemistry Series 251, American Chemical Society, Washington, D.C., 1996, pp. 639–675.
[77] H. Groenzin, O.C. Mullins, (1999) Asphaltene molecular size and structure *J. Phys. Chem.* A 103, 11237–11245.
[78] L.E. Sanchez and J.L. Zakin, (1994) Transport of Viscous Crudes as Concentrated Oil-In-Water Emulsions. *Industrial & Engineering Chemistry Research*, 33, 3256-3261.
[79] R.A. Mohammed, A.I. Bailey, P.F. Luckham, S.E. Taylor, (1993) Dewatering of crude oil emulsions 1. Rheological behaviour of the crude oil–water interface, *Coll. Surf. A* 80: 223–235.
[80] T. Sun, L. Zhang, Y. Wang, S. Zhao, B. Peng, M. Li, J. Yu, (2002) Influence of demulsifiers of different structures on interfacial dilational properties of an oil-water interface containing surface-active fractions from crude oil, *J. Coll. Interf. Sci.* 255:241–247.

[81] S. Singh, J. D. McLean, and P. K. Kilpatrick, (1999) Fused Ring Aromatic Solvency in the Petroleum Emulsions, *J. Dispersion Sci. Tech.*, 20, 279-293 P.

[82] S. Matthew, L. Keith L. B. Gawrys, C. Trail and K. Kilpatrick, (2003) Effects of petroleum resins on asphaltene aggregation and water-in-oil emulsion formation, *Colloids and Surfaces A: Physicochem. Eng. Aspects* 220: 9 -27

[83] A.Yeung, T. Dabros, J. Czarnecki, J. Masliyah, (1999) On the interfacial properties of micrometre-sized water droplets in crude oil, *Proc. Roy. Soc. Lond.* A 455: 3709–3723.

[84] G. Horvath-Szabo, J. Czarnecki, J. Malisyah, Sandwich structures at oil-water interfaces under alkaline conditions, *J. Coll. Interf. Sci.* 253 (2002) 427–434.

[85] D.E. Graham, (1988), Crude oil emulsions: their stability and resolution, in: P.H. Ogden (Ed.), Chemicals in the Oil Industry, *Royal Society of Chemistry,* London, pp. 155–175.

[86] F.E. Bartell, D.O. Neiderhauser, (1949), Film-forming constituents of crude petroleum oils, in: Fundamental Research on Occurrence and Recovery of Petroleum, 1946–1947, American Institute of Petroleum, New York, pp. 57–80.

[87] C.H. Pasquarelli, D.T. Wasan, (1981), The effect of film forming materials on the dynamic interfacial properties in crude oil-aqueous systems, in: D.O. Shah (Ed.), Surface Phenomena in Enhanced Oil Recovery, Plenum Press, New York, pp. 237–248.

[88] O.K. Kimbler, R.L. Reed, I.H. Silberberg, (1966) Physical characteristics of natural films formed at crude oil–water interfaces, *Soc. Petr. Eng. J.* 6:153–165.

[89] C.E. Brown, E.L. Neustadter, K.P. Whittingham, Crude-oil/water interfacial properties—emulsion stability and immiscible displacement, Proceedings of Enhanced Oil Recovery by Displacement with Saline Solutions, May 20, 1977. London, England, pp. 91–122.

[90] C.G. Dodd, The rheological properties of films at crude petroleum–water interfaces, *J. Phys. Chem.* 64 (1960) 544–550.

[91] R.J.R. Cairns, D.M. Grist, E.L. Neustadter, (1976), The effect of crude oil–water interfacial properties on water–crude oil emulsion stability, in: A.L. Smith (Ed.), Theory and Practice of Emulsion Technology, Academic Press, New York, pp. 135–151

[92] H. Groenzin, O.C. Mullins, Molecular size and structure of asphaltenes from various sources, *Energy Fuels* 14 (2000) 677–684

[93] J. G. Speight, (1999). In: J.G. Speight (ed.), Petroleum Analysis and Evaluation, Petroleum Chemistry and Refining. Taylor & Francis Press, Washington, DC.

[94] R. Cimino, S. Correra, A. Del Bianco, and T.P. Lockhart (1995). In: E.Y. Sheu and O.C. Mullins (eds.), Asphaltenes: Fundamentals and Applications. Plenum Press, New York, p. 126.

[95] L. S. Kotlyar, B.D. Sparks, and J.R. Woods (1999). Solids associated with asphaltenes fraction of oil sands bitumen. *Energy & Fuel* 13, 346–350.

[96] E. Y. Sheu, (1998). In: O.C. Mullins and E.Y. Sheu (eds.), Structure and Dynamics of Asphaltenes and other petroleum fractions, Plenum Press, New York. p.115.

[97] B. T. Ellison, C.T. Gallagher, L.M. Frostman, and S.E. Lorimer (2000). The physical chemistry of wax, hydrates and asphaltene. In: *Offshore Technology Conference,* Houston, TX, May 1-4, 2000, OTC 11963.

[98] W. A. Gruse and D.R. Stevens (1960). The Chemical Technology of Petroleum. McGraw-Hill, New York.
[99] R. G. Draper, E. Kowalchuk, and G. Noel (1977). Analyses and characteristics of crude oil samples performed between 1969 and 1976. Report EPR/ERL 77-59, Energy, Mines, and Resources, Canada.
[100] J. G. Speight, (1991). The Chemistry and Technology of Petroleum, 2nd edn. Marcel Dekkar Inc., New York.
[101] W.D. McCain, Jr. (1990). The Properties of Petroleum Fluids, 2nd edn. PennWell Publishing Co., Oklahoma.
[102] W. A. Cruse and D.R. Stevens (1960). In: Chemical Technology of Petroleum. McGraw-Hill Book Publishers Inc., New York City, Chap. XXII.
[103] C. Lira-Galeana and A. Hammami (2000). In: T.F. Yen and G. Chilingarian (eds.), Asphaltenes and Asphalts. Elsevier Science Publishers, Amsterdam. Chap. 21.
[104] H. Warth, (1956). In: The Chemistry and Technology of Waxes, 2nd edn. Reinhold, New York.
[105] O.P. Starusz,, T.W. Mojelski, and E.M. Lown (1992). *Fuel* 71, 1355–1364.
[106] D. Espinat, (1993). In: *SPE International Symposium on Oilfield Chemistry*, New Orleans, LA, March 2–5, 1993, SPE 25187.

11

Tailored Polymer Additives for Wax (Paraffin) Crystal Control

Aurel Radulescu[1], Lewis J. Fetters[2] and Dieter Richter[1]
[1]*Forschungszentrum Jülich GmbH, Jülich Centre for Neutron Science*
[2]*Cornell University, School of Chemical and Biomolecular Engineering*
[1]*Germany*
[2]*USA*

1. Introduction

Crude oils and refined middle distillate products such as diesel fuels, kerosene (jet fuel) or heating oil contain an important fraction of paraffins (alkanes) of high energy content with a broad linear (*n*-paraffins) and branched chain length distribution (Fig.1).

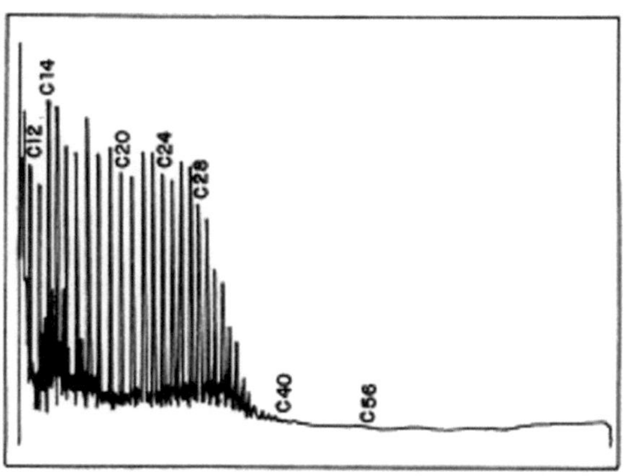

Fig. 1. Gas chromatogram of a virgin crude oil (del Carmen Garcia et al., 2000).

Depending on the type of crude deposits and refined technology applied, this fraction can vary between 10 and 30% (Coutinho et al., 2000). Although they are energetically desirable because of their increased combustion enthalpy with respect to C_5-C_{17} alkanes, the long chain C_{18}-C_{40} *n*-paraffins (waxes) are technically embarrassing when (1) their concentration is too high, (2) crude oils are extracted from deep sea reservoirs and pipeline transported through cold regions or (3) diesel fuels are used during the winter time (Kern & Dassonville, 1992). In these conditions, such fluids undergo dramatic degradation of viscoelastic properties due to precipitation of waxes as a consequence of the temperature drop and reduction of their solubility.

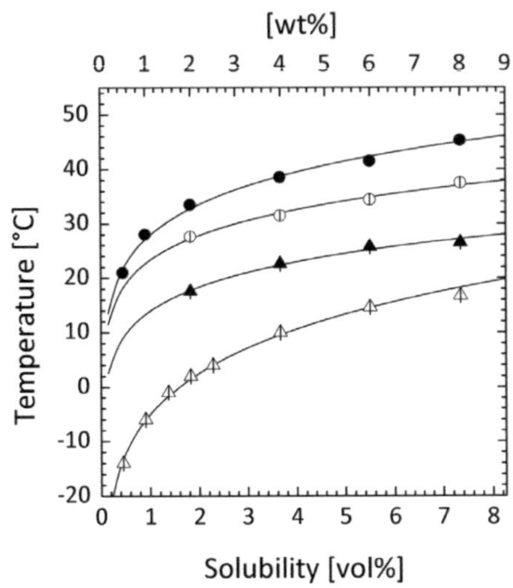

Fig. 2. Observed solubility lines of C_{24}, C_{28}, C_{32} and C_{36} waxes (from bottom to top) in decane (Ashbaugh et al., 2002).

The solubility point of *n*-paraffins (Fig.2) decreases with the increase of the carbon number (Srivastava et al., 1997): the higher the solubility point the more difficult is to keep the wax from precipitating. Wax precipitation is especially problematic during the production phase when the drop in the pressure and temperature of crude oils and the loss of short paraffins (light ends) to the gas phase occur (Pedersen et al., 1991a, 1991b). The waxes form stacked lamellar crystals with sizes of hundreds of micrometers and an overall morphology resembling a "house-of-cards" (Abdallah et al., 2000) that readily entrains liquid oil, primarily through surface tension, and effectively forms an organic gel. The occurrence and deposition of such large wax crystals cause a reduction of the ease of flow of crude oils and a loss of fluidity and filterability of middle distillates (Venkatesan et al., 2005; Singh et al., 1999). Wax deposition affects the storage tanks and conduits (Fig.3).

The pour point (PP) – the temperature at which the system gels and becomes mechanically rigid, - is about 10°C for a typical untreated oil and may vary over a wide temperature range below 0°C in the case of untreated diesel fuels (Claudy et al., 1993). A parameter that directly correlates to the occurrence of low temperature technical problems caused by the wax crystals is the cold filter plugging point (CFPP). This parameter corresponds to the temperature when plugging occurs in a 45μm filter under standardized conditions. For an untreated diesel fuel the CFPP is normally a few degrees higher than the PP.

In order to circumvent these technical problems "pour-point depressants" are added (Giorgio & Kern, 1983; Beiny et al., 1990). Generally, depressants are copolymers consisting of crystalline and amorphous segments that have the capacity to self-assemble in solution without sedimentation even at temperature well below 0°C. They interact favourably with paraffins and moderate the wax crystals morphology so that crude oils and middle distillates remain fluid as their temperature passes through that of PP. The mechanisms by which these polymeric systems modify the wax crystal size and shape are however incompletely understood. As such, the synthesis and choice of additives for crude oils and diesel fuels are largely trial and error rather than based on scientific principles. Moreover, tests to evaluate the effectiveness of wax crystal modifying polymers concern generally the bulk properties and take rarely into consideration flow, cooling rate and composition conditions. A good knowledge of the wax-polymer interaction mechanism at microscopic scale as a function of wax content and temperature variation assesses the ability of a polymeric system to control efficiently the wax crystallization in hydrocarbon solution and enables the tailoring of additives for specific oil and middle-distillate composition conditions. The wax-polymer interaction in solution yields complex aggregates whose morphology strongly depends on the precipitation temperatures of both components. Therefore, besides information on the oil and diesel fuel composition a good characterization of the polymer self-assembling behaviour in solution is a prerequisite condition for the start of a structural study of wax-polymer interaction.

Fig. 3. Plugged pipeline (Banki et al., 2008).

Small-angle neutron scattering (SANS) technique exploiting on one side the strength of contrast matching and, on other side, the wide length scale explored, from 1nm up to 1μm, is a dedicated technique to investigate and elucidate complex morphologies such as those occurring in a wax-polymer-oil system (Richter et al., 1997; Leube et al., 2000; Monkenbusch et al., 2000; Schwahn et al., 2002a, 2002b; Ashbaugh et al, 2002; Radulescu et al., 2003, 2004, 2006, 2008, 2011). An eloquent example is that of diblock polymers (Fig. 4) of semi-crystalline polyethylene coupled to amorphous poly(ethylene-propylene) (PE-PEP) or poly(ethylene-butene) (PE-PEB) which yield in solution plate-like structures. The aggregates consist of a PE core shrouded behind an amorphous brush layer (Richter et al. 1997). The crystalline core serves as a nucleation platform for wax precipitation, while the amorphous brush acts as a steric barrier keeping the aggregates in solution. Rather than forming large platelets themselves, the waxes are sequestered by the diblock lamellar micelles (Leube et al. 2000). The elucidation of this wax modification mechanism by means of contrast matching SANS has found commercial application of PE-PEB (Infineum, Paraflow™) as diesel wax modifiers. More recently, following systematic SANS and optical microscopy studies, it was shown that the poly(ethylene-*co*-butene) random copolymers (Fig. 4), designated PEB-*n*, where *n* is the number of the ethyl side branches / 100 backbone carbons, could assume important oil and refinery applications because of their capability to modify the wax crystals precipitated at low temperature by model oil and diesel fluids (Ashbaugh et al., 2002; Schwahn et al, 2002b; Radulescu et al., 2004).

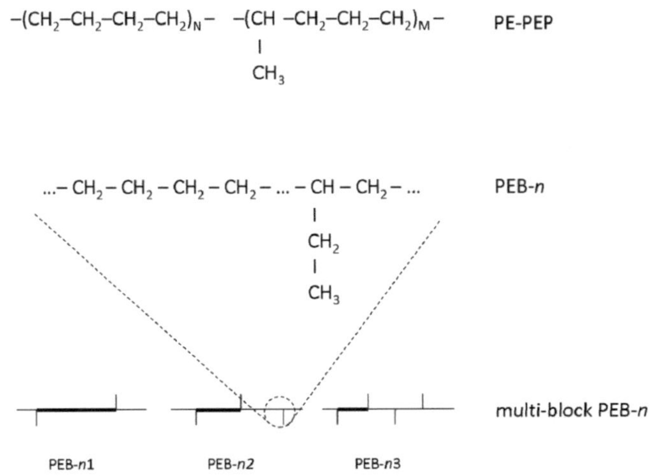

Fig. 4. Polymer structures designed and synthesized as efficient wax crystal modifiers following microstructural studies with SANS: crystalline-amorphous PE-PEP diblocks, semicrystalline PEB-*n* random copolymers with tuned crystallinity (variable number of ethyl side-groups*)* and multi-block PEB-*n* copolymers with graded crystallinity.

The studies on mixed hydrocarbon solutions of PEB-*n* copolymers and single paraffin waxes by contrast matching SANS revealed that these copolymers show selectivity in their wax modification capacity depending on the ethylene content of the backbones (Radulescu et al., 2003, 2004). The more crystalline copolymers show a higher efficiency for longer wax molecules while the less crystalline ones are very efficient for shorter waxes. This suggests that highly efficient PEB-*n* additives for crude oils and middle distillates should consist of segments with graded ethylene content. New polymers presenting variable crystallinity along the chain have been synthesized as multi-block copolymers built from segments representing PEB-*n* random polymers with blocks having different global content of ethyl branches (Fig.4). These new materials exhibit a gradual crystallization tendency of each block with decreasing temperature (Radulescu et al., 2011) which, considering the selectiveness of the PEB-*n* copolymers regarding their wax modification, tunes the crystallization behaviour of a broader distribution of wax molecules like those contained by crude oil and middle distillate fuel systems. In this case a gradual co-crystallization of those wax molecules and polymeric blocks presenting similar precipitation points eventually emerges. SANS and microscopy studies on polymer-wax solutions prepared for conditions closed to realistic ones, for high wax content or mixed wax molecules, have shown that such specifically designed polymers are able to template and control the wax crystallization in moderate size fluffy aggregates, thus arresting the growth of large compact waxy crystals and preventing the wax gelation which otherwise would eventually occur below 10°C.

2. Small-angle neutron scattering

Neutrons interact with matter via the short-range nuclear interactions and hence see the nuclei in a sample rather than the diffuse electrons cloud observed by X-rays. In magnetic samples neutrons are scattered by the magnetic moments associated with unpaired electron spins (dipoles). Unlike the X-rays, the neutrons are able to "see" light atoms in the presence of heavier ones and to distinguish neighbouring elements more easily. Because the cross-section of an atom generally varies between isotopes of the same element, the exploitation of isotopic substitution methods can allow neutrons to highlight structural and dynamic details. Particularly, the strong difference in cross-section between hydrogen and deuterium enabling contrast variation methods is of a great importance for the investigation of synthetic organic compounds. Such complex structures spanning over a length scale from 1nm to 1 μm, can be characterized by SANS method.

2.1 SANS instruments and method

Elastic scattering with neutrons reveals structural information on the arrangement of atoms and magnetic moments in condensed matter systems. The information in such scattering experiments is contained in the neutron intensity measured as a function of the momentum transfer Q

$$\vec{Q} = \vec{k}_i - \vec{k}_f; \quad Q = \frac{4\pi}{\lambda} \sin\theta / 2 \tag{1}$$

where $\vec{k}_i$ and $\vec{k}_f$ are the incoming and outgoing neutron wave vector, λ is the neutron wavelength and θ is the scattering angle (Fig.5). Scattering experiments explore matter in reciprocal space and Q acts as a kind of inverse yardstick: large Q values relate to short distances, while a small Q relates to large objects. Aiming for the mesoscopic scale, SANS is optimized for the observation of small scattering angles using long wavelength (cold) neutrons. Thus, large objects, such as macromolecules, colloids, self-assembled systems, membranes, proteins, polymeric and biomolecular aggregates, etc., can be studied.

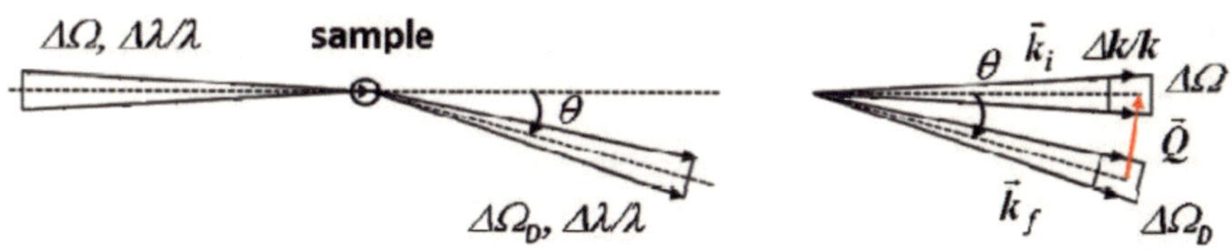

Fig. 5. Scattering process in real (left) and reciprocal (right) space; the divergence of the primary beam is described by the space angle $\Delta\Omega$, while the scattered intensity is measured in a detector element with space angle $\Delta\Omega_D$.

The principle layout of a conventional pinhole SANS instrument is shown in Fig.6. The high intensity requests in the case of SANS can only be achieved by using a large wavelength distribution of the monochromatic beam delivered by the mechanical velocity selector, typically $\Delta\lambda/\lambda$=10%-20%. A resolution-optimized setting of a SANS instrument is achieved if the collimation length L_C equals the sample-to-detector distance L_D. Therefore, the geometrically optimized SANS instruments have a typical length of 40m, reaching 80m for D11 at the ILL, Grenoble. Using neutron wavelengths between 4.5 and 20Å the Q interval between 7×10^{-4}Å^{-1} and 0.5Å^{-1} becomes accessible in the classical pinhole mode.

In order to increase further the Q resolution towards smaller values, focusing optical elements need to be introduced. One possibility are the refractive optical elements - neutron lenses, placed in front of the sample such that a small incoming neutron beam is focused on the detector and thus, smaller scattering angles than for conventional pinhole geometry are accessed (Frielinghaus et al., 2009).

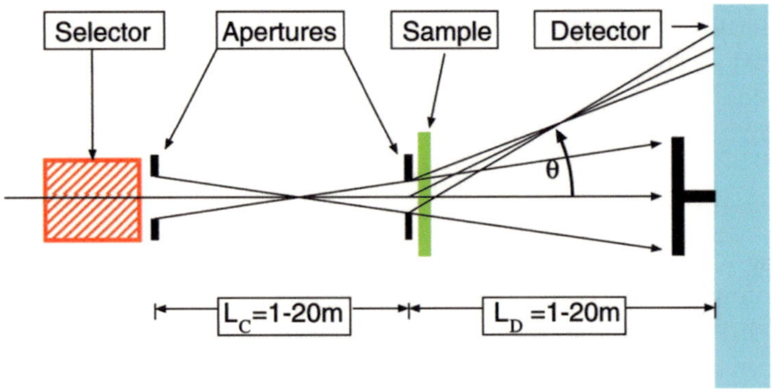

Fig. 6. Principle of pinhole SANS; a neutron beam entering the instrument from the left is the subject of monochromation (done by the velocity selector), collimation over a variable distance (achieved by using a set of adaptive system of apertures), scattering on the sample and detection on a two-dimensional position sensitive detector over long distances; aiming for detecting neutrons at small scattering angles the SANS instruments are very long.

Focusing-SANS using reflective (mirror) systems (Alefeld et al., 2000) was in use since longer time at the KWS-3 instrument operational at the FRM II reactor, Garching-München. In such an instrument the monochromated neutrons enter through a small aperture (typically 1mm) and hit the focusing mirror with the full divergence provided by the neutron guide. The mirror reflects and focuses the neutron beam on a high-resolution position sensitive detector (0.5mm space resolution) placed in the focal plane. The sample position is placed just after the mirror. Both focusing methods extend the pinhole SANS *Q*-range down to $1\text{x}10^{-4}\text{Å}^{-1}$.

SANS is thus a well-established, non-destructive method to examine structure on length scales of 1nm to 1μm.

2.2 SANS cross-section of simple structures

A crucial feature of SANS that makes it particularly useful for synthetic organic macromolecules (soft matter) is the ability to vary the scattering contrasts between different constituents of a hydrocarbon sample over a broad range by H/D substitution. Since the molecules affected by H/D exchange are chemically the same, the physical chemistry of the sample is only marginally modified, if at all. The visibility of an object *M* in solution *S* for example depends on the difference of the solvent/solute scattering length densities:

$$\Delta\rho_M = (\rho_M - \rho_S) = \frac{\sum_i^{atoms\ in\ M} b_i^M}{v_M} - \frac{\sum_j^{atoms\ in\ S} b_j^S}{v_S} \tag{2}$$

where b_i^M denotes the scattering length of the different atoms in *M* and b_j^S those of the atoms in *S*, while v_M and v_S are the effective volumes occupied by the objects composed of the atoms in the respective sums. Since b_H = -3.57x10^{-13} cm, b_D = +6.57x10^{-13} cm and b_C = +6.65x10^{-13} cm, H/D replacement allows for huge variation of $\Delta\rho$ for hydrocarbons. Using a mixture of suitable amounts of deuterated and protonated solvent in general it is possible to achieve "zero contrast" $\Delta\rho = 0$ for one component of the system (Fig.7). The technique of contrast variation allows for highlighting parts of a structure and is of essential importance for the investigation of complex multi-component systems. The exchange of hydrogen with deuterium either in the solvent or explicitly in chemical groups of the organic macromolecules of interest (labeling) is then required. Deuterium labeling of synthetic organic macromolecules (polymers) is easily achieved by special synthesis methods.

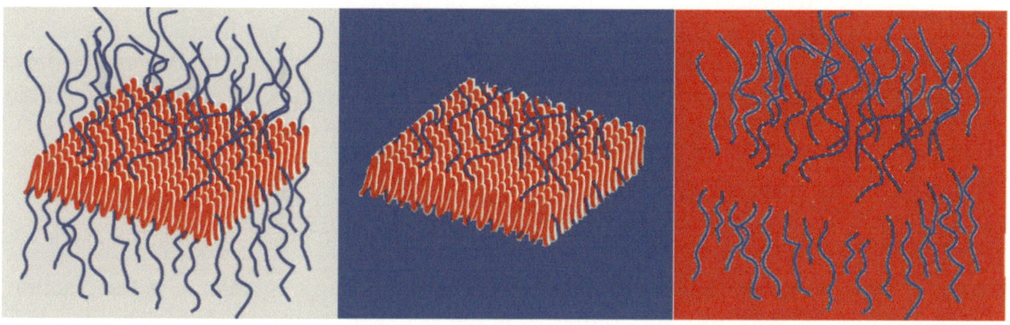

Fig. 7. Example of contrast matching between different domains of a diblock copolymer two-dimensional micelle (crystalline-amorphous core-brush morphology) and solvent: by modification of the solvent scattering length density the full contrast, core contrast and brush contrast (from left to right) can be adjusted.

The contribution to the small angle scattering intensity from some object is contained in the macroscopic cross-section $d\Sigma/d\Omega(Q)$, which is the absolute square of the Fourier transform of its scattering length density distribution $\Delta\rho(\vec{r})$ normalized by the sample volume V_{sample}. In case of N identical particles of volume V_p which are located at random positions and random orientations in a solution (with $N\,V_p = \phi\,V_{sample}$, where ϕ is the volume fraction of the scattering particles in solution) and are homogeneously decorated with a constant contrast factor $\Delta\rho$, the macroscopic scattering cross-section can be written as

$$\frac{d\Sigma}{d\Omega}(Q) = \frac{N}{V_{sample}} \Delta\rho^2\, V_p^2 P(Q)\, S(Q) \quad (3)$$

where $P(Q)$ represents the particle form factor, which relates to intra-particle correlations, and $S(Q)$ the structure factor, which denotes the inter-particle correlation effects. The analysis of the form factor $P(Q)$ and structure factor $S(Q)$ in terms of structural models (Pedersen, 1997) delivers information about the size, shape, number density, and correlation between the scattering particles in the sample. The evaluation of the "forward scattering" $d\Sigma/d\Omega(Q\rightarrow 0)$, which amounts to $\phi\,\Delta\rho^2 V_p$, offers information about the volume fraction of the scattering particles while knowing the contrast from the synthesis procedure and the size of the scattering objects from the scattering data.

For multi-component systems containing for example, oil, polymer and wax, the scattering cross-section in terms of the partial scattering functions assumes the form:

$$\frac{d\Sigma}{d\Omega}(Q)=(\rho_P-\rho_S)^2 P_{PP}(Q)+2(\rho_P-\rho_S)(\rho_W-\rho_S)P_{PW}(Q)+(\rho_W-\rho_S)^2 P_{WW}(Q) \quad (4)$$

where the indexes P and W indicate the polymer and wax, respectively. If the scattering length density of the single components is changed within given limits by hydrogenation or deuteration, either the polymer ($\rho_W=\rho_S$) or the wax ($\rho_P=\rho_S$) can be left visible in the sample.

The wax and semicrystalline polymer molecules yield in solution complex morphologies displaying multiple structural levels that are sometime hierarchically organized on a scale from nanometers to hundreds of microns. Basically, these morphologies evolve from lamellar or rod structural units. Depending on the polymer architecture, a crystalline-amorphous morphology emerges in solution either as core-brush lamellae or as density modulated rods. For a core-brush two-dimensional morphology formed by diblock copolymers in solution (Richter et al., 1997) the cross section is expressed as

$$\frac{d\Sigma}{d\Omega}(Q)=\phi\frac{v_c}{V_c}P(Q)\left(\pi R^2\right)^2\frac{D(QR/2)}{(QR/2)}+I_{blob} \quad (5)$$

Thereby, ϕ is the volume fraction of polymer in solution (as aggregates), v_C is the fraction of the crystallizable segment in the polymer, R is the lateral dimension of the lamellae (discs) and $D(x)$ denotes the Dawson function. The second term in Eq.5 arises from the polymeric structure of the brush (the "blob" scattering). $P(Q)$ is the form factor of the density profile perpendicular to the lamellae surface including the contrast factors of the core and the brush parts and the density profiles of the polymer volume fraction. The form factor of an infinitely large plate of the thickness d considering a simple rectangular density profile is

$$P(Q)=\left[\sin(Qd/2)/(Qd/2)\right]^2 \quad (6)$$

For the amorphous brush different approaches can be adopted varying from a simple rectangular profile to a parabolic or even a more complicated prediction. When the platelets stack the structure factor of one-dimensional paracrystalline order can be used, which in the case of an infinite stack has the form

$$S(Q)=\frac{\sinh\left(Q^2\sigma_D^2/4\right)}{\cosh\left(Q^2\sigma_D^2/4\right)-\cos(QD)} \quad (7)$$

where D is the stacking period and σ_D is its Gaussian smearing.

The form factor appropriate for an ensemble of isotropic oriented homogeneous rods of thickness a and very large length $2H$ can be expressed, for the condition $Q>>\pi H$, $H>>a$, as

$$P(Q)=\frac{\pi}{Q\,2H}\left[\frac{2J_1(Qa)}{(Qa)}\right]^2 \quad (8)$$

where $J_1(x)$ is the first order Bessel function of the first type. More complicated equations requiring numerical calculations are involved for inhomogeneous rods (Radulescu et al., 2004) or for rods with elliptical cross-section (Bergström & Pedersen, 1999).

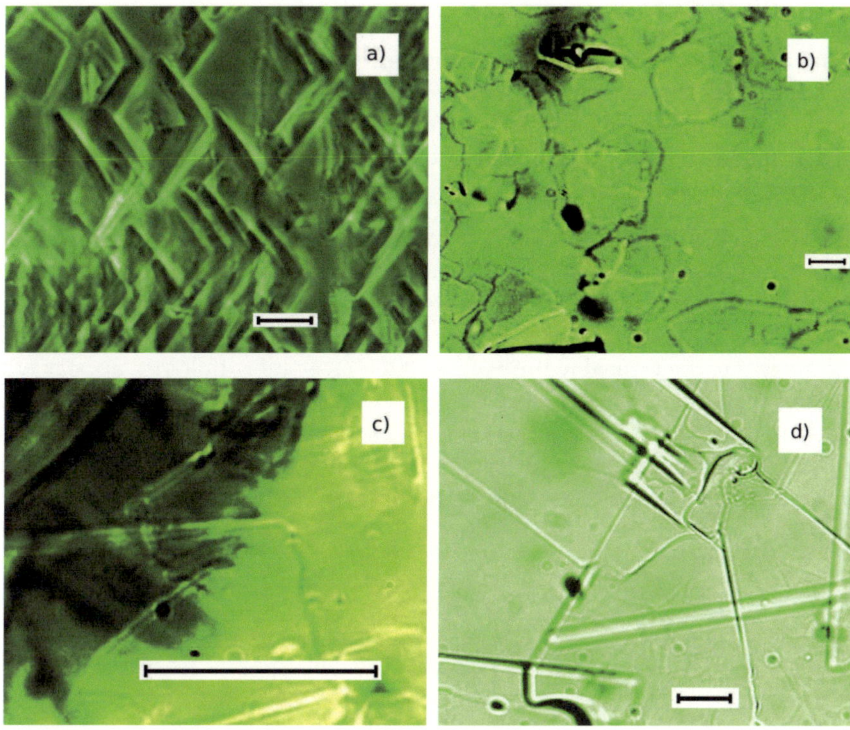

Fig. 8. Micrographs from decane solutions of mixed 1%C_{36}/1%C_{30} at 0°C (a), 0.5%C_{24} at -20°C (b), 1%C_{36} at 20°C (c) and 4%C_{24} at -10°C (d); scale bars 30μm.

The combination of scattering data with fractal geometry concepts recently commenced to become a general path of investigating the complex morphologies displaying multiple structural levels on wide length scales (Beaucage & Schaefer, 1994). Fractal approaches describe power-law regimes often observed in measured scattering profiles with the exponents depending on the geometric structure of the scattering objects, $d\Sigma/d\Omega(Q) \approx Q^{-p}$. If the mass of an object scales with its size according to $M \approx R^D$, one deals with the mass fractals that can be simply characterized by the exponent D, the mass fractal dimension. The scattering cross-section from mass-fractal objects is proportional with Q^{-D} and thus, a power law regime in Q with an exponent $p=D=1$ relates to rod-like structures, $p=D=2$ to platelets and $p=D=3$ to uniformly dense structures. Surface fractals are uniformly dense objects presenting a fractally rough surface that scales with their radius $S \approx r^{D_S}$, where D_S can assume values between 2 (smooth surface) and 3 (extreme roughness). For fractally rough surfaces, the power law exponent becomes $p=D_S$-6 and, if the object possesses a smooth surface, $p=4$ (Porod scattering), while for a rough surface $3 \le p \le 4$. Empirically, one may also find slopes steeper than those corresponding to $p=4$. In this case, the objects have diffuse rather than fractally rough interfaces. An observed structural level and its associated power-law for simple well defined shape objects can be analyzed using the graphical presentation of data in terms of Guinier approximation, ln(Q^p $d\Sigma/d\Omega$) vs. Q^2, which delivers in a direct way the "forward scattering" and the size of the objects (Schwahn et al., 2002a).

3. Wax crystallization from hydrocarbon solution

Below the solubility line (Fig.2) wax precipitates in large crystals. Depending on solution conditions, such as molecule length, wax concentration, temperature, or whether there is a single- or multi-wax solution, the crystal morphology varies to a large extent (Fig.8). Isolated well-defined lozenge-shaped pyramidal plates (Fig.8a), flat conical objects with irregular edges (Fig.8b) or "house-of-cards" morphologies of compact crystals (Fig.8c,d) are observed.

Such large crystals with sizes between several and hundreds of microns provide a strong small-angle scattered intensity (Fig.9). Although in all conditions large crystals are formed the profiles differ significantly and relate to the asymptotic power-law scattering behaviour typical of the differing morphologies formed. The size of the crystals finds itself within micrometer range, i.e. the typical features from such large morphologies appear within very low Q domains, well out of the measured range by classical pinhole SANS. The power-law indicates formation of three-dimensional fractal-like structures having an extremely rough surface (p=3) as in the case of short chain waxes at low content or those where p=4 (typical for high wax contents or long waxes). Mixed wax crystals exhibit surface fractal features.

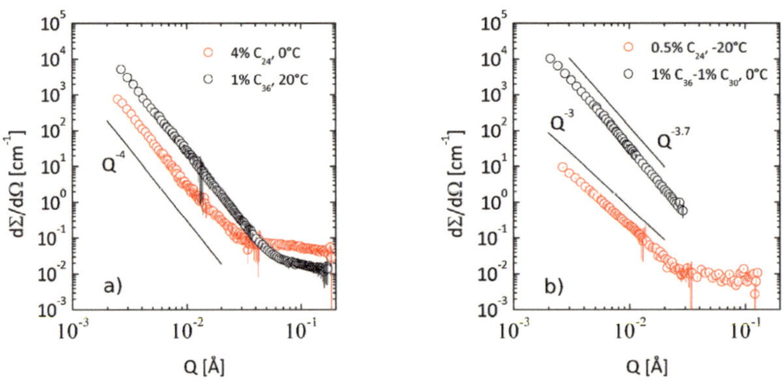

Fig. 9. SANS traces from solutions of different waxes in decane.

4. Crystalline-amorphous PE-PEP diblock copolymers

The crystalline-amorphous poly(ethylene)-poly(ethylene-propylene) (PE-PEP) diblock copolymers yield in decane platelet-like aggregates (Richter et al., 1997; Leube et al., 2000). Such a morphology is adopted as a consequence of the crystallization of the PE-blocks in thin lamellar crystallites (d≈20-50Å), whereas the PEP forms brushes on both sides of the lamellae. Unlike the lower molecular weight M_W compounds that yield isolated platelets, the higher molecular weight copolymers form larger platelets showing stacking tendencies due to mutual van der Waals attraction acting on the large surface area. SANS investigations were performed in order to quantitatively understand the aggregation phenomena occurring in solution and to decipher the influence of PE-PEP additives on the wax crystals formed at low temperatures in common hydrocarbon solutions. Empirically, the PE-PEP diblock systems have proven to exhibit wax crystal modification activity in various diesel fuels. Whether a co-crystallization of wax molecules with the PE-block in the core or another wax-polymer interaction mechanism takes place, was the fundamental question addressed by SANS. For this purpose, a series of well-defined diblocks with different PE-PEP compositions and molecular weights was synthesized by anionic polymerization of polydienes followed by the saturation of the resulting polymer by hydrogenation. This is a convenient method in the case where well-define molecular weights and structures are needed. In a first step polybutadiene-polyisoprene (PB-PI) diblocks were obtained by polymerization of 1,3-butadiene and 2-methyl-1,3-butadiene in nonpolar hexane. The M_W was characterized by low-angle laser-light scattering and by size exclusion chromatography. The thus synthesized polymers were in a second step saturated by hydrogenation or deuteration at 80-100°C under a pressure of 25-30 bar using palladium on barium sulphate as a catalyst. Due to the occasional occurrence of 1,2-addition of butadiene the PE chains in the final polymer contain about 2 ethyl side branches/100 backbone carbons.

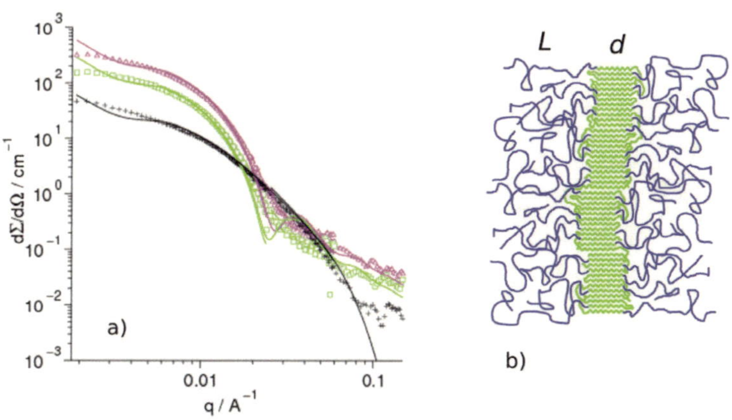

Fig. 10. (a) Example of scattering patterns from PE(1.5kg mol^{-1})-PEP(5kg mol^{-1}) diblock in decane for the common (green), core (black) and brush (pink) contrast conditions (Leube et al., 2000). The solid lines denote the simultaneous fit of the data according to the two-dimensional core-brush model (Eq.5) including a rectangular core profile and a parabolic brush profile; (b) Sketch depicting the PE-PEP structure in solution as emerged from the model interpretation of the SANS data, with *L* the brush length and *d* the core thickness.

To utilize the full capabilities of SANS the contrast was controlled by the variation of the scattering properties of different polymer components by varying the degree of hydrogenation and deuteration in the precursor PB-PI block copolymers and/or during the saturation. Furthermore the wax was contrasted out or made visible depending on the degree of hydrogenation and deuteration of the different components.

Fig.10a presents the scattering data from a low M_W PE-PEP copolymer in decane (ϕ_{pol}=2%) at room temperature for different contrast conditions enabled by the variation of the solvent scattering length density in equal steps between fully deuterated decane (ρ_S=6.2 x 10^{10} cm^{-2}) and fully protonated decane (ρ_S=-0.3 x 10^{10} cm^{-2}). Detailed measurements were performed on decane solutions of largely different M_W copolymers by varying the polymer volume fraction in solution and the scattering length density of polymer blocks and solvent. The fit of such data using Eq.5 with different profiles for the core and the brush morphologies delivered the platelet thickness d and the brush length L, both parameters which vary with M_W. Taking into consideration the structural features revealed by SANS and the thermodynamic aspects of the platelets formation (core thickness and brush length adjust themselves from a balance between the entropy loss in the stretched brush on the one hand and the enthalpy costs of folding the polyethylene chains and defect energies associated with the incorporation of ethyl side branches on the other hand) the structure of the PE-PEP aggregates (Fig. 10b) consisting of a thin crystalline core exhibiting a certain surface roughness and a stretched amorphous brush was fully characterized and understood.

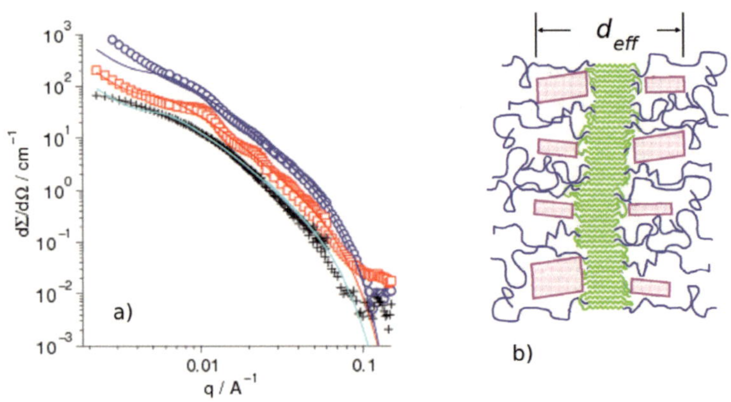

Fig. 11. (a) Example of scattering from PE(1.5 kg mol^{-1})-PEP(5 kg mol^{-1}) with no wax (black), 0.5% C_{36} (blue) and 0.5% C_{30} (red) added, under common core and wax contrast, at 5°C (Leube et al., 2000): the scattering patterns follow a Q^{-2} behaviour; (b) Sketch depicting wax crystallites (pink) nucleated at the crystalline PE core and sequestered by the PEP brush.

The effect of wax on the scattering from diblocks was studied in different contrast conditions (Leube et al., 2000), which allowed the core, the brush or commonly the core and wax visible. A comparative analysis of the polymer core and brush scattering profiles for neat and C_{36} hexatriacontane wax-dopped polymer solutions in decane offered the first insight about the changes occurring at the level of polymer morphology in the presence of wax. The results revealed no change of scattering for core contrast, thus no change in the core thickness, while the brush density profile was found to be shifted outward in the presence of wax compared to the polymer self-assembling case, which indicates more stretched hairs close to the core surface. These findings prove that there is no incorporation of wax into the core. Apparently, the wax enters the brush probably by a nucleation and growth on top of the core. SANS investigations under common core and wax contrast condition revealed an increase in thickness of the two-dimensional aggregates, i.e. the adsorbed crystallized wax enters through the brush and adds to the effective core thickness.

Following detailed contrast matching SANS investigation (Fig. 11a) the increase in the effective core thickness d_{eff} due to addition of wax crystallites was evaluated. Supposing an average surface coverage by wax, from the fit of the scattering patterns the crystallized wax volume fraction was obtained: for 0.5%C_{36} wax dissolved in the initial solution almost 100%

wax aggregation was found at 5°C, while in the case of 0.5%C_{30} solution the wax aggregation fraction is only 57%.

In conclusion, by means of the SANS micro-structural investigation method the structure and morphology of the crystalline-amorphous polymer-aggregates was resolved (Fig.11b). Of greater importance for further design and optimization of efficient polymer additives for wax crystal control was the understanding of the interaction mechanism between the polymers and waxes in this specific case. This mechanism has found commercial application of PE-PEB (Paraflow™) as diesel wax modifiers.

These crystalline-amorphous diblock copolymers are effective at breaking the wax gels at low wax concentration. In applications where the dissolved wax concentration is large, such as crude oils, wax precipitation can bury the PEP or PEB brush layers negating thus the steric stabilization advantages and grow ultimately in large crystals. Other more versatile crystalline–amorphous polymer architectures expressing a higher efficiency in controlling the wax crystallization at high wax content have been designed based on micro-structural information obtained by SANS.

5. Crystalline-amorphous poly(ethylene-butene) copolymers PEB-*n*

The behaviour of PE-PEP and PE-PEB diblock copolymers with respect to their ability to control the wax crystallization invited the evaluation of PEB-based copolymer architectures where semi-crystalline and amorphous segments are combined in an alternative or random manner. One such family is represented by nearly random copolymers of ethylene and 1-butene obtained by the anionic polymerization of butadiene with variable 1,2- and 1,4- modes of addition (Morton & Fetters, 1975). The microcrystallinity of these polymers designated PEB-*n*, where *n* is the number of ethyl side branches/100 backbone carbons, can be tuned by changing the ratio ethylene to butene segments, i.e. by varying *n*. Thus, PEB-0 represents essentially crystalline high density PE, while for $n>13$ crystallinity is absent in the bulk state. The reactivity ratio product obtained from the ^{13}C-NMR evaluation of the sequence distribution characteristics of PEB copolymers suggested their random character. The self-assembling features of PEB-*n* copolymers in decane solution were investigated by SANS. Following carefully observation of the scattering patterns with decreasing temperature it was found that the assembling events appear at a temperature which is the higher the lower *n* (Ashbaugh et al., 2002; Radulescu et al., 2003). Thus, PEB-7.5 shows first aggregation tracks already at 40°C while PEB-11 only at 0°C. Despite this difference the scattering cross-section of copolymers with different number of ethyl side groups shows at low temperature a well-defined Q^{-1} power-law profile which is indicative for rod like aggregates. Fig.12a displays example of SANS patterns from the PEB-7.5 (6kg mol^{-1}) aggregates at -10°C for different polymer volume fraction ϕ_{pol} in solution. (Radulescu et al., 2004). The length of the rods is not accessible by classical SANS since no saturation of scattered intensity towards low Q is observed. The peak-like structure appearing at the same Q position for different ϕ_{pol} denotes intra-particle correlations and arises from longitudinal modulation of the rod density. Following the interpretation of the experimental data by appropriate structural models the polymer rods were characterized with respect to their lateral size and density parameters and a self-assembling mechanism was proposed. The polymer volume fraction inside the rod structures with thickness of 50-60Å and modulation period of 300-350Å is low and could range between 4% and 20% depending on the actual volume fraction of rods in solution.

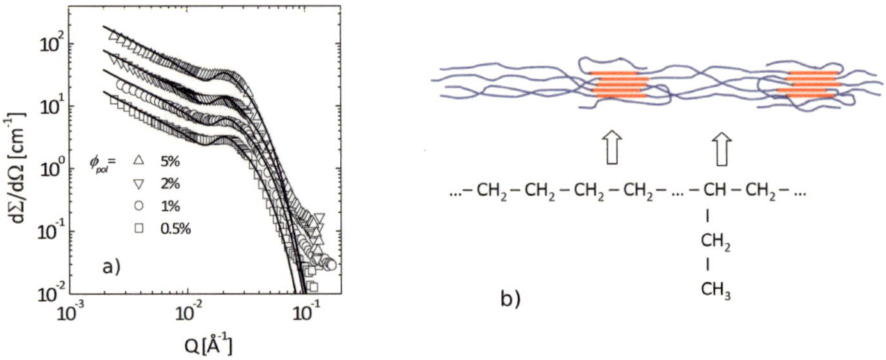

Fig. 12. (a) Examples of SANS patterns from PEB-7.5 copolymer in decane for different ϕ_{pol} at -10°C (Radulescu et al., 2004); (b) Sketch depicting the structure of PEB-*n* random copolymers in solution as emerged from the analysis of SANS data.

This finding shows that the polymer assemblies are solvent swollen open objects. Formation of such morphology was explained by taking into account the molecular architecture of the PEB-*n* random copolymers: the aggregation events occur as a consequence of crystallization of the longer PE sections within one chain while the amorphous sequences containing ethyl side groups form a loose corona around the crystalline nucleus and screen it against the intrusion of other chains by osmotic repulsion. The rarity of co-crystallization events promotes thus a one-dimensional growth (Fig. 12b). For PEB-*n* materials with lesser crystallinity an attenuated or "no peak" feature was noticed in the SANS patterns. That indicates a more homogeneous polymer density inside rods (Schwahn et al., 2002a).

The self-assembly behaviour of the PEB-*n* copolymers with variable crystallinity led to the conjecture that their interaction with waxes may depend on the correlation between the self-assembly temperature and the wax solubility point within the investigated range of wax concentration. The density of the ethyl side groups and the M_W may thus control the efficacy of different PEB-*n* copolymers in modifying the size and the shape of waxy crystals formed in decane. The understanding of their positive effect on the low-temperature viscoelastic properties of waxy fluids was possible following combined rheology, microscopy and SANS studies (Ashbaugh et al, 2002; Schwahn et al., 2002b; Radulescu et al., 2004, 2006). Fig.13a,b presents micrographs collected at different temperatures from decane solutions of 1% C_{36} and 4% C_{24} when 0.6% PEB-7.5 is added (Radulescu et al., 2004, 2006). Marked changes in the wax crystallization habits are observed in the presence of polymer compared with crystallization from pure solutions (Fig.8c,d): instead of big compact plates several hundreds of μm in size smaller morphologies exhibiting a soft texture or thin needles with modulated thickness appear in the presence of polymer.

The formation and hierarchical evolution of such morphologies with decreasing temperatures were characterized by SANS over a wide *Q*-range. Again, the use of contrast allows the identification of wax conformation and the polymer inside the common aggregates. This also aids in the understanding of the microscopic interaction mechanisms.

The assembly events depend on the length of the crystallizable segments that are either the wax molecule or the ethylene sections of the copolymer. Formation of multilevel hierarchical morphologies is a consequence of one component to crystallize prior to its companion and to template the final overall morphology. Conversely, for the case of well-matched self-assembling properties cooperative co-crystallization is allowed. Typical pinhole SANS results are shown in Fig.14a,b, which presents under polymer and wax contrast, respectively, scattering patterns from wax-polymer mixed solutions which are typical for the two mechanisms identified (Radulescu et al., 2003, 2004).

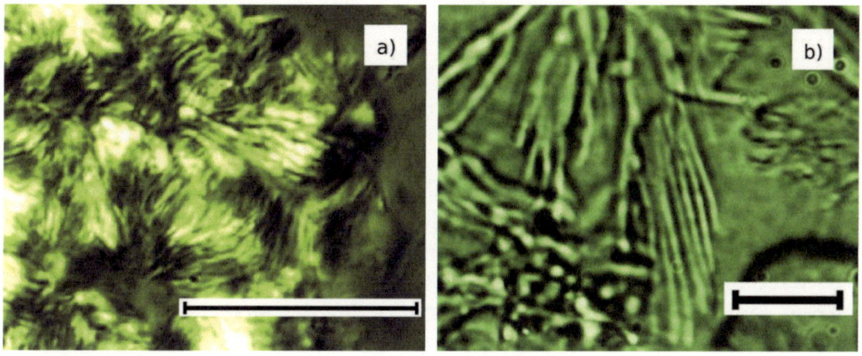

Fig. 13. Micrographs from PEB-7.5 doped decane solutions of 1%C_{36} at 20°C (a) and 4%C_{24} at -10°C (b); scale bars 30μm (Radulescu et al., 2004, 2006).

In the first example, when 0.6%PEB-7.5(6kg mol^{-1}) and 1%C_{24} are mixed in decane, the polymer commences self-assembling in its characteristic density-modulate rod-like structure well before the wax crystals appearance. At temperatures below the wax crystallization line the polymer and wax influences each other's aggregation behaviour (Fig.14a). On the one hand, the wax crystallization influences the polymer structure, which changes from rod-like ($\sim Q^{-1}$) to plate-like ($\sim Q^{-2}$) in the presence of wax. Thus, co-crystallization of the polymer and wax occurs below certain temperatures. On the other hand, the wax scattering patterns reveal the correlation peak at the same Q-position where it is observed for the polymer alone. Thus, it looks like in the case when polymer assemblies are formed at higher temperatures than the wax precipitation point the common aggregation habit is dictated by the primordial polymer structure which serves to template the subsequently formed wax-polymer plates in a shish-kebab-like correlated arrangement. This is demonstrated by the TEM observation of aggregates (Fig.15a) isolated at room temperature from a mixed decane solution of high M_W PEB-7.5 (30kg mol^{-1}) and C_{36}, a ternary system that exhibits at higher temperatures similar scattering features and experiences the same aggregation mechanism like the combination of low M_W PEB-7.5 and C_{24}. The simultaneous fit of the scattering data for different contrast conditions considering correlated wax layers embedded into thicker polymer platelets (Eq.5-7) delivered the geometrical and density parameters of the wax-copolymer aggregates. These consist basically of a single layer of stretched C_{24} molecules (d_{wax}=32Å) embedded into thicker polymer platelets (d_{pol}=100-150Å) which grow around the crystalline nuclei of initial polymer rods and are correlated over distances of D=200-230Å with a smearing of σ_D=100-140Å. This mechanism yields eventually a shish-kebab-like morphology (Fig.15a).

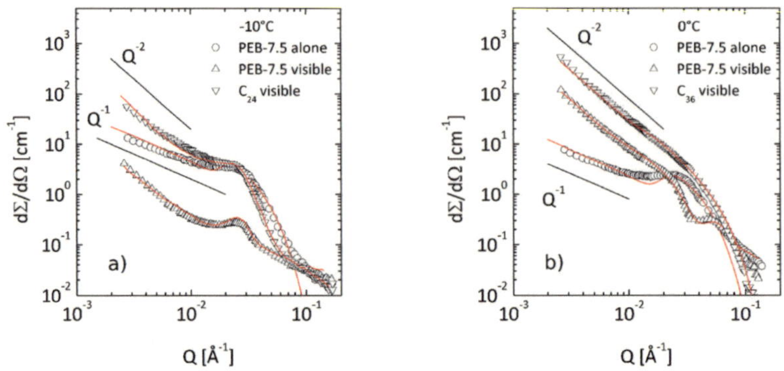

Fig. 14. Examples of wax and polymer SANS profiles (triangles) from PEB-*n*/wax mixed solutions in decane (Radulescu et al., 2004) for 0.6%PEB-7.5/1%C_{24} at -10°C (a) and 0.6%PEB-7.5/0.5%C_{36} at 0°C (b); the polymer self-assembling SANS patterns (circles) are shown in parallel in order to emphasize the structural changes induced by the addition of wax; the curves represent model description of the data (see text).

In the second example, when 0.6%PEB-7.5 (6kg mol^{-1}) and 0.5%C_{36} are mixed in decane, the wax and polymer show quite similar precipitation temperatures and yield extended joint polymer-wax plate-like structures which are loosely correlated due to van der Waals interaction. The co-crystallization of wax and copolymer in thin platelets is again the main mechanism that changes the polymer scattering pattern from ~Q^{-1} into ~Q^{-2} (Fig.14b). A homogeneously distributed polymer profile across well-defined platelets of narrow size distribution was deduced from the observation of the form factor features towards high Q. Again, the simultaneous fit of different contrast scattering patterns according to Eq.5-7 resulted in monolayer of stretched wax molecules embedded within a thicker polymer layer, a morphology which is depicted by Fig.15b.

From the quantitative analysis of the scattering profiles systematically collected over a wide temperature range for different wax-copolymer combinations and volume fractions it was concluded that the co-crystallization mechanism of wax and crystallizing segments of the copolymer emerging as a consequence of the good match between precipitation temperatures of both components corresponds to a high efficiency of the polymer in controlling the size and shape of the wax crystals. In this case, for ϕ_{wax}=0.5% in the initial solution about 80% stays inside polymer-wax common thin platelets at 0°C in the case of C_{36} combined with PEB-7.5(6kg mol^{-1}) and at -22°C in the case of C_{24} combined with PEB-11(6kg mol^{-1}). The amount of polymer contained by the aggregates is very small and corresponds to much less than a half of the volume fraction in the initial solution. The wax volume fraction inside the wax layer is about ϕ_{wax}=70-90%, while the polymer volume fraction inside the polymer layer is about ϕ_{pol}=10-30%. For higher wax contents, a considerable amount of wax is entrapped by the thin platelets, e.g. for ϕ_{wax}=2% C_{36} in the presence of PEB-7.5(6kg mol^{-1}) this amounts at 0°C to about 55%. A lower efficiency was observed in the case of wax crystallization templated by polymer one-dimensional morphologies pre-existing in solution: only a maximum of about 10-15% of ϕ_{wax}=0.5% C_{24} is contained by the correlated polymer-wax plates at -20°C for a corresponding 50% of the polymer consumed from the amount in the initial solution. Lower wax volume fraction, about ϕ_{wax}=30-60% characterizes the wax layer, while the polymer volume fraction inside the polymer layer is about ϕ_{pol}=40-70%. It was shown that, in this case, later crystallization stages or the wax surplus (like for ϕ_{wax}=4%) led to thickening of the platelets anchored at the polymer density-modulated rod-like structures, which ultimately join together and give raise to formation of long needles with modulated thickness like those visible in Fig.13b.

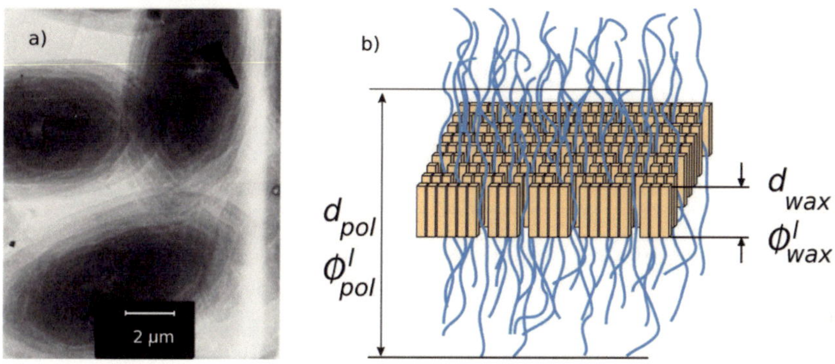

Fig. 15. (a) TEM image of wax-polymer correlated plates resembling a top view of a shish-kebab morphology; (b) Cartoon of the polymer-wax aggregate consisting of monolayer of stretched wax molecules embedded into thicker homogeneous polymer layer; the thickness and material volume fraction characterizing each layer are indicated (Radulescu et al., 2008).

From the micro-structural investigations of mixed solutions of PEB-*n* random copolymers and single waxes it emerges thus the conclusion that these polymers show selectivity in their wax modification capacity depending on the ethylene content of the backbones: the more crystalline copolymers show a higher efficiency for longer wax molecules, whereas the less crystalline ones are very efficient for shorter waxes. This basic result directly suggested that highly active PEB-*n* additives for crude oils and middle distillates should consist of segments with graded ethylene content.

6. Multi-block PEB-*n* copolymers

Two PEB-*n* multi-block copolymers, a tetra-block PEB-2.6/PEB-6.0/PEB-10.9/PEB-13.2 and a tri-block PEB-6.5/PEB-8.9/PEB-10.1, were synthesized via hydrogenation of poly(1,4-1,2) butadiene block random copolymer segments where the 1,4/1,2 ratio decreased as the number of segments increased. For example, the tetra-block material was prepared in cyclohexane using t-butyl-lithium as the initiator (at room temperature). The initial PEB-2.6 block was prepared with no modifier of the butadiene microstructure. On completion of the initial segment, a small amount of solution was removed for micro-structure (H-NMR) and chain molecular weight (H-NMR and GPC). The second segment was carried out with a small concentration of triethyl-amine present. This led to an enhanced vinyl concentration. After hydrogenation, this yielded the PEB-6.0 block. This procedure was repeated to prepare the subsequent PEB-10.9 and PEB-13.2 blocks (Radulescu et al. 2011). The measured averaged molecular weight of the two multi-block materials was 18.9kg mol^{-1} (6.3/6.3/6.3) for the tri-block and 29.7kg mol^{-1} (4.2/8.5/8.5/8.5) for the tetra-block, respectively.

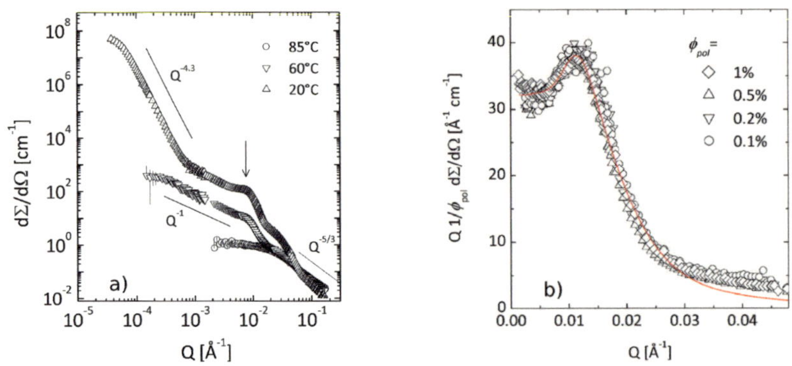

Fig. 16. (a) Selective examples of SANS patterns collected at several temperatures where typical structures are revealed in decane solution of 1% tetra-block copolymer (Radulescu et al., 2011); the power-law scattering behaviour characteristic of different structures is indicated; the arrow denotes the peak arising from intra-particle correlation as concluded following the observation of the constant peak-position for different ϕ_{pol} in solution, which is illustrated in (b) for the case of the tri-block copolymer in decane; the red curve represents the description of data with the density-modulated rod model.

A combination of classical pinhole-, mirror-focusing- and ultra-SANS techniques with microscopy (Radulescu et al., 2011) revealed that decreasing temperature leads to the formation and evolution in solution of multi-sized structural levels showing a hierarchical organization on the length scale from nanometres up to 10 of microns (Fig.16a). The tetra-block and tri-block copolymers show similar aggregation behaviour, and only the temperature at which the self-assembling commences is different for the two materials. This self-assembling behaviour relates to the graded crystallization tendency of the constituent blocks of copolymers upon cooling. In a first aggregation step, the scattering features at 60°C (Q^{-1} power law behaviour and the peak-like structure) revealed rod-like morphologies with modulated density formed as a consequence of the crystallization of the block containing the longer PE sequences following a mechanism described above in the case of PEB-*n* random copolymers. The nature of the aggregates and their structural characterization was established by model interpretation of the results from systematic pinhole-SANS investigations (Fig.16b). Subsequent crystallization events at lower temperature yield new structural features. On the one hand, micellar-like sub-structural morphologies arising from the amorphous segments anchored on growing correlated crystalline nuclei along the rod axis give raise to the occurrence in the scattering profile of a hump-like feature at Q=2-5x10^{-2} Å^{-1}. On the other hand, cross-linking of the one-dimensional aggregates due to co-crystallization of segments with lower crystallinity produced a dramatic increase of the scattered intensity towards low Q. Large-scale macro-aggregates with irregular edges and a diffusive interface (power-law exponent p>4) are ultimately formed as a consequence of these branching and association events. The understanding of the self-association mechanism was completed by the results of microscopy and contrast matching pinhole SANS on mixed solutions of multi-block copolymers and single waxes. These investigations helped elucidate the polymer-wax interaction mechanism at the microscopic level.

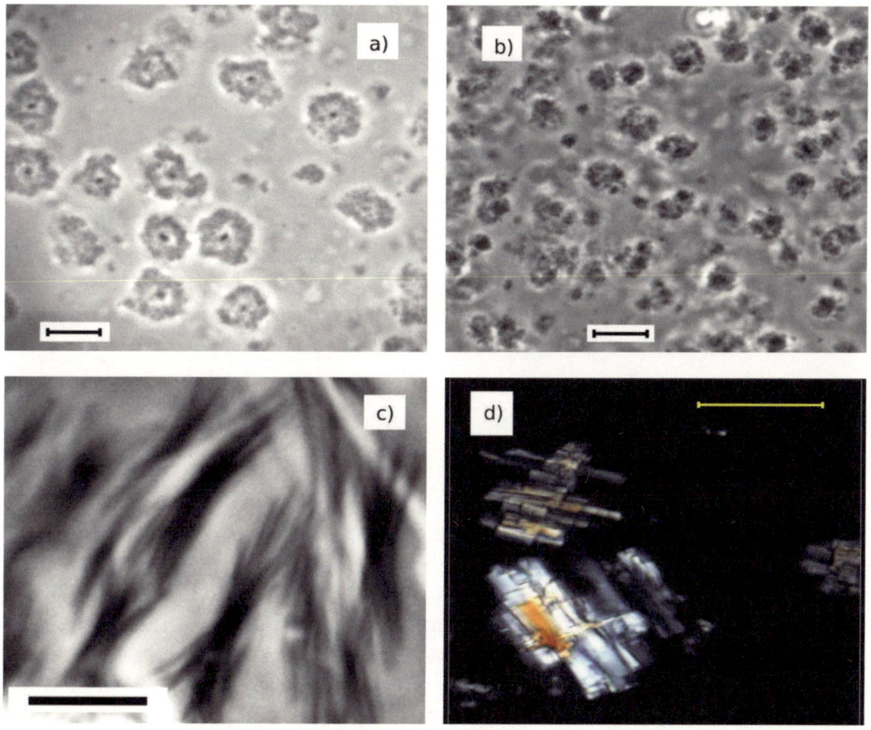

Fig. 17. Aggregates formed at 0°C in decane solution of 1% tetra-block copolymer with no wax (a), 1%C_{36} (b) and 4%C_{36} (c,d) added as revealed by bright field (a-c) and crossed polarizers (d) microscopy (Radulescu et al., 2011); the scale bar: 10μm (a-c), 20μm (d).

Polymer morphologies yielded by the gradual assembling behaviour are emphasized by decoration with wax crystallites (Fig.17b-d) when large wax-entrapping globular morphologies or sheaves and bundles of elongated wax crystals are observed under the microscope. For low wax content, the early formed polymer morphologies incorporate later appearing wax crystallites and hereby, they become more dense and compact. The limited growth of the polymer templates is an indication for less frequent branching events and thus, the incorporation of later crystallizing polymer segments within co-crystals with wax molecules. Fig.17c,d show bundles of fibrils and sheaves of elongated lamellae formed when the wax content is high. These crystal habits seem to be templated and controlled by polymers. After the analysis of systematic SANS data it turned out that the co-crystallization of the wax molecules and copolymers mid-branched crystallizing blocks is the key effect yielding the specific wax-copolymer morphology observed in this case (Radulescu et al., 2011). This emphasizes the specificity of the graded crystalline multi-block copolymers compared to the PEB-*n* random copolymers.

The formation of a primary structural level at certain temperatures is a consequence of crystallization occurring within the highly crystalline segment. Concurrently, cooling promotes a secondary structure which is anchored and hierarchically grows at the initial structure. In this way, the low temperature final morphology, although complex and multi-sized, exhibit smaller dimension and higher compactness compared to those yielded in the case of PEB-*n* random materials.

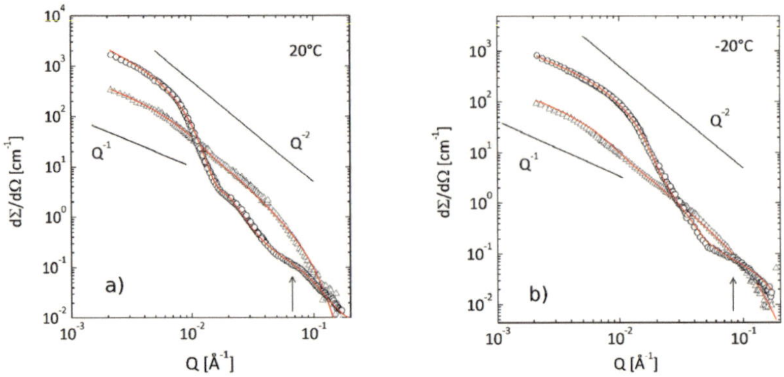

Fig. 18. Examples of SANS patterns measured from mixed 1%tetra-block/1%C_{36} (a) and 1%tri-block/1%C_{24} (b) decane solutions for the polymer (circles) and wax (triangles) visible; the red curves denote the model interpretation of data while the arrows indicate the high-Q scattering detail characteristic for polymer blobs.

The co-crystallization of wax and polymer from common solution in habits that are dictated by the polymer assembling tendency is revealed by a comparative analysis of contrast matched scattering patterns (Fig.18) and those measured from separate wax and polymer solutions (Fig.9, 16). On the one hand, the wax/copolymer common structures are templated by pre-existing polymer rod-like assemblies formed at higher temperatures than the wax crystallization point. This explains the overall one-dimensional morphology of the polymer and wax (Q^{-1} behaviour at low Q). On the other hand, according to the observed polymer scattering features (the disappearance of the intermediate-Q prominent peak and the preservation of the Q^{-1} power law), it seems that co-crystallizing wax and polymer molecules promote the growth of the initial polymer crystalline nuclei along the rod axis, filling eventually the open volume between them and generating more homogeneous aggregates. The polymer patterns were successfully modelled with the homogeneous rod form factor (Eq.8) with an added contribution of the scattering from polymer blobs describing the high Q modulation (Richter et al., 1997). The wax morphology seems to resemble thin and elongated platelets (boards) as revealed by the two characteristic power-laws (Q^{-1} at low Q and Q^{-2} at intermediate Q) and structural levels (kinks of intensity) observed in the scattering profile. The wax scattering patterns were rather well described by the form factor of very long cylinders with elliptical cross-section (Bergström & Pedersen, 1999), which is a good approximation for a tablet- or board-shaped morphology. Again, thin and very long crystals consisting of monolayer of elongated wax molecules and having a width of 500-600Å were revealed in both cases. The lesser extension of common wax-copolymer crystallization in the lateral direction together with the sign of polymer blobs formation is an evidence of aggregates consisting of rather compact very elongated core surrounded by amorphous polymer corona. The core is jointly made by wax and polymer whereas the corona that hinders the growth of the wax crystals into large compact three-dimensional objects consists of amorphous polymer segments. This explains why the correlation effect revealed by the scattering patterns from polymer self-assemblies vanishes in both the polymer and wax scattering patterns at temperatures under the wax crystallization point and why the scattering features of the rods substructure also disappear in the polymer scattering profile. The PEB-*n* multi-block copolymers demonstrate high efficiency in reducing the size and in changing the shape of waxy crystals formed in single wax solutions. Board-like wax crystals that are basically formed (Fig.19) are entrapped by the polymer macro-aggregates or grow in a manner that is controlled by the polymer. The wax-polymer interaction mechanism deciphered with the help of micro-structural investigations is nicely emphasized in the case of high C_{36} wax containing solutions.

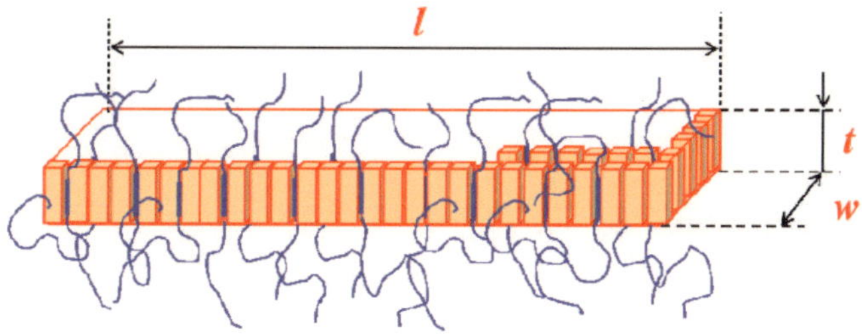

Fig. 19. Sketch depicting the wax board-like morphology controlled by the PEB-*n* multi-block copolymers emerged from the micro-structural studies; the main structural levels characterizing such morphology, the width *w*, the thickness *t*, and the length *l* are indicated.

The wax crystal control capacity of the multi-block PEB-*n* copolymers with graded crystallinity was further explored on more realistic systems with either high content of short *n*-paraffins (4% C_{24}) or mixture of long *n*-paraffin waxes (1%C_{30}+1%C_{36}), close to the conditions encountered in the oil industry applications. Fig.20a,b present the contrast matching SANS results from a 1%tri-block/4%C_{24} decane solution at two temperatures under the wax crystallization point. A co-crystallization of wax and polymer molecules in thin platelets (Q^{-2} power-law behaviour) is again the interaction mechanism revealed by the simultaneous interpretation of the polymer and scattering patterns collected at 0°C (Fig.20a). The plates seem to be weakly correlated as indicated by the broad hump-like feature observed in the wax profile. The modelling of data in terms of two wax and polymer embedded layers of different thickness proves that the wax crystals consist of single layer of stretched wax molecules while the form factor yields details of polymer structure. The analysis of the "forward scattering" resulted in a wax fraction included within the thin lamellar structures of about 15% from the total wax amount dissolved in the initial solution. This value is considerably higher than that obtained in the case of 1%PEB-7.5(6kg mol^{-1})/ 4%C_{24} at 0°C, which amounts to about 6%. A further decrease in temperature (Fig.20b) leaves the polymer conformation almost unchanged as observed in the scattering profile when compare to the 0°C case, but induces a dramatic changes at the level of wax morphology: although a certain amount of wax stays still in a two-dimensional conformation, as proven by the Q^{-2} power-law behaviour observed towards high *Q*, a massive growth of wax in large compact crystals is revealed by the Porod-like scattering observed at low *Q*.

The micrograph from the same sample (Fig.21a) shows the micron size compact aggregates that give rise to such scattering profile. Nevertheless, when compared with the case of the neat wax solution (Fig.8d) the wax crystal control ability of the multi-blocks is striking. The polymer seems to operate by a two-fold mechanism: on one hand, it templates the wax

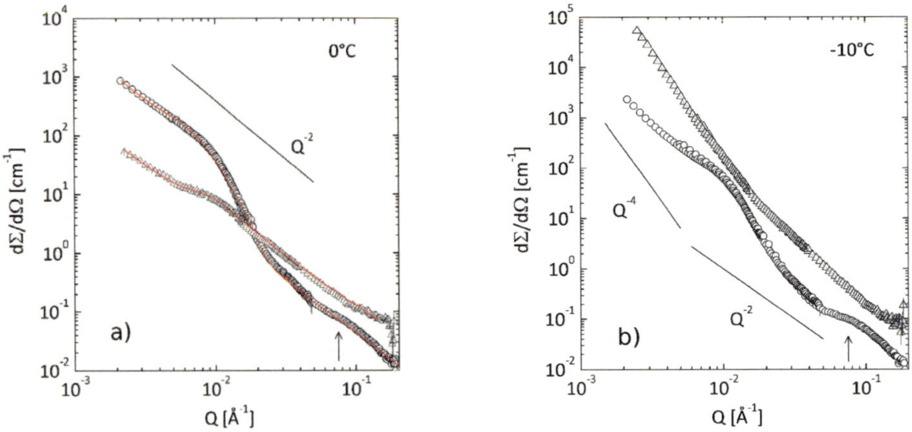

Fig. 20. Examples of contrast matched polymer (circles) and wax (triangles) scattering patterns from 1%tri-block/4%C_{24} decane solution at two different temperatures. The curves and arrows have the same meaning as in Fig.18.

crystallization by means of its early formed primary morphology and, on the other hand, it arrests the growth of large wax crystals by a moderate co-crystallization of wax molecules and mid-branched segments entrapped by the still amorphous segments.

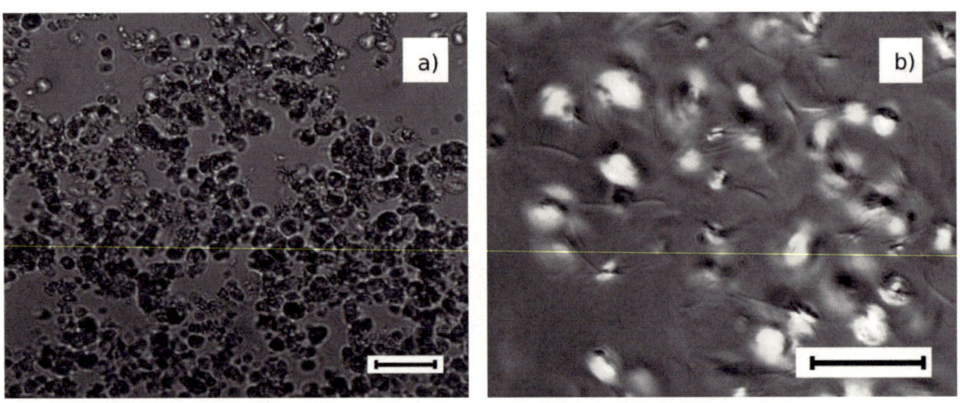

Fig. 21. Micrographs from decane solutions of 1% tri-block and 4%C_{24} at -10°C (a) and 1% tetra-block and 1%C_{30}+1%C_{36} at 0°C; scale bars are 30μm.

In the case of tetra-blocks added to a mixed 1%C_{30}+1%C_{36} decane solution the micrographs collected at 0°C (Fig.21b) also revealed the formation of very small morphologies which differs greatly from the case of polymer undoped solution (Fig.8a). The SANS scattering patterns collected for wax contrast (Fig.22a) show that most of wax stays in a two-dimensional morphology at 10°C and 0°C, under the crystallization point of the two wax molecules. A moderate increasing tendency of the scattered intensity towards low Q with a power law exponent $p{\approx}3$ is evident at 0°C when wax-driven agglomerates showing irregular rough edges are formed.

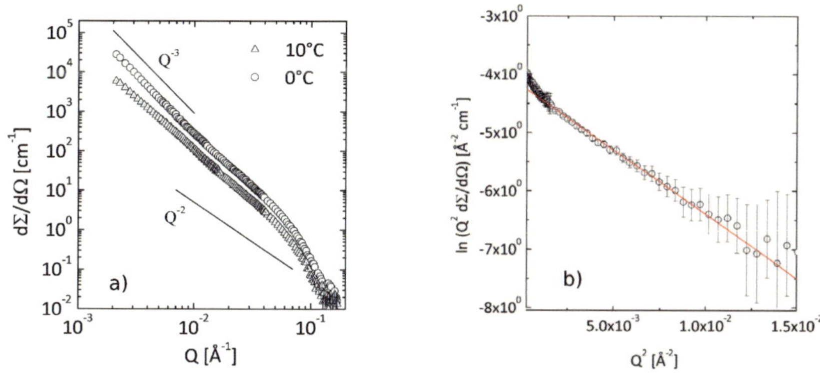

Fig. 22. SANS patterns under wax contrast from a 1% tetra-block and 1%C_{30}+1%C_{36} solution in decane (a) and the Guinier presentation of the large Q data at 0°C, delivering in a direct approach the thickness and the "forward scattering" of the wax lamellae (b).

A close inspection of the micrograph shows besides these aggregates some other elongated objects resembling the one-dimensional crystals observed in the case of tetra-block and single C_{36} wax. These fibrillar-shaped crystals seem to be dominantly wax containing, as in the case of large C_{36} content (Fig.17c), while the larger irregular morphologies must result from waxes entrapped by polymer branched aggregates. A quick quantitative analysis of data in a two-dimensional Guinier approach (Fig.22b) delivered the "forward scattering" from the thin wax layers which revealed that about 50% from the total amount of wax dissolved in the initial solution is contained by the two-dimensional morphologies at 0°C, which again proves the efficiency of these polymers in controlling the wax crystallization from hydrocarbon solutions.

7. Concluding remarks

Hopefully, in this work we have demonstrated the uniqueness and usefulness of small-angle neutron scattering to solve and characterize the complex morphologies formed in common solutions of *n*-paraffin waxes and polymeric systems upon decreasing temperature and, based on the contrast matching method and with the support of the complementary microscopy approach, to understand the subtleties of wax-polymer interaction mechanism at a mesoscopic scale. The elucidation and the knowledge of fundamental aspects concerning the structures and morphologies emerging in such systems as a function of temperature, concentration in solution and, more important, similarities in the assembling and crystallization behaviour between the two systems of interest, namely the wax and the polymer, can lead to direct predictions and immediate applicability with respect to the wax crystal control capacity of distinct polymeric systems. Thus, highly efficient polymer additives can be tailored for acting against specific technical problems experienced by the oil and refinery industry as a consequence of the cold wax crystallization depending on particularities of the systems and conditions involved. As a general conclusion, polymeric systems presenting graded crystallinity have proven a high efficacy and versatility in controlling the crystallization of waxes in various conditions.

8. References

Abdallah, D.J. & Weiss, R.G. (2000). n-Alkanes Gel n-Alkanes (and many other organic liquids). *Langmuir*, Vol.16, No.2, pp.352-355, ISSN 0743-7463

Alefeld, B.; Dohmen, L.; Richter, D. & Brückel, T. (2000). X-Ray Space Technology for Focusing Small-Angle Neutron Scattering and Neutron Reflectometry. *Physica B*, Vol.283, No.4, pp.330-332, ISSN 0921-4526

Ashbaugh, H.S.; Radulescu, A.; Prud'homme, R.K.; Schwahn, D.; Richter, D. & Fetters, L.J. (2002). Interaction of Paraffin Wax Gels with Random Crystalline/Amorphous Hydrocarbon Copolymers. *Macromolecules*, Vol.35, No.18, pp.7044-7053, ISSN 0024-9297

Banki, R.; Hoteit, H. & Firoozabadi, A. (2008). Mathematical Formulation and Numerical Modeling of Wax Deposition in Pipelines from Enthalpy-Porosity Approach and Irreversible Thermodynamics. *International Journal of Heat Mass Transfer*, Vol.51, No.13-14, pp.3387-3398, ISSN 0017-9310

Beaucage, G. & Schaefer, D.W. (1994). Structural Studies of Complex-Systems Using Small-Angle Scattering - A Unified Guinier Power-Law Approach. *Journal of Non-Crystalline Solids*, Vol.172-174, pp.797-805, ISSN 0022-3093

Beiny, D.H.M.; Mulli, J.W. & Lewtas, K. (1990). Crystallization of Normal-Dotriacontane from Hydrocarbon Solution with Polymeric Additives. *Journal of Crystal Growth*, Vol.102, No.4, pp.801-806, ISSN 0022-0248

Bergström, M. & Pedersen, J.S. (1999). A Small-Angle Neutron Scattering (SANS) Study of Tablet-Shaped and Ribbonlike Micelles Formed from Mixtures of an Anionic and a Cationic Surfactant. *Journal of Physical Chemistry B*, Vol.103, No.40, pp.8502-8513, ISSN 1089-5647

Claudy, P.; Letoffe, J.M.; Bonardi, B.; Vassilakis, D. Damin, B. (1993). Interactions between n-Alkanes and Cloud Point-Cold Filter Plugging Point Depressants in a Diesel Fuel. A Thermodynamic Study. *Fuel*, Vol.72, No.6, pp.821-827, ISSN 0016-2361

Coutinho, J.A.P.; Dauphin, C. & Daridon J.L. (2000). Measurements and Modelling of Wax Formation in Diesel Fuels. *Fuel*, Vol.79, No.6, pp.607-616, ISSN 0016-2361

del Carmen Garcia, M.; Carbognani, L.; Orea, M. & Urbina A. (2000). The Influence of Alkane Class-Type on Crude Oil Wax Crystallization and Inhibitors Efficiency. *Journal of Petroleum Science and Engineering*, Vol.25, No.3-4,pp.99-105, ISSN 0920-4105

Frielinghaus, H.; Pipich, V.; Radulescu, A.; Heiderich, M.; Hanslick, R.; Dahlhoff, K.; Iwase, H.; Koizumi, S. & Schwahn, D. (2009). Aspherical Refractive Lenses for Small-Angle Neutron Scattering. *Journal of Applied Crystallography*, Vol.42, No.4, pp.681-690, ISSN 0021-8898

Giorgio, S. & Kern, R. (1983). Filterability, Crystal Morphology and Texture - Paraffin Dewaxing Aids. *Journal of Crystal Growth*, Vol.62, No.2, pp.360-374, ISSN 0022-0248

Kern, R. & Dassonville, R. (1992). Growth Inhibitors and Promoters Exemplified on solutionj Growth of Paraffin Crystals. *Journal of Crystal Growth*, Vol.116, No.1-2, pp.191-203, ISSN 0022-0248

Leube, W.; Monkenbusch, M.; Schneiders, D.; Richter, D.; Adamson, D.; Fetters, L. & Dounis, P. (2000). Wax-Crystal Modification for Fuel Oils by Self-Aggregating Partially Crystallizable Hydrocarbon Block-Copolymers. *Energy Fuels*, Vol.14. No.2, pp.419-430, ISSN 0887-0624

Monkenbusch, M.; Schneiders, D.; Richter, D.; Willner, L.; Leube, W.; Fetters, L.J.: Huang, J.S. and Lin, M. (2000). Aggregation Behaviour of PE-PEP Copolymers and the Winterization of Diesel Fuel. *Physica B*, Vol.276-278, pp.941-943, ISSN 0921-4526

Morton, M. & Fetters, L.J. (1975). Anionic Polymerization of Vinyl Monomers. *Rubber Chemistry and Technology*, Vol.48, No.3, pp.359-409, ISSN 0035-9475

Pedersen, J.S. (1997). Analysis of Small-Angle Scattering Data from Colloids and Polymer Solutions: Modelling and Least-Square Fitting. *Advances in Colloid and Interface Science*, Vol.70, No.,pp.171-210, ISSN 0001-8686

Pedersen, K.S.; Skovbor, P. and Ronningsen H.P. (1991a). Wax Precipitation from North Sea Crude Oils. 4. Thermodynamic Modelling. *Energy Fuels*, Vol.5, No.6, pp.924-932, ISSN 0887-0624

Pedersen, W.B.; Hansen, A.B.; Larsen, E.; Nielsen, A.B. and Ronningsen, H.P. (1991b). Wax Precipitation from North Sea Crude Oils. 2. Solid-Phase Content as a Function of Temperature. *Energy Fuels*, Vol.5, No.6, pp.908-913, ISSN 0887-0624

Radulescu, A.; Schwahn, D.; Richter, D. & Fetters, L.J. (2003). Co-Crystallization of Poly(Ethylene-Butene) Copolymers and Paraffin Molecules in Decane Studied with Small-Angle Neutron Scattering. *Journal of Applied Crystallography*, Vol.36, No.4, pp.995-999, ISSN 0021-8898

Radulescu, A.; Schwahn, D.; Monkenbusch, M.; Fetters, L.J. & Richter, D. (2004). Structural Study of the Influence of Partially Crystalline Poly(Ethylene Butene) Random Copolymers on Paraffin Crystallization in Dilute solutions. *Journal of Polymer Science Part B – Polymer Physics*, Vol.42, No.17, pp.3113-3132, ISSN 0887-6266

Radulescu, A.; Schwahn, D.; Stellbrink, J., Kentzinger, E.; Heiderich, M.; Richter, D. & Fetters, L.J. (2006). Wax Crystallization from Solution in Hierarchical Morphology Templated by Random Poly(Ethylene-co-Butene) Self-Assemblies. *Macromolecules*, Vol.39, No.18, pp.6142-6151, ISSN 0024-9297

Radulescu, A.; Fetters, L.J. & Richter, D. (2008). Polymer-Driven Wax Crystal Control Using Partially Crystalline Polymeric Materials. *Advances in Polymer Science*, Vol.210, pp.1-100, ISSN 0065-3195

Radulescu, A.; Schwahn, D.; Stellbrink, J.; Monkenbusch, M.; Fetters, L.J. & Richter, D. (2011). Microstructure and Morphology of Self-Assembling Multiblock Poly(Ethylene-1-Butene)-n Copolymers in solution Studied by wide-Q Small-Angle Neutron Scattering and Microscopy. *Journal of Polymer Science Part B – Polymer Physics*, Vol.49, No.2, pp.144-158, ISSN 0887-6266

Richter, D.; Schneiders, D.; Monkenbusch, M.; Willner, L.; Fetters, L.J.; Huang, J.S.; Lin, M.; Mortensen, K. & Farago, B. (1997). Polymer Aggregates with Crystalline Cores: the System Poly-Ethylene-Poly(Ethylenepropylene). *Macromolecules*, Vol.30, No.4, pp.1053-1068, ISSN 0024-9297

Schwahn, D.; Richter, D., Wright, P.J.; Symon, C.; Fetters, L.J. & Lin, M. (2002a). Self-Assembling Behaviour in Decane Solution of Potential Wax Crystal Nucleators Based on Poly(co-Olefins). *Macromolecules*, Vol.35, No.3, pp.861-870, ISSN 0024-9297

Schwahn, D.; Richter, D.; Lin, M. & Fetters, L.J. (2002b). Cocrystallization of a Poly(Ethylene-Butene) Random Copolymer with C-24 in n-Decane. *Macromolecules*, Vol.35, No.9, pp.3762-3768, ISSN 0024-9297

Singh, P.; Fogler, H.S. & Nagarajan N. (1999). Prediction of the Wax Content of the Incipient Wax-Oil Gel in a Pipeline: an Application of the controlled Stress Rheometer. *Journal of Rheology*, Vol.43, No.6, pp.1437-1459, ISSN 0148-6055

Srivastava, S.P.; Saxena, A.K.; Tandon, R.S. & Shekher, V. (1997). The Measurement and Description of the Yielding Behavior of Waxy Crude Oil. *Fuel*, Vol.35, No.7, pp.1121-1156, ISSN 0016-2361

Venkatesan, R.; Nagarajan, N.R.; Paso, K.; Yi, Y.B.; Sastray, A.M. & Fogler H.S. (2005). The Strength of Paraffin Gels Formed Tinder Static and Flow Conditions. *Chemical Engineering Science*, Vol.60, No.13, pp.3587-3598, ISSN 0009-2509